Global Cities

Urban and Industrial Environments

Series editor: Robert Gottlieb, Henry R. Luce Professor of Urban and Environmental Policy, Occidental College

For a complete list of books published in this series, please see the back of the book.

Global Cities

Urban Environments in Los Angeles, Hong Kong, and China
Robert Gottlieb and Simon Ng

The MIT Press
Cambridge, Massachusetts
London, England

This book was set in ITC Stone Serif Std by Toppan Best-set Premedia Limited. Printed and bound in the United States of America.

Library of Congress Cataloging-in-Publication Data

Names: Gottlieb, Robert, 1944- author. | Ng, Simon Ka-Wing, author.
Title: Global cities : urban environments in Los Angeles, Hong Kong, and China /
    Robert Gottlieb and Simon Ng.
Description: Cambridge, MA : MIT Press, 2017. | Series: Urban and industrial
    environments | Includes bibliographical references and index.
Identifiers: LCCN 2016034962 | ISBN 9780262035910 (hardcover : alk. paper)
Subjects: LCSH: Economic development--Environmental aspects--California--
    Los Angeles. | Economic development--Environmental aspects--China--
    Hong Kong. | Economic development--Environmental aspects--China. |
    Sustainable urban development--California--Los Angeles. | Sustainable urban
    development--China--Hong Kong. | Sustainable urban development--China. |
    Environmental policy--California--Los Angeles. | Environmental policy--China--
    Hong Kong. | Environmental policy--China.
Classification: LCC HC79.E5 .G6573 2017 | DDC 333.77--dc23 LC record available
    at https://lccn.loc.gov/2016034962

ISBN: 978-0-262-03591-0

10 9 8 7 6 5 4 3 2 1

To our families: Marge Pearson, Casey and Andrea Pearson-Gottlieb, Rebecca Wong and Nathan Ng

Contents

Acknowledgments

The idea for our book originated in talks that we gave in Los Angeles and Hong Kong. As our research progressed, we were able to get support and advice from others who were also engaged as researchers and activists in Los Angeles, Hong Kong, and China. In addition, two foundations provided important financial support: the Henry Luce Foundation through the China-Environment program at Occidental College, and the WYNG Foundation in Hong Kong through Civic Exchange. This support created opportunities for us to further develop our research connections and to identify the issues and policy debates taking place in each of the global cities and regions we profiled.

Both of us have benefited from our work in organizations that nurtured and extended the action-research approach we describe in the book—Bob as a cofounder and long-time executive director of the Urban & Environmental Policy Institute in Los Angeles and Simon through his long-standing research role at Civic Exchange in Hong Kong. To get the book in shape for publication, we had the fortune of working with our editors at the MIT Press—Clay Morgan and Beth Clevenger—who understood the value and the challenges of pursuing this type of project. Clay had worked with Bob for eighteen years on the development of the Urban and Industrial Environments series in which this book is published. After Clay retired in 2014, Beth came in as MIT Press environmental editor and became a champion of the quality environmental books that have been published by the MIT Press. Beth continued to communicate to us the importance of this work, and was supportive of us throughout the process. We also want to thank Miranda Martin, who ably assisted us on many MIT Press publication-related matters.

We especially want to thank the staff and former staff of both UEPI and Civic Exchange for their support and passion for social and environmental justice. Mark Vallianatos, Martha Matsuoka, Rosa Romero, Angelo Logan, and Heng Lam Foong were especially helpful, and Professor Alex ("Sasha") Day ably took over Bob's role in the Occidental China-Environment Program. At Civic Exchange, Christine Loh and Lisa Hopkinson laid the foundation in 2000 as cofounders of the organization, enabling it to prosper as the go-to think tank in Hong Kong. Over the years, Civic Exchange has drawn together like-minded people, including Yanyan Yip, Mike Kilburn, and Veronica Booth, who were attracted to Christine's vision and charisma and worked together with a mission to make Hong Kong a better place to live.

Several colleagues played an important role in helping us move this project forward. We especially want to thank Mark Vallianatos and Carine Lai for their research and writing contributions to the book and Sunny Lam, Leo Chan, Tracy Wong, Kristie Chu, Kristina Kokame, Simiao Lai, and Anne Ewbank for their research assistance. Special thanks to Professor Jim Sadd for his designing our Los Angeles map and to Jianqiang Liu, Rosa Romero, Deborah Murphy, Carla Truax, Wendy Gutschow, and Jim Gauderman for helping us obtain photos and source material. We appreciate the support of Sylvia Chico, Robin Craggs, and Bhavna Shamasunder of Occidental College, and Iris Chan, Michelle Wong, and Rae Leung of Civic Exchange. We want to thank the two anonymous reviewers as well as Professors Yan Hairong, Yuk Wah, and Peter Brimblecombe for their comments and feedback on the manuscript. We also want to thank Alexis Lau of the Hong Kong University of Science and Technology (HKUST) for sharing his air quality data. We are grateful to Patrick Fung of the Clean Air Network, Mathew Fung and Keong Fung of the Hong Kong Institute of Planners, Civic Exchange, Joe Linton of Streetsblog LA, Wendy Gutschow of the University of Southern California, Wendy Ramallo of the Council for Watershed Health, Steven Hines of the Metropolitan Water District of Southern California, Emily Hart, and Meiqin Chen for giving us permission to use their photos and images. It was a pleasure to work with Occidental students Anna ("Skye") Harnsberger, Miranda Chien-Hale, Jennifer Yi, Rachel Young, Tsering Lama, Connie Li, and Marissa Chan, and HKUST students Rachel Tai and Keith Chan as well as Tracy Wong, Leo Chan, and Kristie Chu. We also want to thank our many research and activist colleagues, including those from the Moving Forward Network and THE (Trade, Health, and Environment) Impact Project (especially Andrea Hricko, Carla Truax, Angelo Logan, Mark Lopez, Jesse Marquez, Jessica Tovar, Sylvia Betancourt, and Penny Newman, among many others); from the Los Angeles Food Policy Council (Joann Lo, Alexa Delwiche, and Colleen McKinney); from Kadoorie Farm and Botanic Garden (Idy Wong and May Cheng); Angus Lam from Partnerships for Community Development; Shi Yan from Shared Harvest; and Li Kai Kuen from the Tai Po CSA. All were helpful and provided valuable information, and they are all invaluable participants in the issues we have discussed in the book. HKUST faculty Chak Chan, Arthur Lau, and Jimmy Fung, and Nanjing University faculty Cong Cong, Ma Junya, Fang Hong, and Dean and Associate Vice President Zhou Xian, and the students from the Environmental Club, among others, were gracious hosts, and Occidental Professor Xiao-huang Yin and Nanjing University Professor Cong Cong ably facilitated the arrangements for our connections to Nanjing University. We want to thank Ma Weichun, Zhang Yan, Bao Cunkuan, and students from the Department of Environmental Sciences and Engineering of Fudan University in Shanghai who hosted our visit in May 2014, and City University of Hong Kong faculty Graeme Lang and Robert Gibson for hosting our talks at City University, also in May 2014. Special thanks to Helena Kolenda and Li Ling from the Henry Luce Foundation and Anthony Ng and Belinda Winterbourne from the WYNG Foundation. Bill Barron and the late Anthony Hedley were special friends and long-time collaborators. Our families are very special to us and are always supportive; our love and thanks to Marge Pearson, Casey Pearson-Gottlieb, Andrea Pearson-Gottlieb, and Rebecca Wong and Nathan Ng.

Maps

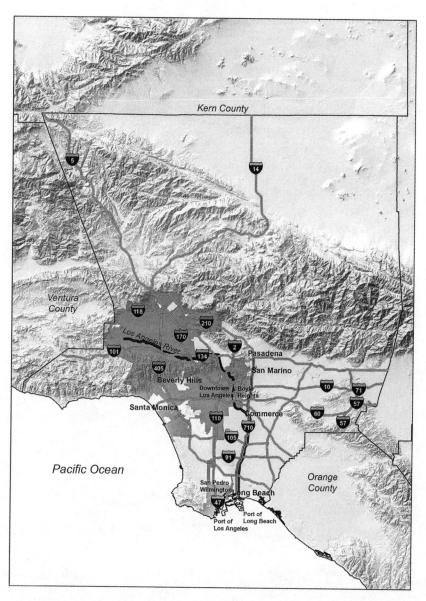

Los Angeles

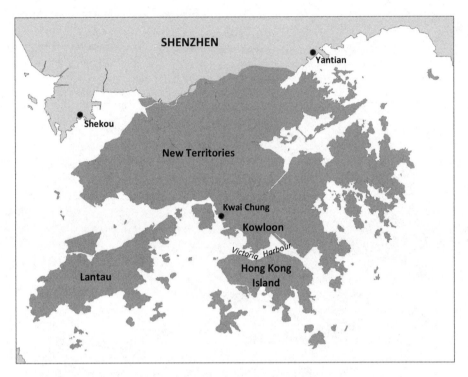

Hong Kong

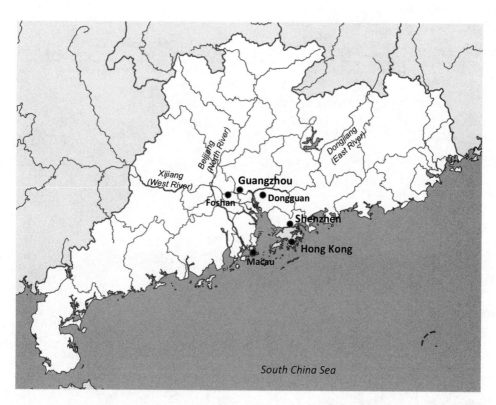

Guangdong Province

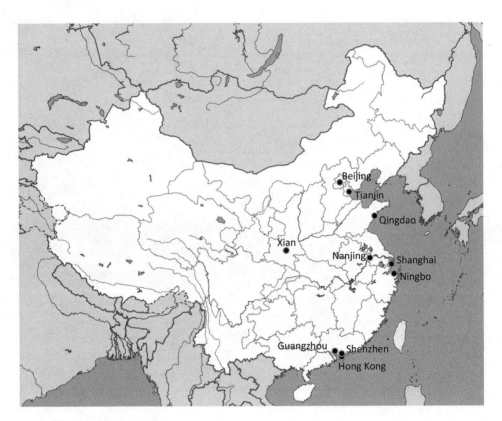

China

1 Moving Forward Together: An Introduction

Global City Connections

In October 2010, six hundred community activists, policymakers, academics, researchers, and participants from outside the United States gathered for a two-day conference in the Los Angeles port community of Wilmington. The theme of the conference was the global trade and goods movement system and its impacts. A few weeks before the Los Angeles gathering, a group of researchers, policy analysts, academics, and shippers had come together in Hong Kong to talk about some of the same issues facing the residents of Hong Kong and port cities in China such as Shanghai and Shenzhen. A speaker at the Los Angeles meeting, Veronica Booth, who was part of the local think tank that had pulled together the Hong Kong event, reported on a crucial outcome of a lengthy engagement process with the shipping industry in Hong Kong: a voluntary agreement called the Fair Winds Charter that was about to be endorsed by seventeen shipping companies. The agreement focused on reducing ships' emissions by substituting low-sulfur marine distillate fuels for the highly polluting bunker fuel source that most shipping lines still used due to its lower costs. Booth told the LA conference of the importance for the Hong Kong groups of Los Angeles's long history of community action and policy change. This included a Clean Air Action Plan that the Hong Kong think tank and academic participants had referenced in their discussions with the shipping companies. The title of the Los Angeles conference, "Moving Forward Together," was in this respect an apt name for the emerging linkages among community players, researchers, and policy advocates who were seeking to change conditions in their communities and to insert themselves as players in a global system that often ignored community voices and environmental concerns.

Figure 1.1
Veronica Booth at the Moving Forward Together conference, October 23, 2010.
Source: THE Impact Project.

Impacts from the movement of goods through the ports to their final destinations are among the most challenging environmental issues in Los Angeles, Hong Kong, and other port cities and regions in the United States and mainland China where the transport of goods over long distances takes place. Two of the groups most directly engaged in "Moving Forward Together" and the Fair Winds Charter initiative were the Los Angeles-based Urban & Environmental Policy Institute (UEPI) and the Hong Kong-based think tank Civic Exchange, groups with which the two of us had been connected for several decades.

A year after the "Moving Forward Together" conference, a December 2011 meeting was held at the Civic Exchange office in Hong Kong. The meeting was organized to discuss potential collaboration between Civic Exchange and UEPI, which is housed at Occidental College in Los Angeles. UEPI and Civic Exchange differ in their history and organizational structure: UEPI is based at an institution of higher education but functions as its own independent center for research and action, while Civic Exchange is an independent policy research think tank with informal connections to higher education institutions in Hong Kong. Both, however, shared the goal of engaging in research and action to bring about policy change. The port and goods movement issues were seen as one key area for collaboration since the changes taking place in both Los Angeles and Hong Kong were directly connected. Parallel connections with other urban environmental issues, such as air quality and transportation, were also noted. The meeting concluded with both groups expressing interest in further exploring a collaborative relationship.

The following year, several exchanges took place between UEPI and Civic Exchange. One involved presentations at Occidental College by Simon Ng in November 2012 that Robert Gottlieb had arranged. Simon talked about Civic Exchange's work related to air quality issues and its research regarding the built environment and public and open space in Hong Kong. After the presentations, Bob and Simon met to discuss additional collaborations. The idea of writing a book was raised as an opportunity to expand the research on the urban environment in Los Angeles and Hong Kong. Such a book could compare and contrast Los Angeles's and Hong Kong's history and experience with such issues as the ports and goods movement, air quality, water quality, water supply, transportation, land use, public space, open space, and the food system. We asked ourselves: What lessons could be derived from those experiences that could inform the policy debates as well as the role of social movements seeking to influence those debates? We also recognized that such a book would need to identify and analyze how urban environmental issues were being addressed in mainland China and why China's connection to Hong Kong and Los Angeles needed to be a critical part of such a research project.

As we began work on the manuscript, we were often asked: "Why Los Angeles and Hong Kong?" As one colleague put it, "It doesn't seem an obvious type of connection." There was a dearth of literature about global cities that compared Los Angeles and Hong Kong. Some colleagues wondered about the China connection as well; perhaps, some said, we should consider a book primarily about China and the vast changes it had made in the transition to a market economy and from a rural to an urban society and the widely discussed environmental challenges it now experienced. Such questions helped us better target what we needed to address and how we wanted to frame the book.

We have focused on how these two cities (and their larger regions, Southern California and the Pearl River Delta) emerged in the latter half of the twentieth century as *global* cities that have played a critical role in the flow of goods, people, and capital; in their patterns of production and consumption; and in the urban environmental issues that have taken root as a result of the changes they have experienced. China looms large in that context, due to its all-encompassing history and connection to Hong Kong, but also (albeit less visibly) due to the ways it has become intertwined with Los Angeles as well. China is becoming a country of megacities, several of which have also become global cities, such as Shenzhen and Guangzhou, that are integral to our discussion of a Los Angeles–Hong Kong–China connection.

This book, documenting the changing urban environments in Los Angeles, Hong Kong, and China, is a byproduct of those discussions and our eventual collaboration. It is a book about contrasting development patterns that have begun to converge. And it is a book that identifies how urban environmental issues have risen to the top of the policy agendas of each of these places, where and how change is possible, and how changes are constrained.

Figure 1.2
Simon Ng hosting a meeting with representatives from the shipping industry and the Hong Kong government, July 2013. Source: Civic Exchange.

World Cities/Global Regions

Fifty years ago, the late British geographer Peter Hall was among the first researchers to write about the concept of world cities, defining them as centers of political and economic power. Hall referenced their role as ports in global trade systems and, unlike some later researchers, included the urban environment in his elaboration of the definition of a world city.[1] The discussion of world cities was expanded two decades later in the 1980s, at the same time that Los Angeles and Hong Kong and several urban regions in China began to restructure and reposition themselves within a global economy.

The world city hypothesis was further elaborated in a 1986 journal article by University of California at Los Angeles (UCLA) Urban Planning Professor John Friedmann that built on an earlier paper he had coauthored with Goetz Wolff. Along with other academic researchers who were part of what was called the Los Angeles School, Friedmann identified several components of the world city phenomenon in a period of global economic and social transformation. These included "the form and extent of a city's integration with the world economy, and the functions assigned to the city in the new spatial division of labor, [which would] be decisive for any structural changes occurring within it." Friedmann's world city hypothesis further referenced a core financial role within the global economy; lengthy, multilocation supply chains connected to global systems of production; and an ever-widening income divide among residents, linked to the expansion of immigration across and within borders.[2]

While the nation-state was locked into an increasingly interdependent global world, it was becoming hard-pressed to contend with the social and environmental consequences related to these changes. This was particularly due to the rise of neoliberal leaders such as Margaret Thatcher and Ronald Reagan who sought to diminish the role of the social welfare functions of the state and to promote the primacy of the market. "There is no such thing as society," Thatcher famously declared, "only individuals and families" (and its corollary, for her, of free markets). The neoliberals dismissed concerns about the huge income disparities, environmental hazards, and urban demographic shifts associated with the expansion of the world cities and the supremacy of the forces of global capital freed from interventions by the state.[3]

Who were the world cities? Much of the literature and the subsequent discussions about global cities focused initially on such metropolitan areas as New York, London, Zurich, and Tokyo. They referenced an evolving spatial division of labor between core and periphery based on financial and production (e.g., supply chain) relationships. Los Angeles was designated a world city as much for its role as a center for information and communication and cross-border migrations as for any dominant financial role. Mostly absent from the discussions in the 1980s and early 1990s were some of the megacities in the developing world. Singapore and São Paulo were included as primary centers functioning in "semiperipheral countries," as distinct from the global cities in the "core countries" (itself a World Bank designation for advanced industrial market economies). Hong Kong tended to be listed as a secondary, semiperipheral city, despite its role in global trade and finance—particularly for the Asia-Pacific region—and its self-characterization as "Asia's World City." China's cities were largely absent from these categorizations since China (along with India) was presumed to be "only weakly integrated with the world market economy."[4]

Yet it was during the 1980s that China entered the world market economy and Hong Kong secured its role as one of the "highly concentrated command points in the organization of the world economy" which included its own shift from manufacturing to finance and specialized service firms.[5] China's major shift toward export-oriented production and the rapid migration of a rural labor force into China's new megacities, which literally sprang up in just a few years, were changes facilitated in part by Hong Kong. Los Angeles would also play a major role as a destination point for China's new export products and market economy restructuring. China's rural-migrant labor force represented a massive source of cheap labor for the new export-oriented industries, many of which were initially located in Guangdong Province and other coastal areas and often financed by Hong Kong capital. As a result, China began to position itself as the "world's factory" for the production of goods, including those destined for the United States through the ports of Los Angeles and Long Beach, California. By 2001, when China finally joined the World Trade Organization (WTO) after seeking membership for sixteen years, another global economy axis had begun to take shape, stretching from mainland China through Hong Kong to the Los Angeles region. The combination of major financial and business centers and locations where the production, export, and distribution of goods took place along with the change in migration established distinctive global city connections. At the same time, it was in those locations that the social, economic, and urban environmental impacts on the global cities loomed largest.

Development and Change: Los Angeles

Urban development patterns can serve as a proxy for how the issues of the urban environment take root in cities and regions. Los Angeles and Hong Kong illustrate that point, as do many of the megacities in China.

Los Angeles had long been seen as the quintessential horizontal city, the capital of sprawl. Its ever-expanding developments along its shifting urban edges were facilitated by the interurban rail and imported water systems, and then were further extended by the automobile and its concrete roadways that shaped the patterns of development. Los Angeles, in British architect Reyner Banham's famous characterization, became known as "autopia," "a single comprehensible place ... a complete way of life."[6]

This car-oriented suburban or horizontal pattern of development also helped earn Los Angeles the dual reputation as the twentieth-century model of a new type of urban region in the United States and the country's most polluted urban environment. Other US cities followed the horizontal expansion model, influenced by car-oriented suburbanization and urban-edge developments. At the same time, Los Angeles's poor air quality, revealed by its constant haze and smog alerts, became that city's most recognizable environmental problem.

Water and fire became additional environmental signposts. Los Angeles gained notoriety as a "water imperialist" due to the city's (and the region's) search for distant imported water supplies; first from the Owens Valley 230 miles to the northeast of the city, then further away and crossing state lines from the Colorado River, and finally from Northern California. The availability of imported water became the added precondition for Los Angeles's horizontal and regional expansion, allowing areas without their own local water supply to annex to the city and subsequently to the regional imported water supplier, the Metropolitan Water District of Southern California.[7]

The horizontal expansion stretched north and east toward the mountains and the high desert and south beyond the county line to reconfigured agricultural land and shrinking open spaces. Seeking a connection to nature as integral to the new land developments, Southern Californians extended new urban-wildland interface boundaries, creating numerous opportunities for fire, flooding, and debris spills to slide down denuded mountain pathways. John McPhee's essay "Los Angeles against the Mountains," describing debris spills in the San Gabriel Mountains and their foothill communities, complemented author Mike Davis's scathing diatribe to "let Malibu burn" from major fires that continually erupted due to that wealthy coastal community's own inappropriate developments cutting into adjacent forest.[8] In a region prone to fire and flooding, horizontal expansion had created an environmentally combustible and hazardous mix of development that worked against rather than with the region's ecology of water and fire. Perhaps most notable in this mix was the fate of the fifty-one-mile Los Angeles River, which was transformed during the 1940s and 1950s from a semidry, meandering waterway, with ever-shifting pathways and occasionally torrential flows of water, into a straightened, off-limits, barbed-wired concrete channel with high walls and assorted trash.

Even as new subdivisions in their quest for pastoral settings sprang up at the urban edge along this shifting urban-wildland interface, Los Angeles found itself among the most park-poor cities in the country, especially in the low-income neighborhoods of South, Central, and East Los Angeles. Its land dedicated to the car and to parking constituted nearly a third of the land area in the city and the county, even causing some policy bodies to declare that streets dedicated primarily if not exclusively to cars could be considered "open space." And Los Angeles, once the largest agricultural producer in the United States, not only saw much of its farm land taken over by new developments at its edge (and within its core), but it also began to encounter enormous problems of food insecurity, including in many of those same neighborhoods without parks and open space (but filled with freeways and truck traffic). This growing problem of food insecurity was related to lack of access to fresh and healthy foods, exacerbated by supermarkets' abandonment of urban-core communities to move to suburban sites, with their ever greater floor space and large parking lots. Like the residential developments, supermarkets followed the horizontal pathways set by the cars and the freeways.

The horizontal pattern of development, fraught with environmental outcomes, was led by Los Angeles's leading industry group and boosters, its land developers. Los Angeles also had a manufacturing base that had developed in the 1920s and expanded during World War II that included aerospace, rubber, automobile, and textiles. The discovery of oil in the 1920s further represented a major factor in the city's and region's development. Many of these industrial land uses left a toxic legacy in Los Angeles. Petroleum extraction, refining, power generation, and much of the manufacturing and distribution infrastructure contributed to groundwater pollution, land contamination, and hazardous pollutants emitted into the ambient environment.

The development of the entertainment industry, so associated with Los Angeles, was clearly a major economic (and cultural) force in helping shape the region. But it was the land development interests—including real estate developers, the banks and other financial players, the top law firms, and, especially in the earlier half of the twentieth century, newspaper publishers and key policy officials who traded in real estate—that primarily constituted the major decision makers in the region, influencing where and how development would take place.

Many of these land development interests were preoccupied with regional policies rather than global economic considerations. A cross section of those regional players was represented by leading behind-the-scenes power brokers, such as the Committee of 25 group, which consisted of financial, law, insurance, and newspaper executives and was prominent from the 1950s into the late 1960s. During that period, Los Angeles's global profile was primarily associated with the entertainment industry. Hollywood films and television productions had long sought a global profile and by the 1950s had come to dominate international markets while exporting American cultural images, often of suburban, middle-class lifestyles. Other Los Angeles-based industries which had an international profile, such as aerospace, were already beginning to contract by the 1970s, a process that included the departure of the headquarters and operations of key companies and the shuttering of plants within the region. Los Angeles retained some manufacturing but experienced substantial deindustrialization. Cities like Los Angeles that experienced both deindustrialization and sprawl often found themselves "at the mercy of finance capital, real estate developers, speculators, and office builders," argues David Harvey.[9]

Yet as early as the 1970s and 1980s, Los Angeles began to take on more of a global profile as new immigration from Mexico, Central America, and a number of Asian countries including China led to rapid demographic changes. Los Angeles came to be recognized as a truly global city because of its population, as well as the rise of immigrant neighborhoods, international cuisine, and, as urban planner James Rojas has argued, the reurbanization of areas in the city where immigrants had grouped and settled.[10] Even some of the areas once considered "suburbs" took on more of an urban identity due to this demographic shift. Though Los Angeles continued to expand horizontally, it also became denser as a region.

As these demographic shifts continued to occur into the 1980s and 1990s, the region began to position itself, through its rapidly modernizing twin ports of Los Angeles and Long Beach, to become a point of destination for an expanding global economy. Policymakers and public officials led by long-time mayor Tom Bradley (1973–1993) promoted global trade and goods movement as essential economic drivers for the region. Major infrastructure programs were pursued at the port, as well as connected projects such as the Alameda Corridor, which facilitated rail transport from the port to complement the hundreds and thousands of trucks departing from the ports.

This new global profile for Los Angeles created by the huge migrations and a logistics-centered economy came with economic and environmental impacts. Economically, a labor pool of undocumented migrants exacerbated the income divide and related forms of exploitation, while air pollution and other environmental impacts were caused by the ships, the ports, the fumigation facilities, and the trucks, railroads, and warehouses.

The election of Mayor Antonio Villaraigosa in 2005, made possible in part by more than a decade of community and environmental activism, promised a new era of environmental change. Villaraigosa spoke of Los Angeles becoming the "greenest and cleanest big city in America" and wished to facilitate a shift toward "elegant density" and away from the horizontal development model.[11] Various community and environmental networks and organizations developed their own agendas for change, some but not all of which was embraced by the mayor and other policymakers. Initiatives involving parks, trees, and open space, a plan to increase alternative energy sources, a Clean Air Action Plan at the port and a shift to cleaner-burning fuel for trucks, and the development of a Food Policy Council were among the changes proposed and implemented. But the changes in this period were uneven and sometimes counter to an avowed "greening" agenda, and Villaraigosa left office somewhat of a disappointment to his one-time community and environmental supporters.

Eric Garcetti, who followed Villaraigosa as Los Angeles's mayor in 2013, also pledged to continue and expand an environmental agenda for the city, with a focus on such goals as "complete streets" (more bikeable and walkable places), greater attention to public space, and alternative food opportunities. But many of the problems endemic to the city and the region, such as the enormous income disparities that were also reflected in widespread urban environmental disparities and a growing homeless population, came to be recognized as requiring a more radical reenvisioning and restructuring agenda in order to make Los Angeles, as Villaraigosa had liked to proclaim, the "greenest city in the world." Today, Los Angeles remains the region with the most pollution, the longest car commutes and most congested traffic, among the greatest income disparities, and a reputation as the most environmentally challenged city in the United States. To create a more livable and just Los Angeles continues to require more of a transformation in policymaking, a major change in the public discourse, and grassroots organizing to develop constituencies prepared to challenge the still dominant forces that shape the region. These are changes that would be at once local and global, and they would position Los Angeles's own social movements as place-based and part of an emerging alternative global politics.

Development and Change: Hong Kong

If Los Angeles became the oft-cited horizontal city even as it sought to become denser and greener, Hong Kong emerged as the quintessential dense, vertical city, while continually remaking and extending its complex and varied relationship with mainland China, especially the industrial and urbanizing centers in Guangdong Province.

Hong Kong's early history is bound up with its role as a British colony and its connecting role with China for the trafficking of goods, people, and money. This history dated back to its establishment as a British colony in 1842 and continued through much of the nineteenth and twentieth centuries as the colony grew and its role expanded. Originally a fishing village with no natural resources but a sheltered deep-water harbor, Hong Kong traditionally served as an entrepôt with China. There was also a vibrant local manufacturing industry, supported by a quarter of Hong Kong's workforce in 1931, but its contribution to the economy was often downplayed by the colonial government.[12] The situation changed drastically in 1951 when the

United Nations imposed a trade embargo on China as a result of China's entry into the Korean War, and Hong Kong's position as an entrepôt took an immediate blow. In response to the changing circumstances, Hong Kong laid out plans to expand and further develop its industrial base. Economic transformation was in full swing, supported by the influx of immigrants from Guangzhou and Shanghai, bringing with them capital, entrepreneurship, and labor supply. During this period of industrial takeoff, Hong Kong became the world's major producer and exporter of textiles and garments, toys, watches, wigs, and high-fashion apparel. During the 1960s and 1970s, export trade continued to grow remarkably, while the volume and value of reexport trade diminished further. Hong Kong was competing with the likes of Singapore, South Korea, and Taiwan—collectively called "Asia's four little dragons"—as the newly industrialized phenomena in East Asia. Analyst Y. C. Jao echoed that "Hong Kong's postwar transformation from an entrepôt into an industrial economy based largely on the export of labor-intensive manufactured goods ranks as one of the most spectacular success stories in recent history."[13]

As Hong Kong's transformation into a manufacturing center gathered pace, urban development also expanded amid an exploding population, rising demand for housing and transport, and the underlying thirst for land. By the onset of World War II, Hong Kong had a population of about one million. After the war, population shrank to 600,000, but within five years it grew to two million. By the end of 1959, the number passed the three-million mark.

A major fire in the temporary squatter area in Shek Kip Mei in 1953 which left tens of thousands of people homeless, coupled with the pressing need for housing from a booming population, pushed the colonial government to roll out a public housing policy with the objective of providing affordable housing for its low-income residents, including the new migrants from China. New public rental housing estates made up of multistory blocks were built on the outskirts of the old urban areas, and later in designated areas further away. The satellite town model, which was later turned into a much more comprehensive program to develop new towns, was a housing-led initiative that diverted forces of urban growth into pockets of undeveloped land outside the urban core—mainly the northern shore of Hong Kong Island, and Kowloon Peninsula south of Boundary Street. Up to the end of the 1970s, over two million people, or roughly 40 percent of the population, were beneficiaries of the government's public housing program.[14]

The transport sector also experienced tremendous growth and attracted substantial investment during the period of economic takeoff and population boom. A number of consultancy studies on long-term transport need and strategy were commissioned by the government in the 1960s, leading to major plans for road improvement, highway construction, public transport expansion, and the building of an underground railway system in Hong Kong. This was a golden period in the history of Hong Kong's transport development, when the foundation of an efficient, affordable, and safe transport system was put in place.

While Hong Kong was slowly emerging as a major manufacturing center, China was facing its own period of turbulence that culminated in the Cultural Revolution of the mid and late 1960s, which included a strong antiurban bias. By the late 1970s, after Mao's death and the full ascendance of Deng Xiaoping in China and the huge economic and geographic shifts toward market-based capitalism and rapid urban expansion that he launched, Hong Kong stood poised to take full advantage of the changes that were going to occur north of its border, drawing upon its ethnic connection and geographical proximity to mainland China.

China's market turn in 1979 marked a new era in the country's economic strategy, moving away from that of import substitution and self-reliance to an economy that encouraged foreign trade and foreign investment. Guangdong and Fujian Provinces were granted special autonomy in handling economic activities with the outside world. In 1980, four special economic zones (SEZs) were established—Shenzhen, Zhuhai, and Shantou in Guangdong, and Xiamen in Fujian—to spearhead external economic cooperation and technical exchange, and to attract foreign capital for producing export-oriented goods. The selection of the four SEZs was based on a number of factors: Shantou and Xiamen are home for many overseas Chinese, whereas Shenzhen and Zhuhai are neighbors to Hong Kong and Macau. In particular, Shenzhen's proximity to Hong Kong, with its capital, know-how, and infrastructure (such as port facilities), was a huge advantage and integral to its early development.

With the changes in China, Hong Kong's economy also changed as it transformed from a major manufacturing center to one that financed and helped facilitate the expansion of manufacturing in Guangdong Province.

Increasingly, Hong Kong became more of a service hub for an industrialized Guangdong, providing such services as insurance, transportation, banking, finance, design, and marketing. Hong Kong's financial role had become central to the burgeoning flow of funds throughout the Asia-Pacific region. The port continued to expand as a transshipment center for the region, despite the diminished role of Hong Kong's own manufacturing sector, and became the busiest container port in the world from 1987 to 1989, and again from 1992 to 1997 and from 1999 to 2004.

Hong Kong's population increased by about one million each decade from 1960 to 1980, and reached 6.6 million at the turn of the century and 7.3 million by 2016. To relieve pressure on the developed urban areas, more new towns in the New Territories began to be planned by the 1980s, with the objective of building public housing estates with less density and better amenities, as space was less of a constraint there. Nevertheless, Hong Kong continued to maintain a strong commercial center in the central business districts. Dispersion of population to new towns and other areas outside the urban core led to a swelling of inbound commuter journeys to the business districts during the morning peak period and likewise a large volume of outbound journeys spreading across the evening rush hours. The need to relieve traffic congestion and to maintain efficient movements of people triggered a continuing wave of new transport projects—to expand the highway system, to add new lines to the rail network, and to enhance public transport services.

The success of Hong Kong's public transport system—featuring mainly rail and buses plus other complementary modes such as minibuses, taxis, ferries, and trams—is to a certain extent sustained by the city's high-density development, with its concentration of people and activities. A compact and vertical urban morphology, however, is the root cause of some of the major environmental challenges in Hong Kong. Urban main roads with tall buildings lining up on both sides, for example, create urban street canyons with very poor wind circulation and ventilation. Air pollutants such as nitrogen oxides and particulate matter emitted from road traffic are then trapped at the roadside in high concentration, putting the health of pedestrians at risk. While Hong Kong has been able to embark on an ambitious plan of rail development and public transport improvement in the last few decades, it still suffers badly from traffic congestion, inhospitable streets for pedestrians, and a general lack of open space for people. Hong Kong also became a huge consumer and importer of energy, further contributing to air pollution and carbon emissions.[15] Hong Kong's air pollution problems today have even surpassed those of Los Angeles, and, like Los Angeles, its water supply problems became magnified due to its dependence on imported water. As early as the 1960s, Hong Kong experienced massive cutbacks and rationing of water, subsequently forcing it to rely on an imported water supply from Guangdong's Pearl River Basin for as much as 70 to 80 percent of its water.

The Hong Kong government's first attempt to study the need for environmental control was to commission a consultancy study in 1974. The Environment Branch was established in 1975, and upon completion of the study in 1977, the Environmental Protection Unit was set up within the Environment Branch to formulate environmental protection policy and legislation. During the 1980s, a systematic approach was taken to handle environmental issues, and the Environmental Protection Department was established in 1986, putting almost all pollution control and prevention works under one roof. In 1989, the White Paper on pollution further explained the scale of the environmental challenge, and put forward a ten-year plan to tackle different types of pollution.[16]

Despite the plan, progress to improve the environment has been slow, and in some aspects stalled, much to the frustration of academics, scientists, environmentalists, some businesses, and the general public. The situation remained very much the same after the 1997 departure of the British, the creation of Hong Kong's Special Administrative Region (HKSAR), and the establishment of the one country, two systems framework (with China overseeing and placing political limits on Hong Kong's role). Since 1997, the frequent restructuring of the policy bureau responsible for environmental protection within the Hong Kong government disrupted any significant environmental policy development. A key barrier to effective environmental governance continued to reside with the decision makers at the very top. Environmental protection was never given top policy priority in the colonial government and was largely overlooked by the first two chief executives of the HKSAR in favor of economic development. The lack of leadership and interest filtered all the way down to the different levels of the civil service. While the administration of Chief Executive Leung Chun-ying (2012–) was criticized for its economic policies and failure to address the growing income divide as well as its resistance to greater democracy in the political sphere, it did achieve some prominence in the area of the environment. This was largely due to the appointment of environmental leaders Wong Kam-sing and Christine Loh as secretary and undersecretary of the Environment Bureau. Through Wong's and Loh's roles, the government put more emphasis on environmental issues, notably on reducing air pollution and protecting public health.

Hong Kong's social movements have also contributed to the greater focus on the environment as well as a range of other social issues, notably democracy and transparency as witnessed by the 2014 Occupy Central events. Researchers and academic participants have played a role in raising greater awareness of the scale of environmental problems, including air pollution, lack of open space, and food safety and food system problems, among others. For Hong Kong, the tension between economic development, real estate forces, and the environment remains prominent, even as the powerful issues of climate change and resource dependence loom over such policy debates.

The Connection to China

While Los Angeles and Hong Kong experienced major changes during the 1980s and 1990s, the urban, industrial, and environmental changes taking place in China during this period were even more extensive. A huge migration from rural to urban and a multifaceted pattern of urban and periurban growth complemented the explosion in industrial development, including the production of manufactured goods for export. Massive infrastructure projects were designed and completed that were breathtaking in their reach and in the disruptions of existing settlements and the environmental consequences that resulted from their implementation. These changes further had a palpable effect on agriculture and food production, resulting in an enormous loss of farmland and the uprooting of traditional agricultural practices and widespread adoption of industrial agriculture methods. Massive water and energy projects helped fuel the changes in production and agriculture, while the use of pesticides skyrocketed to increase productive capacities but also generated enormous health and environmental impacts. In the cities, whose population continued to double and then triple in just a few years, the car became an increasingly popular symbol of wealth status and an extension of the drive toward modernization while, at the same time, China's cities witnessed an explosion in rail development. And with the focus on export production (and subsequently of beefing up domestic consumption) as central to economic policy, several of China's ports quickly climbed to the very top rankings in the world as measured by the number of container boxes being loaded and unloaded and with Los Angeles a major destination point for the massive increase in US consumption of those cheap Chinese goods.[17]

Table 1.1
Top 10 container ports of the world, 2001 and 2015

	2001	2015
1	Hong Kong	Shanghai
2	Singapore	Singapore
3	Busan	Shenzhen
4	Kaohsiung	Ningbo-Zhoushan
5	Shanghai	Hong Kong
6	Rotterdam	Busan
7	Los Angeles	Guangzhou
8	Shenzhen	Qingdao
9	Hamburg	Dubai
10	Long Beach	Tianjin

The social and environmental impacts from these changes were immediate and unprecedented in the size and speed in which they occurred. While millions of rural villagers began to achieve greater food security and reduced poverty, gaps between rich and poor simultaneously increased at a far greater rate than even in Los Angeles and Hong Kong where wealth disparities had become such an important issue. The environmental outcomes, increasingly visible both inside and outside China, were both an embarrassment to the State, a call to action, and a major health and economic concern. These changes were particularly pronounced in the Pearl River Delta area which linked Hong Kong to Guangdong Province and its major cities and megacities such as Guangzhou, Foshan, Dongguan, and Shenzhen.

The severity of the environmental issues in China have had a direct impact on people's health and the conditions of daily life, particularly in such areas as air and water quality, and food safety and food production. Climate change has also become increasingly prominent, as China's thirst for energy resources, its massive use of coal, and its status as the leading country in the world of carbon emissions (by volume, though not per capita), has indicated how quickly China's environmental footprint has changed, itself a byproduct of its rush toward marketization and modernization. China has truly become a global player in so many different arenas, as cities like Shanghai, Beijing, Guangzhou, and even Shenzhen have achieved their own global city status.

As China has changed, so too has its connection to Hong Kong—and to Los Angeles. Hong Kong, of course, has always been connected to China, from the British seizure in 1842 to the 1997 handover, the establishment and implementation of the Basic Law regarding Hong Kong's governance, and the 50 year one country, two systems framework, with all its unsettled outcomes. Chinese migration to Hong Kong has a long history from the very origins of its colonial past, though the uptick of Chinese migration in recent years to Hong Kong (and to Los Angeles) has been an even greater influence regarding the class and demographic nature of those cities. For Hong Kong, mainland China events immediately resonate, whether the 1989 Tiananmen events (where one million people demonstrated in Hong Kong—its largest demonstration ever—in support of the Tiananmen demonstrators) or the SARS outbreak in Guangdong in 2003 that also directly affected Hong Kong. Today, migration from mainland China—and its huge tourist population—draws both rich and poor, even as Hong Kong's wealthiest individuals and companies have developed their own mainland China investment portfolios and political ties. Hong Kong is informally— and directly—connected to the Pearl River Delta, emblematic for example, of the Hong Kong to Guangzhou high speed train now under construction. Yet despite this apparent seamless border and complex, intertwined relationship, Hong Kong and its counterparts in China, such as neighboring Shenzhen, represent distinctive places—and challenges—in confronting their own environmental, social, and community issues.

The China (and Hong Kong) connection to Los Angeles also takes multiple forms.[18] It includes the greatly expanded investments by Chinese companies (some state-owned) in Los Angeles real estate and a wide range of businesses. It includes China achieving its world's factory designation in the 1980s and 1990s that ran through Hong Kong and the ports of Los

Angeles and Long Beach. It includes Chinese migration to Los Angeles and California that has a nearly 150 year history. Los Angeles has become a leading immigrant city—upward of 35% of its population are immigrants—and its Asian immigrant population is a key part of the city's demographic change. Its Chinese population numbers more than 500,000 and whole new areas in Southern California, from Monterey Park to San Marino, have become the new destination points for migrants and tourists alike. Monterey Park, in fact, has the largest percentage of Asian residents (66%, most of whom are Chinese) of any city in the United States. The Chinese tourist population has also skyrocketed, from 158,000 visitors in 2009 to 779,000 in 2015, becoming in the process the second biggest source of international visitors (after Mexico) to Los Angeles.[19]

Today, the relationships between China, Hong Kong, and Los Angeles have undergone both subtle and profound changes. They have contributed to the reconfiguration of global city networks. And they have led to China's emergence as a leading global economic player and as a primary contributor (along with Hong Kong and Los Angeles) to the environmental impacts that also define the contemporary global city.

Comparisons and Contrasts

Whether in relation to migrations, economic ties, environmental impacts, political interdependencies, the movement of goods, or their global city roles, Los Angeles and Hong Kong and their megacity counterparts in China are connected. The flow of goods and its environmental impacts provide a starting point. The Los Angeles-Hong Kong-China ties around port, global trade, manufacturing shifts, and the movement of goods have deepened the connections, expanding them to a point where a change in output or other economic (or political) variables can influence all three areas. When strikes shut down the ports in Los Angeles (as they did in 2002 and again in 2012) the impacts could be felt by the retailers and goods movement industries in the United States and the suppliers and manufacturers and shippers and port operations in Hong Kong and China. Similarly, when China began to put in place new policies designed to expand its domestic markets for the goods manufactured in China that had been primarily produced for export (including through the ports of Hong Kong or Shenzhen to the Ports of Los Angeles and Long Beach), the LA ports and goods movement industries sought to calibrate what that might mean for their own operations. And when China's economy finally began to slow down in 2014–2015, that change also impacted Los Angeles and Hong Kong, not just in relation to their ports but also with respect to migration and investment.

At the same time, the emerging community and environmental advocacy around port and goods movement impacts in Southern California and the policy debates and changes that resulted have been followed attentively in Hong Kong and China. The potential for connections between community and environmental social movement actors (at least at the level of what changes are being advocated or implemented) have become more apparent, albeit often in complex and less direct ways. Not yet a form of "globalization from below," nevertheless the connections around community and environmental advocacy, whether through NGO relationships, shared research and policy innovations, or budding ties among activists, have begun to emerge. Moreover, the differences in the organizational and political roles and influence of social movements, NGOs or other community and environmental actors in Los Angeles, Hong Kong, and China, are also important and noteworthy, creating limits to the opportunities for connections, even as the interest in such connections grows.

The areas of the urban environment discussed in the book—water, air, food, transportation, and public and open space and the built environment—all provide a mix of comparative and contrasting histories and current impacts. Each of the places have experienced major environmental changes in a relatively short period of time: Los Angeles especially after World War II; Hong Kong through the 1950s to the 1970s, and then again in the 1980s and after the transfer from Britain to China in 1997; and China after the introduction of market economies and an export-oriented industrialization and the policies promoting urban expansion since 1979. Some of the environmental problems now endemic to those places such as air pollution, the restructuring of their urban form, water imbalances and quality problems, the transformation of agriculture and food security and food safety concerns provide important lessons for each other about the environmental impacts and the limits and opportunities as well as the strategies for change.

Yet it is important to not overstate the parallels and comparisons. Each of the places—Los Angeles, Hong Kong, or the global cities in China such as

Shanghai and Shenzhen and Guangzhou—are distinctive places, with their own particular histories and changing urban and environmental dynamics. What becomes valuable about identifying those histories and the environmental changes taking place is not to ignore the place-based nature of those histories and changes, but to learn from and appreciate what makes them distinctive as well as connected. It also provides valuable knowledge in a globalized world and the increasingly dominant role of major global cities and regions. These urban places are the primary actors today in influencing and impacting local and global environments. And they are the central locations in identifying the challenges and opportunities and the linkages to bring about environmental change.

Regional Geographies

The regional geographies of these places help situate the issues of the urban environment. The Los Angeles region is at once connected to a regional watershed, a regional airshed, a regional (and global) food system, regional transportation systems, and regional land use patterns. Los Angeles has long experienced boom and bust population cycles, but since its rapid territorial expansion in the 1920s after its first major imported water project came on line, it has witnessed continual population growth. From a population of more than half a million in 1920, the city doubled its population by 1930, continued to grow modestly during the Depression years of the 1930s (thanks in part to immigrations from the depressed regions of the Southwest and Midwestern US), and then saw its population take off after World War II. The city's population reached nearly two million in 1950, three million in 1980, four million today. The city's boundaries stretch outward from the Pacific Ocean to the mountains and beyond to the San Fernando Valley, and to the south through a strip of land to the Los Angeles port area at San Pedro. Los Angeles County includes the City of Los Angeles but also 87 other cities and large unincorporated areas. The county population is more than double the population of the city, having reached nearly ten million today. The larger Los Angeles or Southern California metropolitan region, represented for example by the boundaries of its regional water wholesaler, the Metropolitan Water District of Southern California, extends 5,200 square miles and includes six counties, dozens more cities, and as many as 19 million people.

Geographically, Hong Kong is located immediately south of Guangdong Province, on the eastern side of the Pearl River estuary, with Macau located on the western side, and Guangzhou at the top of the river mouth. This is commonly known as the Pearl River Delta region (PRD), one of the most economically advanced and developed areas in China, and also one of the most populous with over 63 million inhabitants (including Hong Kong's 7.3 million).[20] Unlike Los Angeles, Hong Kong is administratively cut off from its immediate region and hinterland in southern China since the colonial days, and now as the Special Administrative Region of China under the one country, two systems framework. However, Hong Kong is very much part of a regional watershed, a regional airshed, a regional (and global) food system, and increasingly, a regional transportation network.

In China, the uprooting of what had been predominantly a rural society took place after 1978–1979. This is the primary time frame for our discussion of urban environmental change in China, as it increasingly became urban-oriented, with explosive growth and huge migrations to its Tier 1, Tier 2, and Tier 3 cities. These informal classifications of urban areas into different Tiers have become a tool used substantially by economic interests and have included such criteria as population, gross domestic product, housing sales, and income levels, among other factors. The Tier 1 cities (and regions) are still considered the economic drivers linked in part to urban expansion and industrial activity, although central government policies have begun to focus on the development and expansion of Tier 2 cities (including certain provincial capitals) and Tier 3 cities where the rural-urban divide, while changing, is not yet as pronounced.

The Tier 1 cities of Guangzhou and Shenzhen are located within the PRD region where rapid urban growth has been connected to the major manufacturing and export industry centers that have contributed to vast migrations from rural areas and other less developed regions. The fast growing cities such as Dongguan and Foshan in the PRD, while considered to be Tier 3, have nevertheless emerged as prime destinations for new capital investments (including foreign investment) and industrial activity, due to their more recent spurts in population and economic growth. The Beijing and Shanghai regions constitute the two other major urban regional constellations associated with Tier 1 developments. While the book primarily discusses the PRD and South China with their long standing connections to Hong Kong as well as to Los Angeles, the issues of the urban environment and environmental problems more broadly are experienced throughout mainland China, albeit in some places more than others.

These geographies—of Los Angeles, Hong Kong, the PRD, and mainland China—situate the book as a discussion of regions: regional population and economic development centers; regional watersheds and airsheds; regional food systems; and regional land uses. But the book also describes these places as part of a global nexus—global cities and regions that are interconnected—even as those problems of the urban environment impact daily life at the most local level. Those impacts are magnified by the global flows of capital as well as of people via new global migration routes.

The Structure of the Book

The chapters for the book are organized by issue area—the critical urban environmental problems and strategies for change in Los Angeles, Hong Kong, and China. Each chapter provides a discussion of the particular histories and issues for Los Angeles and Hong Kong and how they resonate and how they differ from each other. It discusses some of those same issues for China and the ways that China is connected to Los Angeles and Hong Kong. This chapter (chapter 1) provides the background discussion for the book. Chapter 2 describes the history and the growth of the ports, the supply chain infrastructures, and the logistics industries that constitute the goods movement system. It identifies the community and environmental impacts they have generated as well as the struggles to green the system. Chapter 3 discusses the history and current challenges that air pollution poses. Both Hong Kong and Los Angeles—and many of the cities of China—have had and continue to experience the most polluted and hazardous ambient environments in the world. How they have responded, whether by community mobilization, policy initiatives, or by the changing dynamics around such areas as transportation, energy, and land use, suggest whether and how much of a change toward cleaning the air can be accomplished. Chapter 4 focuses on water, including water supply and water quality, as well as what has been called the water quality-driving-water supply relationship. All three places have utilized and become dependent on massive transfers of water. For China, especially, water quality is a major concern, but it also continues to affect Los Angeles and Hong Kong, with Hong Kong dependent on water from China. Chapter 5 looks at the food system, from seed to table, whether local or global, industrial and/or organic. It focuses on how food is grown and produced, where it is sold, and where and how it is consumed, as well as the efforts to construct alternatives and elevate the diverse cultures around food that can counter an increasingly globalized, industrialized food system. Chapter 6 addresses the multiple urban transportation issues all three places confront, whether the rise of the automobile as a dominant or emerging transportation mode and the non-auto-based systems such as rail and their impacts. Chapter 7 discusses the questions of public and private space, land use, and the built environment, including the diminishing nature of open space. The final chapter, chapter 8, discusses the social movements and policy initiatives for change and the barriers to change. Where and how social movements and policy initiatives can affect change provide lessons not only for Los Angeles and Hong Kong and China, but more broadly for changes in the urban environment in global cities and the other places around the world that they influence.

The book has been organized to identify and evaluate the complex interplay between core and periphery, global city/region and local places, and to elaborate the shared and distinctive experiences of Los Angeles and Hong Kong and their connections to China and its cities. As awareness of these connections increases, it helps identify the need to think and act globally and locally, and to find out where and how environmental change can take place in these stressed and environmentally challenged places.

2   The Global Goods Movement System

A Doll's Journey[1]

Everyday products, whether an Apple computer or an iPhone, a cotton shirt or denim jeans, a clove of garlic or a bottle of wine, dog food or toothpaste, or a doll for the holiday season, are today produced, transported, and sold through an intricate weave of producers, suppliers, shippers, trucks, trains, warehouses, and retail outlets. At each stage of the journey of these products, the different components of a goods movement system has taken root. The major players shaping the economics and structure of a system that connects Los Angeles with Hong Kong and China are led by the ship owners (who transport the goods across borders and oceans) and the large end users (who sell the goods). They have developed a strategic connection to the producers (those who make the goods), the port owners and managers (who bring in and send out the goods), the rail, truck, and warehouse companies (who take the goods to and from the ports, repackage them, and send them to their retail destination), and the consumers who have long been hooked on high-turnover, cheap goods for sale. Hong Kong export trading executive Victor Fung has called the front end of this journey "dispersed [and borderless] manufacturing," while community and environmental groups at the back end see it as creating environmental sacrifice zones.[2]

A China–Hong Kong–Los Angeles nexus has come to play an outsized role in this goods movement system. Take a doll's journey, which is emblematic of how the system operates. It may begin at a factory in Shenzhen where different parts have been assembled from other factories and subcontractors in Guangdong Province in southern China (where 70 percent of all dolls in the world are produced) or from other countries in Asia with even lower labor and manufacturing costs. The factories producing the dolls may be owned or financed by Hong Kong investors or entrepreneurs, such as Early Light Industrial Co., Ltd., controlled by the politically well-connected manufacturer and real estate developer, Hong Kong billionaire Francis Choi, who has called himself "the toy king."[3]

Once assembled, the doll continues its journey from the manufacturing facility to the ports, perhaps by truck or barge to Shenzhen, the third largest container port in the world, or to Hong Kong, the fifth largest in the world. Packed in twenty-foot containers, the dolls, along with the computers, iPads, T-shirts, and hundreds of other goods, are loaded onto giant container vessels—some even four football fields in length—owned by some of the largest ocean vessel companies in the world. Most of the ships still run on bunker fuel, the grade of fuel lower than diesel and more polluting, although some companies like Maersk, signatories to the Fair Winds Charter, have switched to a cleaner-burning fuel since 2011 while berthing in Hong Kong. The doll may take up to a few days to arrive at the ships from the manufacturing plant, depending on the plant's location and the domestic transport or feeder shipping method. Loading and unloading all the containers for the ships that leave Shenzhen, Hong Kong, or Shanghai could then take another couple of days, depending on the sailing schedule for arrival at either the Port of Los Angeles or the adjacent Port of Long Beach. More than 86 percent of all the toys entering the US ports come from China and Hong Kong, with the Los Angeles and Long Beach ports serving as the dominant gateway for the nearly $10 billion worth of toy imports that are then transported throughout the United States.[4]

The unloading process involves a crane that lifts each container off the ship, placing it onto the chassis of a truck. The truck may haul the container (with the dolls inside it) to a rail facility where it is reloaded onto a train, powered by two to four diesel-fueled locomotives. The container from the train may then be moved to a truck that takes it to a huge distribution center where it may be further reloaded to the truck that eventually departs for the big box store where it will be sold.

The retailers, such as Walmart, are key players in the goods movement/ supply chain system. They are deeply connected to the manufacturing and supply chain relationships in China, working with global brands (such as Mattel, in the case of the dolls) and operating through subcontractors and various other Chinese (and Hong Kong) suppliers and middle men. The retailers and brand-name companies have fueled the massive consumption patterns (and continuing search for cheap goods) in the Los Angeles region and throughout the United States that have in turn transformed the United States into a consuming rather than a producing country. The retailers also operate their own fleets of trucks and have relationships, primarily through subcontractors, with the massive warehouses in the inland port areas.

Thousands of trucks, barges, or rail lines now operate along this supply chain from manufacturer to ports and to retail stores and consumers. The huge international container ships, some reaching more than 400 meters long and 54 meters wide, can now carry upward of 19,000 TEUs (the twenty-foot equivalent units), with plans to increase that capacity to 21,000 TEUs in the near future. Warehouses which receive and then reload containers may themselves be as large as two million square feet. Environmental, public health, workplace, and community impacts arise along each of these pathways. Public health concerns arise primarily from the diesel pollution as well as other environmental, community, and labor impacts resulting from this system of producing, moving, and consuming goods along the entire supply chain.

Recognition of these impacts and concerns have led to more than a decade of research, voluntary agreements, regulatory and policy action, and community and environmental mobilization in the Hong Kong, southern China, and Los Angeles regions. These have resulted in some important changes, including new regulatory and policy requirements, legal action, and community agreements. Defenders of the system, such as the shipping lines, manufacturers, port operators, railroad and trucking firms, and the big retailers, have disregarded or minimized its impacts. They have argued that the goods movement system provides multiple advantages: job creation, regional and global economic development, and an expanding flow of inexpensive goods for consumers—such as this doll, another new model for the holiday season, which may sell briskly because of its low price. These differing perspectives have resulted in numerous protracted battles, with some important environmental victories contending with plans for expansion that create further health and environmental impacts.

Even as awareness of the impacts of the goods movement system increases, the expansion continues apace. According to Los Angeles port officials, for example, international trade increased nearly thirty times at US West Coast ports since 1970, with Los Angeles and Long Beach ports far outpacing any US counterpart.[5] At the same time, ports in Hong Kong and China have until recently witnessed phenomenal growth to become among the largest ports in the world. These global trade and goods movement connections have become a crucial part of regional and global economies *and* a major source of regional—and global—pollution. In the twenty-first century, the links between Los Angeles, Hong Kong, and China have made earlier dreams of establishing a Pacific littoral that stretches from LA to Asia more of a reality, while also providing the basis for "a maritime world economy."[6] With those global connections have also come the challenges to local communities, regions, public health, and the environment that had neither been anticipated nor incorporated into the planning process for global trade and goods movement.

The Goods Movement System

Hong Kong
In Chinese, Hong Kong means "fragrant harbor." Today, the port of Hong Kong is one of the busiest in the world, with more than 30,000 oceangoing

Figure 2.1
Cranes and containers at the Port of Los Angeles. Source: THE Impact Project.

vessel calls and about 160,000 river vessel calls each year.[7] Hong Kong handles over 22 million container boxes each year, and 90 percent of Hong Kong's cargo is handled through the port.[8] In terms of their contribution to its economy, the port and related sectors account for about 2.3 percent of Hong Kong's gross domestic product and 2.7 percent of its total workforce.[9] The shipping and port industry has always been regarded as one of the key pillars of Hong Kong's economy. Since its inception as a British colony, the city and its harbor have been seen as indivisible and symbiotic.

From a relatively isolated small fishing village and a refuge for pirates in southern China, Hong Kong emerged as a valuable trade outpost with China after it became a British colony in 1842.[10] Hong Kong's only natural endowment has been its deepwater harbor. Entrepôt trade blossomed due to the harbor and Hong Kong's strategic location. Until the end of the nineteenth century, trade mainly consisted of British exports to China passing through Hong Kong, such as textiles from Britain, cotton from India, and pepper from the Straits. In return, Chinese goods such as tea, silk, and porcelain were shipped out.[11] Trade activities also included what came to be known as the coolie trade, "a traffic in human labor that resembled a commodity trade," as David Meyer put it in his book on the global metropolis. As Meyer noted, huge numbers of coolies went abroad (or came back) through Hong Kong, with the United States, including California, the largest destination. Hong Kong firms in turn became the primary supplier of Chinese goods to the coolie migrants and their families.[12]

By the twentieth century, British dominance in trade began to be gradually reduced, and Hong Kong's entrepôt trade partners became more diversified.[13] Nevertheless, entrepôt trade remained the bloodstream of Hong Kong's economy up to World War II, and Hong Kong reliance on China's trade continued to remain very strong. After World War II, Hong Kong was in an even better position to handle China's trade with other countries, as it became the only remaining foreign concession in China. Postwar recovery was subsequently disrupted by China's entry into the Korean War in the early 1950s as well as a series of trade embargoes and retaliation between China and the western world.[14] But as entrepôt trade declined, Hong Kong sought to expand its domestic industrial base. The foundation for this change had already been laid by the entrepôt economy, including the commercial and financial framework as well as the shipping network and port facilities which allowed Hong Kong to quickly adapt in order to support its domestic industries. The shift to local manufacturing, such as textiles and garments, plastic products, toys, electronics, and watches and clocks, also led to a change in Hong Kong's trade patterns. As trade with China declined during this period, a new "seaward hinterland" identity emerged through new trade partnerships and shipping routes.[15] Export-driven external trade in the 1960s and 1970s clearly replaced reexport as the driver of growth and modernization of the port of Hong Kong.[16]

This same period also marked another phase of port expansion and improvement. Apart from the government adding deepwater mooring buoys and private companies adding their berthing and warehousing facilities, British Colony Governor David Trench appointed a Container Committee in 1966 to consider the need for container handling facilities in Hong Kong. The committee's recommendations were positive, and eventually Kwai Chung was chosen as the site to develop the terminal. The first berth of the Kwai Chung Container Terminals was opened for operation in 1972. By 1976, the six-berth terminal had an annual capacity of up to 1.5 million TEUs. Although a relative latecomer (globally) to containerization, by 1979, 54 percent of the general cargo handled at the port of Hong Kong was containerized.[17]

Hong Kong's port expansion was facilitated by a combination of government subsidies and private investment. The government held title to the land where the different port facilities were developed, and leased the land to various private interests. These private entities in turn came to own the ports and establish the linkages with the shippers and the goods manufacturers and also participated in the management structures that provided oversight of the port operations. The largest of the port operators, Hutchison Port Holdings Limited (HPHL), itself a subsidiary of Hutchison Whampoa controlled by Hong Kong's richest investor, Li Ka-shing, emerged as a port power broker in Hong Kong; it also came to play a major investor role in the eventual development of Shenzhen and several other China ports as well as ports in the Americas, Europe, Australia, the Middle East, and Africa. HPHL further established multiple relationships and partnerships with both private and state-owned port companies in China, and by the 1990s had secured its role as the largest port operator in the world.[18]

It was China's shift to a market-oriented economy in the late 1970s and the establishment of special economic zones in the provinces of Guangdong and Fujian that created another shift in Hong Kong's economic orientation and inevitably affected the fortune of the port sector. The special economic zones set up next door were able to attract foreign investment and establish new bases for industrial production with abundant land and labor and other types of preferential treatment. Hong Kong's proximity to southern China and its already established linkages placed it in a favorable position to support the modernization drive in southern China and to trigger its own economic restructuring, from a manufacturing to a service economy. Spatial relocation of Hong Kong's domestic industries into Guangdong got under way in the 1980s and 1990s, and by the turn of the century there were already over 50,000 Hong Kong-owned factories in Guangdong (a number that continued to increase into the twenty-first century). Despite the relocation, most of the finished or semifinished products made in Guangdong were shipped back to Hong Kong via land-based or river-based transportation for final processing, labeling, packaging, and ultimately for export. Cross-border container trucking was at its peak during the 1990s, with drivers dashing between factory locations in different parts of Guangdong and Hong Kong on a daily basis. Traditional border checkpoints at Man Kam To and Sha Tau Kok were congested all the time, and a new checkpoint at Lok Ma Chau was opened to ease the traffic.[19]

The growth in cross-border road freight traffic was expected, not only because these products were connected with Hong Kong-based investment, but also because there was a lack of port infrastructure and capacity in Guangdong to handle the shipments at that time. During the first two decades of economic liberalization in southern China, the port of Hong Kong was the best choice in the region, given the number of shipping routes available, the frequency of shipping services, the efficient turnaround time at the terminal and even at midstream, and its highly developed ancillary services. Its efficient performance explained a period of high growth in port traffic and throughput in Hong Kong. During the eighteen years from 1987 to 2004, Hong Kong was crowned the busiest container port in the world on fifteen occasions. To handle additional throughput, the Kwai Chung Container Terminals expanded into nearby areas in Tsing Yi and Stonecutters Island. At present, the entire terminal complex has nine container terminals operated by five private companies, with twenty-four berths of about 7,700 meters of deepwater frontage. It covers a total area of about 279 hectares, including container yards and container freight stations, with a total handling capacity of over 20 million TEUs a year.[20]

However, with the construction and expansion of container port facilities in Shenzhen (both in Yantian and Shekou), Guangzhou (including Huangpu and Nansha), Zhuhai (including Gaolan), and other smaller ports in the last fifteen to twenty years, Hong Kong's position as the hub port in southern China has been strongly challenged. Finished products can be easily shipped to Yantian or Shekou via road or rail, and often at a lower terminal handling charge relative to that at Hong Kong. As a consequence, cross-border road freightage and the value of domestic exports (i.e., finished products made in Hong Kong) have declined significantly. Moreover, since 2013, Shenzhen has surpassed Hong Kong in terms of annual container throughput, and in 2015, Hong Kong was also overtaken by Ningbo-Zhoushan, another Chinese port, and dropped to fifth place among the world's largest ports.[21]

Despite the rise of the Chinese ports, Hong Kong continues to be a major global port operator. It has transitioned from a gateway to a transshipment port and now faces challenges as it has inadequate facilities to accommodate the new generation of megaships. The 1997 transition to the one country, two systems framework failed to undercut Hong Kong's historic role due in part to a provision included in the Basic Law that allowed Hong Kong to remain an independent customs territory, underlining its importance to China. In the last few years, with China's own lower rate of growth, Hong Kong's shipment of exportable goods and container throughput have declined as well. Despite its challenges, Hong Kong nevertheless remains a major player in the global goods movement system.[22]

Los Angeles

Today, the Port of Los Angeles remains the busiest container port in the United States, with its neighbor just to its south, the Port of Long Beach, the nation's second busiest. Together the two ports make up the largest port facility complex in the country. In 2015 about 37 percent of the nation's imports came through these two ports, making Los Angeles and Long Beach critical to the system of trade and goods movement traffic in the United States. The ports are also the primary destination of the goods that come from China and Hong Kong into the United States, establishing a new version of the old Pacific trade routes.

The Port of Los Angeles encompasses 7,500 acres along forty-three miles of waterfront, including twenty-seven passenger and cargo terminals,

Figure 2.2
Hong Kong's Kwai Chung Container Terminals with residential buildings in the background. Source: Civic Exchange.

on-dock intermodal facilities, and railyards. The Port of Long Beach resides on 3,200 acres, and comprises twenty-two terminals, ten piers, and eighty cargo berths that handle nearly 5,000 vessel calls a year, as well as on-dock rail facilities, with more than six million TEUs passing through the port. Plans to accommodate the huge megaships are now in place at both ports, with the first of the megaships, one that carried 19,000 TEUs, having arrived at the Port of Los Angeles in December 2015. These huge ships have become partly responsible for increased congestion and long delays that take place at each of the ports' operations. Port officials anticipate that the megaships will continue to increase in size, as shipping companies consolidate and look for greater efficiencies in scale. Those trends represent numerous challenges for port operations, which already include massive tie-ups from unloading to greater environmental and labor impacts at the ports and additional community impacts along the supply chain.[23]

A focus on the Pacific has long been a dominant interest of the various advocates and boosters of the ports of Los Angeles and Long Beach, even in their earliest years as the two port developments took shape. For example, in a 1935 book published by the LA Chamber of Commerce about the history of the Port of Los Angeles, the long-term head of the Los Angeles Harbor Department, Clarence Matson, spoke of the "westward march of empire and civilization which is now reaching its climax on the eastern shores of the Pacific with Los Angeles as its apex."[24]

The development of the Port of Los Angeles was first proposed in the 1890s. It proceeded at a modest pace until new oil fields were discovered in the region in the 1920s. After the Panama Canal had opened in 1914, the port was able to expand by shipping the oil through the canal to East Coast refineries, a necessity since Los Angeles lacked oil storage and refining capacity.[25] Increased agricultural production, including citrus products, also led to the increased movement of goods through the port. But it was after the introduction and widespread adoption of containerization in the late 1950s and early 1960s that the Los Angeles and Long Beach ports began to position themselves as major destination points for international shippers. A greater number of imports could be taken by truck and rail to locations throughout the United States, a process also made possible with the development of the US interstate highway system.[26]

After World War II, US businesses' and government officials' interest in new global trade opportunities and investments, sparked by talk of a new Pacific Rim constellation of players and by US economic and political influences, led to renewed focus on Pacific trade. In the late 1950s and early 1960s, delegations from Los Angeles would continually travel to various Asian destinations, including Hong Kong, Taiwan, Manila, Tokyo, and Singapore, to explore increasing their port's capacity for imports and to look for new trade opportunities for US exports. By that point, the Panama Canal's channels were already too small to handle the larger ships that had been built thanks to the revolution in containerization. The Panama Canal's limitations had helped fuel subsequent expansions to the Los Angeles and Long Beach ports in the 1980s and 1990s, which enabled the two ports to accommodate larger ships and increased traffic. As the ports expanded, Los Angeles mayor Tom Bradley boasted that Los Angeles was becoming the "gateway city for the Pacific Rim."[27]

This growth in international trade from the Pacific exploded even further in the new century with the huge volume of exports entering the United States from China. In response, both the Los Angeles and Long Beach ports embarked on massive expansions of their own facilities. Alongside their continuing competition, the ports began to cooperate in anticipation of the increased trade and the need to accommodate the largest container ships that had to bypass the Panama Canal trade routes due to their size.[28] Periodic efforts were made to anticipate future growth, leading to further changes in port operations and facility developments, with the hope that Los Angeles could become "the trading center of the world," as one official put it.[29]

During this period, new terminals were built on existing vacant land and existing container terminals were redeveloped and expanded. Waterside berths were also deepened and a bridge replacement was built to accommodate more goods movement traffic. To further accommodate the growth of the Port of Los Angeles, the Alameda rail corridor was constructed. This $2.4 billion, twenty-mile rail link, which opened in 2002, went from the port to the huge railyards and intermodal facilities situated in the low-income communities to the south and east of downtown Los Angeles. Plans were also made to expand the Interstate 710 (I-710) freeway, the primary route for the thousands of trucks driving to and from both ports, which passes through the same neighborhoods located next to the large intermodal facilities. At the eastern edge of the Southern California region, in an area known as the Inland Empire (or the Inland Valley, as some residents preferred to call it), massive new warehouses and intermodal facilities were constructed as inland ports where the goods could be repackaged and then transferred to their final destinations. The huge retailers like Walmart, working in tandem with the major rail companies such as BNSF and establishing supplier relationships with the warehouse operators, further extended the goods movement system in Southern California and around the United States, where other inland ports took root. "This is a big-box market," one inland port official characterized the rise of the inland ports.[30]

These developments led policymakers to talk of a logistics industry revolution in the Los Angeles region, while its land use and environmental impacts were largely ignored. Supporters of this goods movement sector characterized it as a win-win for the region—good for the ports and the logistics industries such as warehousing, trucking, and the railroads; good for certain sectors of the Los Angeles regional economy and for the global economy; good for consumers seeking cheap imported products; good for a US economy relying on imports to fuel consumer spending; and good for US exporters that could ship their goods to overseas markets. It was seen as beneficial for Hong Kong and China as well, for both applauded the journey of goods across the Pacific. In an interesting 2007 presentation at Occidental College on the benefits of global trade, a visiting delegation from China's Anhui Province responded to an overview of the system's environmental and community impacts by saying that the students should focus instead on how China's export industries serve as a major benefit for US consumers. "Your shoppers will want these goods; we can make them more cheaply, which is what your consumer wants," the delegation leader asserted—a comment often repeated in both China and the United States.[31]

If the boosters of Los Angeles's goods movement had concerns, these had more to do with competition from the newly expanding ports in the southern and eastern United States that eagerly awaited a new enlargement of the Panama Canal's capacity. With the Panama Canal's projected expansion to accommodate larger ships and the anticipation of reduced shipping costs that would result, a frenzy of new port expansion occurred in eastern US seaboard places such as Jacksonville, Savannah, Miami, and Gulfport, Mississippi, and even in older port complexes like New York and New Jersey. In addition, the Panama Canal expansion was seen as a major boost for the export of natural gas from the United States, which had seen huge increases in production due to advanced technologies such as hydraulic fracturing, or fracking.[32]

Even before the Panama Canal expansion could be completed, there were already concerns, expressed primarily by China, that it would not be big enough to accommodate the ultralarge crude carriers and even the largest gas carriers, among the massive new ships that had entered the global goods movement system. In response to such concerns, in June 2013 the Nicaraguan Congress granted the HKND (Hong Kong Nicaragua Canal Development) Corporation, a Chinese investment company based in Hong Kong, exclusive rights to construct a huge canal project in Nicaragua; according to the investors, the canal would accommodate those largest ships and help further China's quest for new sources for its huge energy needs, including, potentially, US gas exports. While the HKND investors claimed that the Nicaragua Canal project was ready to break ground (with an anticipated completion date in 2020), industry analysts were skeptical whether the project would ever be completed, given its enormous community dislocations and environmental impacts (for example, threatening Lake Nicaragua's ecosystem that the project needed to pass through) and its huge price tag ($50 billion, nearly five times Nicaragua's own annual economic output).[33]

Aside from the murky Nicaragua Canal prospects, the Panama Canal expansion seemed more pressing to the Los Angeles and Long Beach ports. As a result, the focus of the LA port boosters became "Beat the Canal." While there were hopes that the "greening" of the LA/Long Beach ports (forced in part by environmental advocacy and policy mandates) could help modernize and meet regulatory barriers, the greater concern was that greening the port operations would slow down port expansion and reduce their competitive edge threatened by the enlarged Panama Canal, which finally opened in 2016, and (the elusive) Nicaragua Canal expansion.[34]

Even as the Los Angeles and Long Beach port officials worried about shipper opposition and the costs associated with the new greening policies, the community and environmental groups who had been the key backers of greening the ports argued that greener ports could potentially become the trend around the world. Prospective greening initiatives at the Port of Hong Kong, influenced by the changes in Los Angeles, could inspire changes at Shenzhen, Shanghai, and other Chinese ports and help create further momentum for a more expansive greening strategy on both sides of the Pacific. Despite these breakthroughs, it remained to be seen whether the negative environmental and community impacts of the global trade and goods movement system would remain deeply embedded in that system's very nature and operations. New research only reinforced those concerns, as the question of the environment and of community health, far from disappearing or becoming marginalized, emerged at the heart of the debates about the system's future.

Impacts

Hong Kong

Compared to Los Angeles, Hong Kong's goods movement system is relatively simple. International movement of freight is mostly carried by water (over 90 percent in tonnage in 2014, including via ocean and river), with only a small fraction moved by road (less than 7 percent).[35] Rail freight across the border with Guangdong had never been of any significance in the past, and the service was terminated after June 2010. Internally, most freight is distributed by trucks, and there are also barges that operate as feeders as an alternative for moving cargo inside the harbor area, especially between cargo working areas, wharves, and terminal facilities. As such, most environmental impacts are associated with ship and port operations as well as road freightage.

However, compared to Los Angeles, Hong Kong has experienced less of a government policy focus or public awareness of the environmental impacts of its goods movement system, even as it has expanded in scale as a result of economic growth and burgeoning external trade. Construction and operation of the infrastructure and the moving components, including port and terminal facilities, oceangoing vessels, harbor craft, and the trucking industry, are major economic drivers for Hong Kong, and any adverse impact on the environment is usually accepted as an inconvenient consequence of trade and economic development. Identifying a connection between environmental impact and health outcomes is almost nonexistent in this context. For example, when cross-border road freight was flourishing in the 1990s and early 2000s due to growing economic ties with Guangdong, the main issue that concerned people was traffic congestion at the border checkpoints, which caused delivery delays and economic problems related to the traffic. The air pollution and noise at the border area and along major trucking corridors was perceived at most as a secondary concern.[36] During that period, government projections of cargo movements were optimistic based on past trends, and the idea of building a Port Rail Line (PRL) connecting the Kwai Chung Container Terminals with the rail network in mainland China was rekindled in 2000.[37] About the same time, the Port and Maritime Board commissioned a consultancy study to look into ways to make Hong Kong the preferred logistics hub in southern China.[38] These projections and studies demonstrated the government's desire to strengthen Hong Kong's role as a transshipment center in the region. Yet the assumptions behind these projections began to be questioned, and the need for a thorough assessment of the economic benefits and environmental costs of moving increasing amounts of freight through Hong Kong was also raised.[39]

These questions led various groups such as Civic Exchange to inquire about the contribution of ship and port operations to Hong Kong's serious air pollution problems, given its position as a leading seaport in the world and the intensity of shipping activities in the port area. Although air pollution was getting worse, public perception, backed up by government figures, suggested that power plants were a main source of local air pollution, followed by the road transport sector. The contribution from ships and port activities was considered insignificant, while power generation and roadside emissions were higher on the government's policy agenda. It was in this context that Civic Exchange completed its first background paper on ship emissions in Hong Kong and the Pearl River Delta in 2006, including a review of international regulations and best practices in ship and port emissions control.[40] Their review included the research, community action, and policy initiatives taking place in Los Angeles during this period which demonstrated the importance and feasibility of port and ship emission reduction initiatives. Recognition of these efforts in Los Angeles helped strengthen the emerging arguments in Hong Kong about the importance of focusing on air pollution from ships and the port sector; in turn, the Los Angeles approach provided a potential model for the research and policy analysis then being conducted in Hong Kong.[41]

Background research findings were shared with the shipping industry and agents in other related sectors, such as container terminal operators, truck owners and drivers, shippers and logistics companies, as well as government officials representing the Environmental Protection Department and the Marine Department. This led to a multiyear engagement process that was launched in 2008, with regular roundtable discussions and meetings set up with different stakeholders. These sessions led the government in 2008 to commission a study to compile an emission inventory for marine vessels operating in Hong Kong waters. The study was completed in 2012.

With updated assumptions and a new, activity-based methodology, the new marine vessels emission inventory showed that the contribution of ships to air pollution was higher than previously estimated. By 2013, ships were identified as contributing 50 percent of SO_2, 31 percent of NO_x, and 36 percent of PM_{10} emissions in Hong Kong. It was estimated that almost 70 percent of ship emissions were contributed by oceangoing vessels, and among those, container vessels were major emitters, with 80 percent. It was also found that 30 to 40 percent of oceangoing vessel emissions were produced while the vessels were hoteling at berth or anchorage. As would be expected, Kwai Chung Container Terminals and other major berthing locations were identified as emission hotspots.[42]

In another 2012 Civic Exchange report that calculated ship emissions in the Pearl River Delta region, the impact of these emissions on public health was also assessed. It was estimated that ship emissions of SO_2 alone (hence providing a conservative estimate, as other pollutants were not included) accounted for 519 cases of premature deaths each year in the PRD. Four control scenarios were assessed, and the report concluded that establishing an emission control area in the PRD would bring about a 91 percent reduction in premature deaths.[43]

Research on port and ship emissions led the Hong Kong government to begin to identify whether and how new policies could be established to reduce the air pollution impacts from the ships entering, docking at, and leaving its ports. But even as those efforts began to be explored, Hong Kong's port traffic, along with that at several other ports in China including neighboring Shenzhen, continued to be identified as serious polluters, representing what one journal article characterized as "the Dirty Ten." These were the ports whose ships still utilized the most polluting heavy fuel oil and were responsible for the elevated levels of such pollutants as $PM_{2.5}$. As more studies continued to be published, they further expanded on the link between ship emissions and air pollution and heightened the concerns of the public health and environmental groups, and, increasingly, of policy officials as well.[44]

Los Angeles

During the first years of the twenty-first century, the question of the health and environmental impacts from the expansion of global trade and the goods movement system began to more urgently enter the debates in Southern California. As identified in the doll's journey, this included emissions from the ships (as they entered the ports and idled, awaiting unloading); emissions from the cranes that took containers from the ships onto the trucks; emissions from the trucks and railroads transporting the goods; emissions and the "around-the-clock" bright lights, noise pollution, and extended use of land in the heart of dense urban communities by the intermodal railyards; emissions from the highways and high-traffic roadways used for freight transport; and emissions and land use impacts at the inland port warehouses.

Community, health, and environmental concerns have especially focused on the diesel emissions that occur along each of these pathways. This concern has led some community residents to describe their neighborhoods adjacent to the goods movement operations as "diesel death zones."

The concerns are warranted. Diesel is considered a mobile source air toxic (MSAT) by the US Environmental Protection Agency (EPA). In California, it is regulated as a toxic air contaminant (TAC). The TAC designation for diesel was made in 1998, based on more than thirty studies showing that worker exposure to diesel exhaust is linked to lung cancer and other health effects. In 2012, pointing to cumulative research, the International Agency for Cancer Research, a part of the World Health Organization, identified diesel as carcinogenic to humans (Group 1), changing its earlier 1988 classification of diesel as a probable carcinogen (Group 2A) based on sufficient evidence that exposure is associated with an increased risk of lung cancer. Research studies have also identified such diesel-related health impacts as asthma, reduced lung development in children, cardiovascular disease, lung cancer, and premature death, among numerous other community and health impacts.[45]

The plans to expand the I-710 freeway as the primary goods movement corridor for trucks illustrate the environmental and public health battles that are being waged. Leaving the ports traveling on the I-710 in a car is a frightening experience. The roadway is almost entirely filled with trucks throughout the day. As many as 43,000 truck trips to and from the ports occur daily, with estimates suggesting that the number could increase to 80,000 truck trips per day by the year 2035. Serious accidents happen periodically; the roads are severely congested, which only increases emissions; the roadways are subject to major wear and tear, which is an expense for the truckers as well as a public expense; and serious impacts from emissions are a constant problem.[46]

The trucking industry, the transportation agencies, and various regional bodies such as the Southern California Association of Governments (SCAG) have concluded that the way to address these issues is to expand an eighteen-mile stretch of the I-710. At one point, that included the possibility of constructing a multibillion-dollar second level to the freeway. This double decker of exposures and pollutant impacts was promoted by goods movement interests as a way to limit exposures by adding lanes to reduce traffic congestion, despite the anticipated increase in truck traffic. Community and environmental critics countered that a tripling of the freeway's capacities could enormously increase emissions, pointing to research studies that identified how increased capacity generates "induced traffic"; that is, if they built it, more traffic would come, negating any lessening of congestion. Alternatives that have been identified by the community groups include expanding public transit, mandating the use of zero-emission technologies, providing community benefits such as pedestrian and bike improvements, and establishing green belts, open space, and trail improvements along the LA River which abuts the I-710 corridor.[47]

Contentious debates about the community and environmental impacts from air, noise, and light pollution have also centered on rail transport, including the rail corridors, the huge railyards, and the intermodal facilities. Some of the largest railroad companies such as the Union Pacific and BNSF Railway have sought to significantly increase their own freight traffic capacities to coincide with the flow of goods coming from China and other Asian countries and ports. The first major expansion of the rail freight system in Southern California involved the construction in the 1980s and 1990s of the Alameda Corridor, designed to overcome right-of-way bottlenecks and speed the rail traffic coming out of the port. Linked to this opportunity for faster rail freight, the Union Pacific and BNSF developed ambitious plans to increase their role. A proposed expansion to more than double the size of an intermodal facility (the Intermodal Container Transfer Facility, or ICTF) was developed, as well as the Southern California International Gateway (SCIG) facility, the proposed new railyards next to the ICTF. The expanded ICTF and SCIG facilities were designed to handle as many as three million containers a year. These "off-dock" facilities are directly across from an elementary school, a park, residential homes, and other community places.[48]

The intermodal railyards in the city of Commerce southeast of downtown Los Angeles, which include four railyards and high-traffic corridors within a small community of working-class immigrants, as well as another huge rail facility at the eastern end of the region in San Bernardino have also become flashpoints for concern over community, health, and environmental impacts. These intermodal rail facilities generate substantial sources of diesel exposure from the trains, trucks, and cargo-handling equipment and off-road equipment that moves the bulk cargo. Health risk assessments of the facilities in Commerce and San Bernardino by the California Air Resources Board have found extremely high cancer risks (as much as 1,000 times higher than what is often identified as a threshold for cancer risk), with elevated cancer risks extending as far as eight miles from the yards.[49]

Figure 2.3
Trucks leaving the Port of Los Angeles across from a neighboring park. Source: UEPI.

Beyond air quality concerns, light and noise pollution from the intermodal rail facilities and the trucks are a constant factor affecting the health and quality of life of nearby residents, primarily in low-income communities. The bright lights, often resembling stadium lights, remain on twenty-four hours a day, which can cause sleep disorders and sexual dysfunction. The noise levels are also constant, causing major health impacts such as psychological effects (e.g., elevated stress levels), physiological effects (e.g., hearing loss and increase in blood pressure), and mental health effects (e.g., increased anxiety). In Commerce, the railyards are adjacent to several freeways (including I-710) and other heavy-traffic roadways that overwhelm the community with emissions and the stress from the noise and traffic.[50]

Figure 2.4
Line of trucks along a main boulevard in the city of Commerce. Source: East Yard
Communities for Environmental Justice.

Meanwhile, the San Bernardino railyards and the huge warehouses that
have sprung up in the inland counties of Riverside and San Bernardino
have generated substantial pollution and land use impacts. Along with the
intermodal yards, warehousing is an integral link in the goods movement
chain. Approximately a quarter of the thirteen to fifteen million shipping
containers entering the region's ports each year are transloaded in warehouse storage facilities before leaving greater Los Angeles. The LA region
has approximately 700 to 800 million square feet of warehouse facilities of
which about 25 percent are port-related—with future port-related warehouse development plans poised to significantly expand those numbers.[51]
As close as one can come to "sprawl in a box," these structures are spartanly
built, lined with dozens of truck bays. Warehouse operations are connected
to distant locations—factories in China, railyards near the port, Walmarts
in Illinois—rather than connected to the fabric of life in the communities
in which they are located. Walmart alone has 160 distribution centers,

including 42 regional centers around the United States that are larger than one million square feet each.[52]

The fastest growing cluster of warehouses serving the goods movement and logistics industry are in the Inland Valley. In 2013, 23.3 million square feet of industrial space was picked up for warehouse and logistics industry purposes in the Inland Empire, while the next year as much as 22 million square feet of warehouse space was leased.[53] The unchecked growth of warehouses and the roads and railyards that serve them has dramatically transformed the landscape of the area and the health of its residents. Farmland has been converted to immense, windowless warehouses surrounded by asphalt and chain link fencing—and some of the worst air pollution in the nation has been measured there. Mira Loma Village, a low-income Latino community in Riverside County, which is surrounded by warehouses and distribution centers, has been identified as having some of the highest levels of particulate pollution in the nation, while Mira Loma children have the slowest lung growth and weakest lung capacity of all children studied in southern California.[54]

Due to this web of impacts caused by the goods movement system, opposition emerged at the ports and along the regional goods movement corridor in the Los Angeles region. Such opposition has since spread to other port and goods movement communities and regions across the United States, and connections have formed with environmental critics and researchers in Hong Kong and China as well as other places throughout the world. Community, health, and environmental advocates, along with new research and policy initiatives, have forced a shift in how the system is characterized and have helped change the debate to what has been causing those impacts and how they can best be addressed.

The Connection to China

The contemporary rise of China's ports and shipping industry coincides with China's post-Maoist evolution into a globally oriented market economy. In 1978, the total value of China's imports and exports was US $20.6 billion (or 35.5 billion yuan). It ranked thirty-second in world trade and accounted for less than 1 percent of the world's total trade volume. Although Hong Kong had ranked among the top two or three largest port cities in the world, China's ports—even its largest and historically rooted ports in

Shanghai and Guangzhou—were small in relation to its Asian counterparts such as Singapore and Hong Kong. However, with the shift toward export production and its role as the world's factory, China's ports grew rapidly, particularly after the mid-1980s when China initiated negotiations with the World Trade Organization (WTO). The mainland government made huge investments in port infrastructures, and over the next twenty years those investments surpassed all the other investments in the world combined. By 2001 when China formally entered the WTO as its 143rd member, its total value of imports and exports had jumped to US $509.6 billion (4,218.3 billion yuan) and then grew again in the next dozen years to US $4,158.99 billion.[55]

The trade in goods through seaports has had a long and complex history in China. Shipping to and from Guangzhou (formerly Canton) can be traced back to the Eastern Han period around the first century AD, when merchants from the Roman Empire explored opportunities for obtaining goods through sea transport via a marine Silk Road that stretched from Canton to the European coastal cities. The development of the treaty ports in the nineteenth century, including Canton, created strong negative associations about imperial control and the carving up of port territories, concessions, and enclaves as places reserved for foreigners. By the Maoist period, those historically rooted negative associations about the global designs of foreign powers as reflected in global trade lasted well into the mid-1980s even as China began to construct its export-based manufacturing and transport infrastructure to accommodate its own global trade designs.

The negotiations to incorporate China into global trade systems (initially as part of GATT, the General Agreement on Tariffs and Trade, and subsequently with its successor, the WTO) changed that dynamic from one of mistrust of foreign manipulation to expectations of expanded trade as an economic driver. When an agreement was reached in December 2001 for China to join the WTO, China's role in global trade became even more explosive. From 2000 to 2009, the average annual increase in growth rates of China's exports and imports reached 17 percent and 15 percent, respectively; figures much higher than the 3 percent annual growth rate of world trade as a whole. At the same time, foreign direct investment, including in China's export industries and its ports, grew enormously as well.[56]

The tenth anniversary of China's entry into the World Trade Organization in 2011 became an occasion to celebrate China's new global role.

Chinese Premier Wen Jiabao enthusiastically characterized this change as "a momentous event in China's opening-up to the outside world." New private investors, including from Hong Kong, became major international port operators, and, since 2012, as many as six Chinese ports (in addition to Hong Kong) have continued to be listed among the ten largest container port operations in the world. State-run and financed companies have either joined or established partnerships with huge private (and global) operators such as HPHL to undertake the rapid expansion of port operations in China and subsequently to expand investments in port operations around the world.[57]

The rush to expand was influenced in part by local governments seeking an economic bonanza with the development of their new megaports, not unlike the rush to develop other huge transportation projects in China's cities. As a result, new problems of overcapacity emerged due to a more chaotic development process. Those problems were reinforced by the global economic slowdown of 2008–2009, increased labor costs due in part to protest actions, the rise of the megaships (which also experienced their own overcapacity problems), and eventually, recognition of the environmental problems associated with the port developments. The rate of increase in the value of imports and exports in China peaked in 2002 and began to slow by 2009; the total value declined over the next three years. The mainland government sought to address some of these trends by pushing for consolidation among the largest Chinese shipping companies (for example, combining the state-owned China Shipping with another huge state-run shipping firm) and by developing a more robust "hub and spoke" framework that favored key regions such as the Yangtze and Pearl River Deltas. This shift was part of the government's emerging strategy to increase domestic consumption of goods and to reduce the economy's long-standing reliance on export production. Slogans that were introduced, such as "Go West" (into the interior) and "Strategy for developing the Yangtze River" (establishing a goods movement link, for example, between Shanghai and Chongqing), typified the new focus on domestic goods movement.[58]

Missing in this change in orientation until recently has been an environmental focus. While Hong Kong began to conduct research and take action to reduce and control ship emissions in the last few years, the focus in China remained on port capacities and cargo throughputs, whether domestic or foreign. Private investors looked at environmental questions from a bottom-line perspective while government agencies, increasingly absorbed by issues like air pollution, looked more toward problems related to power generation, industrial production, urbanization, and motorization.

One impetus for change in China has been Hong Kong's own experience in ship emissions control. This has led to some pressures on Guangdong, especially on Shenzhen, given the recognition that Guangdong's ship- and port-related emissions have impacted both China and Hong Kong through cross-border pollution. Those pollution sources cause government intervention in Hong Kong to become less effective when the neighboring ports in Shenzhen fail to take similar actions. As a result, the Hong Kong government began to seek more regional collaboration with Guangdong and Shenzhen in such areas as air pollution monitoring and ship emission controls. At the same time, once Hong Kong sought to move toward a regulatory approach, the shipping industry began to lobby for a uniform standard within the region to reduce the difficulties of compliance and to establish a level playing field within the sector. Each of these pressures eventually translated into a modest but important regulatory approach for Shenzhen and Guangdong. Constructive dialogue and information sharing also occurred between the governments of Hong Kong, Guangdong, and Shenzhen and, at the nongovernment level, at academic conferences and technical exchanges between research institutes and universities.

Beyond the pressures coming from Hong Kong, the growing concerns about air pollution in mainland China have established some momentum to move toward tighter air quality standards and comprehensive air pollution control strategies. Among other areas, controlling ship emissions has begun to be identified as one of the means to help reduce air pollution. New information from a 2013 assessment provided by China's Ministry of Environmental Protection, for example, identified the shipping sector as responsible for 8.4 percent of China's SO_x and 11.3 percent of its NO_x emissions. Until 2015, China's ports, including its six largest, had not been required to meet the air pollution standards mandated in a number of port communities around the world, including Los Angeles and Hong Kong. In June 2015, the Chinese government initiated a public consultation on new requirements on ship emissions. Six months later, a series of plans were put in place, including requirements for the Pearl River Delta, Yangtze River Delta, and the northeastern Bohai Bay rim, as designated domestic emission control areas, to use fuel with a sulfur content of no more than 0.5 percent starting in January 2016, first on a voluntary basis and subsequently as a mandatory requirement. In April 2016, Shanghai moved nine months ahead of schedule with other key ports in Yangtze River Delta by making the 0.5 percent sulfur fuel rule mandatory.[59] According to China's Ministry of Transport, the new rules are focused on reducing SO_2's contribution to acid rain as well as its major health impacts such as respiratory conditions and premature death.[60]

As these regulations have begun to be put in place, the impacts of ship emissions on public health for China have been highlighted, given the size of its major port clusters in the Pearl River Delta, the Yangtze River Delta, and the Bohai area, all situated in the most populated coastal regions in China. The evidence from Los Angeles and Hong Kong has also made it clear that emission control and pollution prevention measures for ships and goods movement are both doable and could lead to major health and environmental improvement. The question remains whether the rapidity with which China has increased its economic development, whether in relation to its ports, global trade, or overall push for industrial and urban development, will either continue to overwhelm any environmental initiative or will instead elevate the types of environmental approaches needed to address the impacts already identified from those developments.

Strategies for Change

In Hong Kong, on the morning of July 9, 2010, a meeting took place on the thirty-ninth floor of a commercial building typical of the Hong Kong skyline. It was the Hong Kong head office of Maersk Line, the largest container shipping company in the world. The meeting was attended by representatives from Maersk Line, the Hong Kong Environmental Protection Department, and Civic Exchange. The meeting, which included Tim Smith, chief executive of the North Asia business of Maersk Line and an active player in the Hong Kong Liner Shipping Association (HKLSA), Billy Cheung and Tony Lee of the Hong Kong Environmental Protection Department, and Simon Ng of Civic Exchange, covered issues like fuel switching and the additional cost that would be incurred by the shipping companies, the viability of voluntary action, and whether it could lead to a pathway to future regulation, and the need for communications and collaboration with Guangdong authorities. It was hoped that a new initiative could be forged involving the shippers, the government, and civil society groups.

Preceding this meeting had been a workshop organized by Civic Exchange in mid-June 2010, one of a series of stakeholder engagement activities since 2008 on the topic of ship emissions and the need for prevention and control measures. Shipping lines, container terminal operators, fuel suppliers, government representatives, and academics were present for the June workshop, which highlighted fuel switching as a potential option in Hong Kong for reducing ship emissions. After a number of presentations and group deliberations, major shipping lines attending the workshop verbally agreed to draft a voluntary agreement that ships berthing in Hong Kong waters would use distillate fuel instead of bunker fuel, and that they would be willing to take the proposal back to their members for internal discussion. This proposal was particularly championed by Maersk Line and a few other key members of the HKLSA, as well as the Hong Kong Shipowners Association. The workshop participants recognized that these types of changes had already been instituted in a few other port communities around the world, including Los Angeles and Long Beach, thus putting additional pressures on the ship owners. After the workshop, everyone agreed to reconvene after summer vacation, since it was not clear whether the other HKLSA members would be willing to switch to the more expensive distillate fuel.[61]

As these internal discussions were taking place, preliminary findings of a marine vessels emission inventory commissioned by the Hong Kong Environmental Protection Department were being shared with members of the HKLSA in another meeting held on August 18, 2010. The inventory demonstrated that container vessels contributed a substantial share of emissions, especially when the vessels were at berth or anchorage. Switching to clean distillate fuel, particularly with fuel sulfur content of 0.5 percent or lower, would potentially reduce emissions of SO_2 and PM_{10} by roughly 70 to 80 percent at the berthing locations. After this meeting, the language for the Fair Winds Charter was discussed and finalized in early September 2010. Maersk Line then took the lead by switching its fuel that month, followed by American President Line (APL) in October. The Fair Winds Charter officially started on January 1, 2011, for a two-year period, with seventeen shipping lines signing on to the agreement.

Figure 2.5
Fair Winds Charter. Source: Civic Exchange.

In Los Angeles, in 2001, scientists at the University of Southern California (USC) hosted a conference on air pollution and invited a number of environmental justice groups, including those located in heavily impacted neighborhoods adjacent to goods movement corridors, to hear their findings. The USC scientists were in the midst of a longitudinal study of air pollution and had begun to map air pollution hot spots. During the discussion of exposure to particulate matter and diesel exhaust, several community members who lived adjacent to the ports, truck routes, or the railyards spoke of the health and environmental issues in their neighborhoods.

Toward the end of the conference, Jesse Marquez, a resident of the low-income community of Wilmington adjacent to the Port of Los Angeles, rattled off a series of anecdotes about people in his neighborhood with serious health problems. "The ships at the port are not even regulated for their impacts, yet we face the consequences every day," Marquez told the assembled scientists. "Of course they are regulated," the scientists replied, less knowledgeable about how the goods movement system operated. The scientists soon discovered that Marquez was accurate, suggesting that community experiences and knowledge were also critical factors in helping frame the research. The combination of research, community action, and litigation would soon begin to change the policy landscape.[62]

Six years after the USC conference, a new network of community groups, policy researchers, and scientists that called itself THE (Trade, Health, and Environment) Impact project hosted their own conference on the goods movement system and its impacts in Southern California. THE Impact Project community groups were linked to an evolving environmental justice framework. Based in low-income, predominantly Latino and immigrant neighborhoods, the focus on goods movement represented not just the classic environmental justice argument about the toxic burdens in such communities, but a strong desire among residents to create more livable places, underlining the environmental justice argument that environmental advocacy needs to focus on the places where people live, work, play, eat, and go to school. What became especially compelling about this place-based focus was the understanding that the goods movement system connected the places at the local *and* the regional level as well as within a national and global context where the system operates.[63]

The development of these community-based environmental justice groups, along with other academic/research, health, environmental, and labor networks, has played a role in some important policy changes, community benefit agreements, and environmental changes at the ports in Los Angeles and Long Beach and throughout the goods movement system. While these changes have been substantial—and have been considered a model for other port and goods movement communities throughout the United States as well as in Hong Kong and China—they have also continued to be challenged and at times undermined by the shippers, big-box retailers, and railroad and trucking companies. Along with the development of such policies and new greening initiatives, the question of implementation has been critical.

In China, in the port city of Tianjin, in August 2015 two huge blasts decimated a warehouse as well as the adjoining neighborhood in the industrial Binhai New Area. The explosions killed 173 people and injured many more, although it took several months to identify the number of victims, which included workers, residents, firefighters, and police. A large number of buildings, apartments, vehicles, and other infrastructure were burned or destroyed. The state media reported that a shipment of explosive materials had ignited at a port warehouse run by Tianjin Dongjiang Port Ruihai International Logistics, which stores such substances as sodium cyanide and the toxic substance toluene diisocyanate. The two explosions occurred within seconds of each other. The National Earthquake Bureau said the first blast's strength was equivalent to three tons of TNT and the second was the equivalent of twenty-one tons.[64]

A week after the blast, the Chinese news agency Xinhua published an investigative report that quoted company executives who had been detained by the police as saying they had "good connections with government officials." The article was surprising since Xinhua had not been known for aggressive reporting on major accidents in China. The top party leaders, including President Xi Jinping, criticized the way Tianjin officials, including those at the port, had addressed the causes and consequences of the explosions. Ultimately several officials were arrested, including those at the Ministry of Transport, Tianjin's municipal government, and its port authority. The Work Safety Bureau of Tianjin's Binhai New Area where several port-related warehouses and chemical plants were located, some of them next to new residential developments, was instructed to review the conditions at 583 chemical companies in the area. Their review found problems at eighty-five of them. Ten of those plants were subsequently ordered to relocate because they were seen to be too close to residential areas.[65]

The Tianjin events occurred at a time when port and goods movement issues had begun to receive attention in China, as its December 2015 policy on fuel oil demonstrated. Moreover, one possible outcome of the Tianjin explosions was an immediate drop in throughput numbers at Chinese ports. While such a drop was thought to be temporary and not necessarily directly connected to the Tianjin situation, it likely contributed to China's broader assessment of its goods movement system. That same assessment was taking place in Hong Kong and Los Angeles, where regulating the port and shipping sector's environmental and health impacts was now on policy agendas, including whether and how the implementation of such policies would take place.[66]

Comparing the policy changes in Los Angeles, Hong Kong, and China reveals that some of the more extensive changes have taken place in Los Angeles. One of the first policy changes was a 2003 settlement agreement on plans for a major Chinese state-run shipping company to occupy a $650 million, 174-acre terminal to be built by the Port of Los Angeles. Two wharves would be built at this terminal, larger than any other at the time, to house as many as 200 to 300 container vessels a year per wharf for the China Shipping Holding Company (China Shipping). The terminal would also include ten massive cranes, up to sixteen stories high, that would unload the containers. Yet the site was also just 500 feet from residential homes and would involve new roads to accommodate the anticipated huge increase in truck traffic.[67]

These plans immediately generated community opposition. The ships docked at the port were likely to keep their engines running, potentially days at a time, until containers could be unloaded. Dockside diesel-related emissions for just a single vessel at berth could include as much as one ton of NO_x and nearly 100 pounds of particulates each day before the unloading took place. A lawsuit against the port was filed to stop the completion of the new terminal until an agreement could be reached with the community groups. After losing in district court, opponents of the China Shipping plans achieved a major victory at the appellate court level. As a result, the port (and the City of Los Angeles) decided to settle in order to avoid an additional lengthy court battle whose outcome was not assured.[68]

The results of the agreement were impressive. China Shipping agreed that its ships would use "cold ironing," a long-standing technology used by naval vessels and ferries but never before by container ships. Identified by the port as an "alternative power source for oceangoing vessels," cold ironing allowed the ships to plug into an electric source while at berth. Such a shift represented an expected elimination of more than three tons of nitrogen oxides (NO_x) and 350 pounds of diesel particulate matter for each ship that plugged in.[69] The settlement also called for other environmental changes at the terminal, including the use of dock tractors to run on alternative fuels instead of diesel, shorter cranes, and a shift toward cleaner marine fuels once their feasibility had been evaluated. A community mitigation fund was also established, including incentives to replace diesel-powered trucks, air quality mitigation measures, and community improvements.[70]

The China Shipping agreement turned out to be the opening effort in Los Angeles for addressing future expansion plans and producing environmental changes related to port operations. In 2007, a new proposal was introduced to expand the TraPac facilities from 176 to 243 acres and to reconfigure roadways to accommodate the anticipated increase in traffic. Once again a lawsuit was filed and a settlement was reached out of court the next year. The settlement included a $50 million Port Community Mitigation Fund to be run by a nonprofit and administered by a community and environmental board; $3.5 million for parks and open space; installation of air purification and sound proofing in the nearby public elementary schools and residents' homes; new health services resources and research on health and land use impacts; and potential wetlands restoration projects in the Wilmington and San Pedro areas that neighbored the port.[71]

The TraPac dispute was taking place just prior to a lengthy and contentious policy process that resulted in the San Pedro Bay Clean Air Action Plan (CAAP), adopted in 2006, and a subsequent Clean Trucks Plan the next year. The two ports signed on, although the Long Beach port eliminated one key provision of the truck plan. In many ways, the CAAP and the Clean Trucks Plan provided the most substantial changes up to that point of any port in the United States and for many ports worldwide. The CAAP set significant emission reduction targets by the end of 2011 from the baseline year of 2007—a 45 percent emissions reduction in diesel particulate matter (DPM), nitrogen oxides (NO_x), and sulfur oxides (SO_x). That goal was achieved and even exceeded, in part due to reduced ship traffic related to the Great Recession of 2008–2009. Other changes that had predated the adoption of the CAAP or were put in place subsequently included a ship speed reduction plan, the change to electric shore power, a shift to alternative fuels for cargo equipment including the cranes, and future changes in the fuel sources for the incoming ships. In addition, the state of California began to require oceangoing vessels to use lower sulfur fuels, slow down as they approached shore, and turn off their engines for shore power (or adopt equivalent controls) when docked.[72]

Perhaps the most contentious of the changes involved the transition toward replacing dirty diesel trucks with less polluting trucks at the port. Among other goals, this Clean Trucks Plan included replacing and retrofitting approximately 16,000 trucks in order to meet federal EPA emissions standards by 2012. To achieve these goals, the program featured a $1.6 billion concessionaire model that would require trucking companies that serviced the port to hire truck drivers as employees in return for securing transport contracts with the port. These new employees would replace the heavily exploited system of independent contractors, or *tranqueros*, who resided at the economic margins and would find the truck replacement costs nearly impossible to meet.[73]

Aside from the economic squeeze, the immigration status of the individual trucker contractors further reinforced the potential for exploitation.

Immigration issues were a major factor for the port trucking and goods movement sector. For example, on May 1, 2006, a year prior to the Clean Trucks Plan agreement, 90 percent of all the truck drivers serving the Port of Los Angeles refused to make or pick up their deliveries, in solidarity with the massive Immigration Rights rally taking place that day. Participation in this labor action was significantly influenced by an immigration raid at the port two weeks prior that had targeted the immigrant truck drivers and had caused several drivers to be hauled out of their trucks and detained, with their trucks towed away.[74]

The Clean Trucks Plan was immediately met with opposition from the American Trucking Association as well as by other players in the goods movement system industry, including multinational retail companies such as Walmart and Target which filed an injunction blocking the implementation of the employee concession element of the program from moving forward. A series of court battles subsequently ensued which eventually gutted the concessionaire model (which the Long Beach port had already abandoned) but kept the policy of truck replacement and retrofit. Many of the older diesel trucks were replaced and emissions were further reduced. But the vulnerabilities and poor working conditions for the independent truckers, the *tranqueros*, still largely remained, even as a few victories subsequently took place after the truckers mobilized and a handful of companies changed their status from contractors to employees, as had been required in the concessionaire agreement.

Other changes also failed to proceed smoothly. In September 2015, the Port of Los Angeles revealed that about 20 percent of the environmental mitigation measures for the groundbreaking China Shipping agreement had not been implemented. Moreover, when port officials had provided an additional $5 million to equip its vessels with the technology to meet the shore power rules, those funds were also used for some of China Shipping's fleet even though those ships had abandoned the China to LA route and been substituted with ones that were not all equipped to meet the rules. Port officials blamed these problems and the lack of implementation on inadequate technologies and on the Great Recession, which they felt had warranted a more lax approach. But China Shipping's motivation to escape any additional costs arising from complying with the mitigation measures could also be seen in the context of its own problems of overcapacity, declining freight rates, and higher operating costs, which had factored into the push for a merger by the Chinese government with one of China Shipping's main state-run Chinese shipping rivals.[75]

The community groups that had long been engaged in the issue of pollution at the port were outraged that information had been withheld about these implementation failures and that the port had allowed China Shipping to violate a key statewide rule on the dockside use of shore power. To make matters worse, a few months after the China Shipping implementation problems became public, a second revelation was made that the TraPac terminal agreement had identified only a 50 percent compliance rate instead of the mandated 80 percent use of shore power.[76]

Despite these implementation problems, the community groups and their environmental allies were able to score a major victory in the courts when a Superior Court judge ruled that the proposed SCIG facility, a primary battleground over port expansion in an area already experiencing severe pollution impacts, failed to meet the requirements of the California Environmental Quality Act (CEQA). "Let this victory be a message to the polluters and the policymakers that perpetuate environmental racism," organizer Angelo Logan told the *Los Angeles Times*. Logan continued that the community groups "look forward to working with the Port of Los Angeles to develop a long-range plan that puts people's health at the center." Logan's comment—citing opposition to particular plans that increased health impacts while working with policy officials to develop long-range greening and health-related change—was emblematic of the Los Angeles experience. The changes that had taken place in LA in turn represented a starting point for its counterparts in Hong Kong and China.[77]

In Hong Kong, the long-standing resistance by the government and its various business sectors to regulatory mechanisms often meant more limited change without the power of regulatory enforcement. But when policies were adopted and regulations instituted, the potential for environmental change was substantial. This had been the case with the 2011 Fair Winds Charter agreement.

The Fair Winds Charter had been perhaps the first industry-led, voluntary initiative in the world to seek to reduce ship emissions. One of the charter's strengths had been the buy-in of Maersk Line and some other shipping lines not just to agree to voluntary measures but also to push for government regulation in two years' time in order to create a level playing field within the industry. The Fair Winds Charter participants argued that it was unfair to penalize those who were beginning to meet environmental goals and paying the extra fuel cost to cut emissions through a voluntary scheme. The participants were also conscious that an agreement in Hong Kong alone was not sufficient, and that such an agreement needed to be extended to Guangdong, including the port of Shenzhen. Bucking the trend toward a minimal government role, the Fair Winds Charter further called for the Hong Kong government to actively work for an agreement with Guangdong so that in the long run regional standards could be created on controlling ship emissions in line with international requirements.

While Hong Kong government officials had been part of the Fair Winds Charter discussions and had endorsed the voluntary agreement, the government had initially been a reluctant participant. If the Fair Winds Charter had had to rely on an exclusively voluntary approach, it would never have changed fuel sources and achieved other environmental measures. Arthur Bowring, the managing director of the Hong Kong Shipowners Association, pointed out that the shipping companies were initially concerned that if they publicly supported and followed through with the Fair Winds Charter, "then other places might well demand they do the same thing, and that could really affect their bottom line tremendously."[78]

New research identifying shipping as a primary source for several pollutants, as well as other long-standing air pollution problems related to road vehicles and power plants, led to renewed pressure to address port-related emissions. The government finally interceded with the development of the Clean Air Plan in Hong Kong in 2013. The plan, which laid out a trajectory for various control measures, including direct regulatory mechanisms, to reduce emissions from different sources, became the blueprint of air quality management in Hong Kong. In July 2015, the Hong Kong government moved ahead by mandating all oceangoing vessels at it berths to burn marine fuel with 0.5 percent or less sulfur content, or other compliant fuel like liquefied natural gas (LNG). The government's earlier market-based incentive program, initiated in September 2012 for a three-year period to facilitate ships' switching to low-sulfur fuels, had come up short, with only 12 percent of the ships registered at the port making the switch. The July 2015 regulation would now make the switch mandatory—and in accord with regulations at Los Angeles and several other international ports.

In China, the rollout of new air pollution policies combined with the growing collaboration with Hong Kong and California ports, including Los Angeles and Oakland, has led to policy statements about the need to reduce emissions. The arrangement with Los Angeles included a September 2015 memorandum of understanding that was signed during the US-China Climate Leadership Summit to "share best practices and lessons learned in reducing emissions." The memorandum of understanding also focused on reducing the emissions from vessels "visiting their major respective ports."[79]

By 2016, the greening of the ports had thus made headway in Los Angeles, Hong Kong, and China in important ways. Policies were established to ensure the use of lower-emission technologies and cleaner fuels. Broad air quality plans for particular ports and their regional impacts have been developed and have begun to be implemented. Greater awareness of the community, health, and environmental risks at each stage of the goods movement system has been strengthened due to science-based research, more comprehensive monitoring, and community action. The concept of a "green" (or at least a greener) port and goods movement system has been embraced, or at least referenced.

But despite these changes, the need for environmental change is still at an early stage. Port and goods movement expansion projects continue to be promoted and then challenged through new research on health impacts, new policy initiatives, community mobilizations, legal actions, and the continuing push for less polluting and even zero emission technologies. Beyond the specific battles resides an implicit—and at times explicit—push for a more fundamental change that would also require a rethinking of a global trade system that has only just begun to address the impacts that are integral to how the system functions.

3 Breathing Air

Selling Air

In October 1954, elevated air pollution levels that lasted nine days laid siege to Southern California. As the air quality worsened, a women's group called Smog-A-Tears (a takeoff on the Mouseketeers of the popular 1950s Disney television show, *The Mickey Mouse Club*) donned gas masks and marched through Pasadena, northeast of downtown Los Angeles, carrying signs decrying the "smog attacks" plaguing the region. The night before the Smog-A-Tears demonstration, Pasadena mayor Clarence Winder led a group of six thousand assembled protestors in a group prayer "to deliver us from this scourge." Soon after, the Optimists Club, a group of Los Angeles businessmen and professionals, put on their gas masks at their regular meeting and hoisted a banner that read, "Why Wait until 1955, We Might Not Even Be Alive." As the air pollution continued to blanket the region, bright yellow containers filled with "LA smog" began to be sold to tourists. The containers had the message: "When suffering from an over-abundance of happiness, puncture can and place directly under nose and get depressed."[1]

More than fifty-five years later in Hong Kong in 2010, the environmental group Clean Air Network (CAN) enlisted actor Daniel Wu to do a short video clip promoting the bottling and sale of "fresh air." Hong Kong had been suffering through numerous air pollution episodes that had enveloped the city, reducing visibility and causing a wide range of health concerns. Hong Kong activists led by the NGO group Clean Air Asia took the canned air spoof to another level two years later when they launched a "hairy nose" campaign. If air pollution continued to get worse, the group suggested that one could adapt by relying on hairier noses to filter out the pollution as a final line of defense.[2]

Figure 3.1
Smog-A-Tears demonstration, 1954. Source: *Los Angeles Times* (Common Use).

Figure 3.2
Optimists Club meeting, 1954. Source: Common Use.

Figure 3.3
Actor Daniel Wu promoting "fresh air." Source: Clean Air Network.

In China, the increasing presence of gas masks in places like Beijing and Shanghai has led various entrepreneurs and artists to develop and market visually attractive and fashionable masks for residents to wear—if they can pay for it! One e-commerce website reported that upward of $140 million had been spent on gas masks on its site in just one year. Further capitalizing on worries about health risks from air pollution, Chinese businessman and philanthropist Chen Guangbiao introduced a "canned fresh air" marketing campaign in September 2012. Displaying a bottle that looked like a soft drink can that was filled with "Grade 2 quality air" imported from southwest China, Chen suggested that buyers could open the can, put their nose near the top, and then breathe in. After retail stores would be established, the air could be trucked in and buyers could return for refills! The journalists who wrote about Chen's plan could not decide whether it was a "green initiative" or "performance art."[3]

While Chen's marketing scheme failed to catch on, another idea was hatched from a more powerful source. In March 2014, China President Xi Jinping, at a meeting of the National People's Congress, jokingly suggested to Guizhou Province Governor Chen Min'er that Guizhou might have a competitive advantage because of its relatively better air quality. According to newspaper reports, Xi commented that "air quality is now a deciding factor in people's perception of happiness," and, as a result, the province could "sell air cans in the future." President Xi's comment was embraced as a genuine commercial opportunity by the Guizhou Tourism Bureau to sell bottles of local clean air as a souvenir for tourists. "Once the canned air concept is commercialized," commented Tourism Bureau head Fu Yingchun, "it could become a unique product and a bestseller for the province."[4]

Figure 3.4
Chinese businessman Chen Guangbiao (center) gives cans of fresh air produced by his factory to passersby for free in a financial district in Beijing. Source: Mark Wong/ EPA /LANDOV.

This use of ironic commentary, whether in Los Angeles, Hong Kong, or China, had become a part of the discourse about an air pollution problem that continues to plague each of those places. Groups like CAN in Hong Kong and Clean Air Asia in Asia, parallel groups in Beijing and other cities in China, and groups in Los Angeles have all used humor to complement their community education and social mobilization goals. Such tools as community-based air monitoring and the translation of scientific research into accessible public information have become part of the strategy to "engage the general public," as Fan Xiaqiu of China's Green Beagle environmental organization put it. Groups have also mobilized to push for policies and their effective implementation rather than simply relying on the market, because market forces seem more focused on capitalizing on people's fears.[5]

Highlighting an apparently absurd response (selling clean air), developing a tool to raise awareness through community monitoring, translating and making accessible scientific information, and pushing for policies to address the sources of air pollution have each contributed to the struggle to improve air quality. In Los Angeles, the shift from satire and community action to policy change has been a long, difficult, and sometimes tortuous process. But policy change has occurred, and air quality in Los Angeles has significantly improved in the past three decades, as indicated by several research findings. Despite these gains, the air in Los Angeles still represents a major health burden, and efforts to improve the air continue to be contested.

In Hong Kong, the combination of research findings, monitoring tools, and community action has led to a significant shift in the development of new policies that have situated air pollution as a major arena for government action and new approaches. Action is taking place even as Hong Kong's residents have continued to be plagued by air quality problems, which now extend even to the summer months when air pollution levels usually decline.

And in China, government action, first instituted during the 2008 Olympics as a temporary measure, has led to broad government policy statements and specific interventions, such as limiting the number of annual car registrations and relocating factories away from air pollution hot spots. Yet air pollution episodes remain endemic and, thanks to increased monitoring, the recognition of the extent of the problem has led to more draconian crisis interventions, such as the closing of schools and workplaces during the most severe air pollution episodes.

While there are greater awareness and transparency about the importance of the air pollution issue in each of these places and even some important policy initiatives, they are still just a prelude to addressing structural issues such as Los Angeles's dependence on cars, Hong Kong's port and roadway pollution, or China's mix of rapid urbanization, coal-fueled industrial development, and marketization. Without such major changes, the idea of bottling and selling air, once considered a spoof, runs the risk of becoming the norm, albeit for marketing rather than policy purposes.

Discovering Air Pollution

Los Angeles

On the morning of July 26, 1943, Los Angeles residents, particularly those working in the downtown area, woke up to a dense haze. This was a "gas attack" whose "noxious fumes [were] almost unbearable," according to a front-page *Los Angeles Times* story. It was not the first of such episodes— concerns about air quality had dated back to 1940. But the severity of Black Monday, as the July 26 episode was labeled, threatened not only visibility but the very reputation of Los Angeles as the "smokeless city." Los Angeles had long promoted itself as a place where people from around the United States could come to breathe the fresh air, take in the sunshine, and help alleviate such health concerns as asthma and tuberculosis. The fight against smog, the *Times* would later declare, was nothing less than the fight to "bring the sun back to the city."[6]

The summer of 1943 proved to be a pivotal moment in the Los Angeles experience with air pollution. After the July 26 episode, the immediate focus was on finding a culprit. A war-related synthetic rubber plant in the downtown area was first identified. But even when that plant was shut down, the haze continued. The search for the cause consumed Los Angeles residents and policymakers alike. Multiple explanations were proposed and then replaced by others. Maybe it was the oil refineries, or the backyard incinerators, or the chemical plants. Perhaps it was caused by sewage disposal, or diesel exhaust from trucks, or exhaust from the soap factories, or perhaps auto emissions as Los Angeles's car use continued to climb. The list grew longer with each episode. Whether to regulate or not, or whether to develop public policies and new public agencies, or whether to call for voluntary efforts consumed the public discourse, even as public discontent intensified. Los Angeles County Supervisor John Anson Ford characterized the situation as "the most baffling, the most unprecedented, and the most widely distributed affliction ever to harass [Los Angeles] county."[7]

Through the late 1940s and 1950s, the decline in air quality became increasingly severe, with each new devastating episode appearing to be longer lasting and more threatening. A series of smog episodes in 1949 led to the closure of the LA airport due to poor visibility and fears that the school year could be impacted, as well as increased reports of deaths due to respiratory and cardiac illnesses. In 1953, a group of USC scientists stated that if dangerous air pollution levels continued to increase, a smog evacuation plan for parts of the region would need to be developed. And the nine-day air pollution episode during October 1954 when the Smog-A-Tears demonstrations took place led once again to the closure of the Los Angeles airport and brought about a port closure for incoming vessels. It also created a political firestorm that impacted the California governor's race.[8]

Already by the early 1950s, important and groundbreaking research by scientist Arie Haagen-Smit identified automobile exhaust as the single largest contributor to Los Angeles's decline in air quality. Such research made it increasingly clear to policymakers that the automobile industry needed to better control emissions. Auto industry representatives at first refused to acknowledge that auto exhaust was the primary problem, asserting that such emissions dissipated quickly into the ambient environment and thus did not represent an air pollution problem. Moreover, the auto executives argued, Los Angeles represented a unique case due to its multiple sources of pollution. Correspondence between Los Angeles County Supervisor Kenneth Hahn with auto executives over a nearly twenty-year period beginning in the early 1950s underscored the unwillingness of the auto executives to introduce the control technologies that were becoming available.[9]

By the 1960s, air pollution had turned into a major environmental issue, not just in Los Angeles but throughout the United States. An urban-focused environmental movement was beginning to take shape with groups in local communities like Stamp Out Smog (SOS) and scientist and lawyer groups at the national level such as the Environmental Defense Fund and the Natural Resources Defense Council. On the policy front, the federal Air Quality Act was first passed in 1963 and revised in 1967 to focus on sources and control measures and to provide resources to the states to develop their own approaches, including emission control regulations. Auto manufacturers were increasingly squeezed, with Los Angeles and California leading the way by requiring exhaust control devices and establishing new regulatory bodies with more powers. But even though some emissions were reduced by the late 1960s (particularly carbon monoxide and hydrocarbons), other pollutant levels increased (notably nitrogen oxide).

With air pollution reaching the top of policy agendas during the late 1960s and early 1970s, protest movements were proliferating and air pollution promised to be a major issue in the upcoming 1972 presidential election. In response, Congress passed the 1970 federal Clean Air Act, a far more expansive effort at direct federal government efforts to set regulatory standards that included new clean air requirements to be met at the state level.

In many ways, the early 1970s represented the peak of the air pollution problem in Los Angeles. The region continued to experience intense air pollution episodes that led to school closures and warnings about outdoor activities. The last of such episodes defined as a Stage 3 alert occurred in 1974 in the city of Upland in the Inland Empire area. Media stories intensified public concern, such as when a smoking automobile pulled into a gas station and exploded, killing a fourteen-year-old girl while severely burning a nine-year-old boy and his grandfather. Yet the early 1970s also experienced the first signs of a modest reversal, with new technology-forcing regulations and air pollution standards developed at the state and federal levels and implemented through more empowered regulatory bodies such as the federal Environmental Protection Agency, established in 1970, and California's regional-based air districts, including the South Coast Air Pollution Control District in the Los Angeles region, which was established in 1976.[10]

The regulations on lead in gasoline and implementation of a requirement that automobiles have a catalytic converter to reduce ozone emissions were important examples of the shift taking place. The introduction of lead in gasoline dated back to the early 1920s, when it was discovered that an additive to gasoline, tetraethyl lead (TEL), could significantly reduce the "knock" in gasoline, improve its efficiency, and significantly increase the power of the automobile. This furthered the marketability (and eventual dominance) of automobiles for their "comfort, convenience, power, and style," as General Motors head Alfred Sloan characterized its products.[11]

Lead, however, was already known as far back as the Roman period two thousand years earlier to be highly toxic. Within months after production of TEL commenced, serious health problems erupted at its manufacturing facilities, leading to a campaign in 1924 to ban the additive in gasoline. However, the debates at the time focused on occupational hazards. Once workplace controls began to be introduced a few years later, leaded gasoline proponents successfully argued that the emissions were too small (from a dose-response perspective) to represent a problem for the ambient environment.[12] Yet the subsequent magnitude of vehicle sales and the widespread use of leaded gasoline enormously increased the amount of lead in the air, particularly in the areas of greatest exposure, such as near freeways or high-traffic roads. Lead in paint and in older housing (which heavily impacted low-income residents and people of color, turning lead exposures into a civil rights issue) was also highly contested. The two struggles—lead in paint and lead in gasoline—highlighted the importance of policy interventions, with the 1970 Clean Air Act beginning the process of phasing out its use in gasoline.[13]

While the battles over leaded gasoline were taking place, new efforts were made to target emission reductions of the six common or "criteria pollutants" (CO, NO_x, SO_2, ozone, particulate matter, and lead) targeted by the 1970 Clean Air Act through enforceable standards (for example, a 90 percent reduction of emissions in new automobiles by 1975) and technology-forcing tools. The most immediate impactful tool for certain pollutants was the catalytic converter, which was designed to reduce tailpipe emissions in automobiles and whose first-generation technology was ready for use by 1975. Meanwhile, auto industry figures—including the president of General Motors, whose company had been in partnership with the manufacturer of the tetraethyl lead additive—recognized that leaded gasoline would eventually have to be phased out due to its incompatibility with the catalytic converter. When GM's president, Edward Cole, announced his company's intention to do so at a meeting of the Society of Automotive Engineers, the announcement came as a "bombshell." "Here was General Motors, which had fathered the [tetraethyl lead] additive, calling for its demise," the biographer of the Ethyl Corporation (the manufacturer of TEL) wrote of the astounded reaction at the meeting. By 1975, that combination of growing community opposition to leaded gasoline with the policy-driven use of the catalytic converter technology had caused the auto industry to finally reverse itself, and its long-standing advocacy of the critical importance of leaded gasoline for the automotive experience dissipated as well.[14]

Like the leaded gasoline situation, regulatory changes did not always proceed smoothly. Soon after the passage of the 1970 Clean Air Act, a young attorney named Mary Nichols, recently hired by an NGO, the Center for Law in the Public Interest, brought suit against the EPA to implement provisions that directly targeted automobile *use*. The lawsuit focused on one aspect of the Clean Air Act, the State Implementation Plan (SIP) provision, to argue that regulation of automobile use was an appropriate method to help meet California's SIP mandate to reduce its high levels of air pollution. Such an approach would have been a revolutionary change by establishing new types of policy tools for transportation and land use, but the automobile and oil industry interests fought ferociously and successfully against that approach.[15]

Despite continuing opposition from various industry interests and increased numbers of vehicles on the road, air quality in Los Angeles nevertheless gradually improved. This improvement has primarily been due to the expansion of a policy and regulatory infrastructure that has continued to set standards, explore new technologies, and support research that has identified the multiple hazards of various air pollutants and their "hot spot" locations. Such impacts have been most extensive in these hot spots, which are often low-income communities subject to high-traffic roads and polluting industries. The push for such policy interventions has been led by a continuing grassroots environmental mobilization and heightened (though not always continuous) media attention to the air pollution issue in Los Angeles.[16]

A good news/bad news dialectic has emerged in recent years: new and strengthened policy interventions have been reducing air pollution levels, but they have not been enough to eliminate the range and extent of health hazards and daily-life burdens that air pollution still represents for the region. Los Angeles continues to rank first among cities in the US for the highest concentrations of ozone and second (to cities in California's Central Valley) for particulates, while also continuing to receive an F rating from the American Lung Association's State of the Air Annual Report Card.[17]

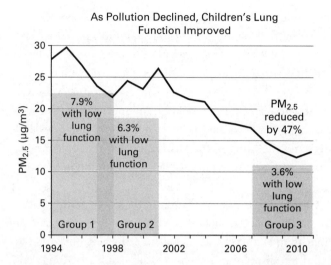

Figure 3.5
Improved air quality in Los Angeles: concentration of PM$_{2.5}$. Gauderman et al., 2015.
Source: Wendy Gutschow.

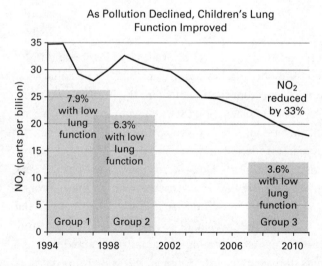

Figure 3.6
Improved air quality in Los Angeles: concentration of NO$_2$. Gauderman et al., 2015.
Source: Wendy Gutschow.

The region also continues to exceed the federal standard for ozone emissions by as many as ninety-two days during the year and more than one hundred days in the Inland Empire area, even prior to the Obama administration's 2015 rule tightening the federal ozone standard, which made compliance even tougher.[18] Moreover, in 2016 a conservative takeover of the board of Los Angeles's leading air regulator, the South Coast Air Quality Management District, led to weaker rules governing oil refineries and other large polluters and the firing of the District's chief executive officer, who had worked closely with community and environmental justice groups.[19]

How then Los Angeles continues to respond to its long-standing battle to reduce the sources and impacts from the "scourge" that Pasadena's mayor had identified more than sixty years earlier is an issue not only for Los Angeles residents but for air-pollution-plagued cities like Hong Kong, Beijing, Shenzhen, and other cities around the world suffering from poor air quality and looking to see how Los Angeles can maintain its momentum in the face of setbacks and eventually shed its long-standing "smog capital" reputation.

Hong Kong

In her 1988 book on Hong Kong, the prolific Welsh travel writer and historian Jan Morris described Victoria Peak as a place where one could view "terrific vistas." Looking out from the Peak toward the Pearl River estuary, Morris wrote that "the hills of Guangdong stand blue in the distance," and with "the city itself precipitously below you ... the early sun catches the windows of Kowloon across the water."[20]

Morris's iconic description of the Hong Kong viewscape from the Peak was shared by many Hong Kong Peak enthusiasts for whom the view of an unobscured Hong Kong skyline was considered a wonder to behold. But during the next two decades, the problem of poor visibility, initially considered a nuisance, became more concerning. In 2004, an Asian edition of a *Time* magazine cover story juxtaposed Morris's 1988 view from the Peak with the view on September 14, 2004, when Hong Kong experienced a record-breaking day of air pollution. That September view from the Peak, Bryan Walsh wrote in the *Time* cover story, "was little more than a smudge. The skyscrapers could be glimpsed only through a veil of noxious smog [and] sunlight did not glint from windows."[21]

It was during the 1990s and 2000s that Hong Kong's visibility began to worsen dramatically. Visibility became a commonly perceived indicator of poor air quality. Prior to the 1990s, low visibility had mostly been caused by fog due to high humidity which was common in March and April. But as the air quality declined, low visibility was strongly associated with smog. While both poor visibility and poor air quality occurred throughout the year, they were especially severe in the winter, due to northerly and north-easterly winds. The number of hazy days, defined as days with visibility less than or equal to 8 kilometers and relative humidity less than or equal to 80 percent, was as high as twenty-eight days a month for a while in 2004. As noted in figure 3.7, that year was one of the worst years for air pollution in the Hong Kong and Pearl River Delta regions.

Air pollution first emerged as a problem during the late 1950s and 1960s when Hong Kong's industrial and economic growth began to take off. The primary concern was at first the dark smoke emissions from power stations, combustion plants, and other industrial installations burning low-quality, high-sulfur fossil fuel. Prior to that point, efforts to address what were con-sidered "smoke" issues were, as in Los Angeles's earlier history, based on

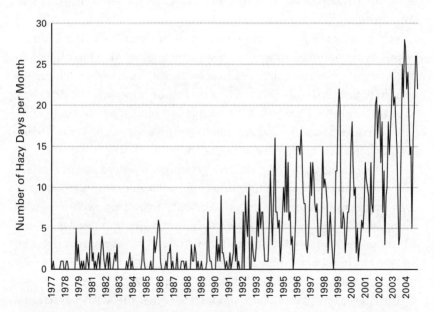

Figure 3.7
Number of hazy days per month in Hong Kong, 1977–2004. Source: Alexis Lau, Hong Kong University of Science and Technology.

nuisance ordinances. A smoke control area was designated first in Shatin New Town in 1962. By 1974 the whole of Hong Kong had become a smoke control area. That year, the colonial government commissioned a study on the need for control measures. Based on the study's recommendations, an Environmental Protection Unit was created within the Environment Branch of the Government Secretariat to help develop policy and legislation in this area. This development helped formalize within the government opportunities for environmental action covering noise, water, air, and waste pollution as well as environmental impact assessments. More substantial efforts through government action took shape with the 1986 creation of an Environmental Protection Department that placed within one agency all major responsibilities in addressing the growing number of environmental problems, including air quality.[22]

Initial air pollution control efforts were established by the 1959 passage of a Clean Air Ordinance. They were then significantly extended when an Air Pollution Control Ordinance (APCO) was enacted in 1983. APCO in turn became the principal legislation addressing air pollution in Hong Kong. In 1987, statutory air quality objectives (AQOs) were further established, which stipulated the concentration levels for seven major pollutants, including SO_2, NO_2, and PM_{10}.[23]

As the authority for air pollution control in Hong Kong, the Environmental Protection Department has been tasked under APCO to achieve its AQOs as soon as is reasonably practicable. The AQOs, however, have never been fully achieved since their 1987 introduction, especially the AQO limits at the roadside. Moreover, AQO limits have generally been lax in comparison to international standards, nor have they been updated or tightened for over a quarter of a century since their adoption.[24]

With worsening air pollution and growing discontent at government inaction among public health professionals, academics, green groups, and even some business leaders, the government finally proposed a review of Hong Kong's AQOs in July 2006 and declared its intent to develop a long-term strategy on air quality management. The review process that followed included a consultancy study commissioned in 2007, and a public consultation that took place between July and November 2009.[25]

The Hong Kong government's proposal to review the AQOs was released around the same time that the World Health Organization (WHO) announced new Air Quality Guidelines (AQGs). AQGs are health-based air quality standards to protect human health from air pollution and were recommended by the WHO working group of international experts based on the most up-to-date peer-reviewed scientific research findings on the health impacts of air pollutants. After the WHO AQGs were made public, the Hong Kong government argued that they might be too stringent for Hong Kong and that economic or social impacts needed to be taken into account before making any changes to Hong Kong's air quality standards. Instead, the government commissioned a study for the review and the development of an air management strategy, rather than directly adopting the WHO AQGs. A group of university-based air pollution researchers in turn criticized the government's approach for effectively delaying a public-health-based response. "If the same level of harm to the public health from our pollution came from an infectious disease," declared University of Hong Kong Professor Anthony Hedley, "[the] government presumably would not be proposing this unnecessary delay in admitting the problem and undertaking whatever measures were necessary to address it."[26]

Yet the Hong Kong government's response continued to be limited. In a speech delivered at the "Business for Clean Air Joint Conference" on November 27, 2006, then Chief Executive Donald Tsang claimed that Hong Kong's air pollution was at "a level comparable with such cities as Tokyo, Seoul, Barcelona, and Los Angeles." Dismissing the concerns of critics, Tsang argued that "foreign investment continues to pour into our city," and that life expectancy in Hong Kong was among the highest in the world. "We [Hong Kong] have the most environmentally friendly place for people, for executives, for Hong Kong people, to live," Tsang concluded.[27]

Tsang's remarks were criticized for understating or ignoring the public health issues associated with air pollution. As concerns about air pollution intensified and new groups such as the Clean Air Network kept the issue prominent, pressure on the government to initiate stronger action continued to mount. Local and regional research studies combined with the growing evidence from researchers around the world on the health and environmental impacts from air pollution made it difficult for the government to ignore the issue. Eventually in June 2010, a list of measures was recommended to improve air quality in Hong Kong. In January 2012, the government proposed new AQOs for Hong Kong, to be adopted together with a package of air quality improvement measures. The new AQOs became effective on January 1, 2014. Critics, however, pointed out that the process of engagement had been too slow and that the government's actions on public health protection were still too limited.[28]

Information communicated to the public has also been seen as lacking. Hong Kong's Air Pollution Index (API) has been communicating the state of air pollution to the general public since 1995 by translating air pollution levels into a scaling system. Every day, API is reported as a number ranging from 0 to 500, with little explanation of the health impact behind the numbers.[29] Members of the general public may be misled to a certain extent by assuming that any numbers below 100 are safe. Yet research has identified that persistent exposure to air pollution levels from 51 to 100 could actually lead to chronic health effects.

In an attempt to better inform policymakers and the general public, Professor Hedley and his team at the School of Public Health of the University of Hong Kong, with the support of Civic Exchange, developed and launched the Hedley Environmental Index (HEI) in late 2008, the world's first index that quantifies the public health and monetary costs of air pollution. Using the image of ticking meters, the HEI website shows real-time, cumulative costs of air pollution since the start of each day in terms of the number of premature deaths, hospital bed-days, doctor visits, as well as tangible and intangible economic loss. HEI emerged as an effective tool to help reframe the public perception of air pollution and facilitate an informed discussion in society about the real cost of air pollution and the consequences of inaction. This additional source of information helped establish a critical mass of research and advocacy that would eventually culminate in new environmental leadership within the government and the push to make air pollution more central to the policy debates across sectors and institutions, whether the ports and the shipping industry, motor vehicles and the transportation system, or land use, real estate development, and the built environment.[30]

Sources and Impacts

Los Angeles

The history of air pollution in Los Angeles has been reflected in the search for the sources of the pollution and how impacts have been identified. During the 1940s and early 1950s, the debates about air pollution sources remained intense. "Many sources have been pointed out as contributors of atmospheric pollution," two officials with the local Air Pollution Control District wrote in 1955. As part of their list, they included "automobile exhausts, incineration of combustible refuse, metallurgical fumes, oil refineries, chemical plants, burning of fuel oil, hot mix paving plants, asphalt roofing manufacturing, railroad locomotives, ships in the harbor, and others."[31] Air pollution control measures varied widely. While the residential (and popular) "Smokey Joe" backyard incinerators came to be prohibited during the 1950s, and various industries, from powerful oil refineries to small dry cleaners, experienced local, state, and federal regulations over the next several decades, auto emissions remained a difficult target despite its recognized role as a primary polluter. Particular automotive pollution control strategies were instituted that clearly reduced some of the pollutant exposures. But the auto-based regulations and greater efficiencies that helped improve air quality were partially offset by the vast highway and road infrastructure developed to accommodate a continuing increase in automobile ownership and use. In addition, new sources have been identified, notably the pollution stemming from the trucks, ships, air traffic, and diesel-powered locomotives and railyards that have become part of LA's growth industry around goods movement and global trade, described in chapter 2.

Despite the improved air quality, health and environmental air pollution impacts continue to beset the region. In the city of Los Angeles alone, air-pollution-related health impacts cost $22 billion annually, with more than two thousand premature deaths per year attributed to air pollution from vehicles. The health impacts due to traffic-related pollution, are extensive and include exposures to ultrafine particles, nitrogen dioxide, and elemental carbon, a marker for diesel. Women experience severe problems such as higher incidences of breast cancer, premature births, low-birth-weight babies, pregnancy complications, and miscarriages. Other health impacts identified for all adults include atherosclerosis, cognitive impairment, diabetes, heart and lung disease, and emphysema. New research on health impacts from particle exposure, such as connections to Parkinson's disease and psychosocial stress, have furthered the understanding of air pollution's extensive impact on public health.[32]

Air pollution due to heavy truck traffic and congested high-traffic surface roads as well as freeways have also been shown to impact nearby residences, schools, playgrounds, and other community places. As in many

other places in the United States, Southern California land use decisions
have resulted in homes, schools, and even parks being located near highways, and subsequent highway construction has ended up even closer to
homes. For example, sixty-five schools are located within one mile of the
I-710 freeway with its huge goods movement-related truck traffic, and there
are more than 600,000 residents (including 212,000 under age eighteen)
who live within 1,500 meters of that freeway. Those living within 1,500
meters also have higher poverty rates and represent a larger proportion of
people of color than Los Angeles County as a whole. In the United States,
59.5 million people live within 500 meters of high-volume roads, including
40 percent of California's population, while 10 percent of the population in
California's major metropolitan areas live in very close proximity to the
highest traffic volume. However, very few air quality monitors are located
in these areas, leaving residents without basic information about their
exposures—and risks.[33]

Children are especially at risk from air pollution impacts when they live
close to freeways and heavy road traffic. One longitudinal study by investi-
gators at USC found that children living near freeways had substantial defi-
cits in lung function development between the ages of ten and eighteen
years, compared with children living farther away. Children also face
increased risk of asthma, and, as a result of asthma exacerbation, increased
wheezing and the need for medication, with potential side effects and quality of life concerns.[34]

Where air quality monitoring has taken place, recognition of the breadth
and extent of air pollution risks indicates its widespread nature as well as
where those risks are most concentrated. The South Coast Air Quality Management District Multiple Air Toxics Exposure Study (MATES), first initiated
in 1986 partly in response to the growing environmental justice movement's focus on toxics hot spots, has become an important source of information about emission sources and air toxics impacts. There have been four
MATES studies, with the latest MATES-IV study released in 2015. MATES-IV
included the latest updated emissions inventory of toxic air contaminants
and a modeling effort to characterize risk across the South Coast Air Basin
(which largely coincides with the Southern California region, including
Los Angeles). Also in response to the growing research and community-
organizing focus on ultrafine particle exposures, MATES-IV added measure-
ments of particle concentrations, which further reinforced the understanding

of the importance of pollution from heavy road traffic, diesel trucks, and ships, among other sources.[35]

The MATES studies and other monitoring data complemented and helped frame important research on air pollution sources and impacts conducted by a team of researchers at USC and UCLA. Their research was based in part on a twenty-year Children's Health Study which measured the lung function of three separate cohorts of children over a four-year age range and in five study communities (Long Beach, Mira Loma, Riverside, San Dimas, and Upland). These communities had some of the highest concentrations of nitrogen dioxide, ozone, and particulate matter.[36]

The link between exposure levels and health outcomes that the USC-led teams identified became an important component of the advocacy work of the environmental justice/community-based groups associated with THE (Trade, Health, and Environment) Impact Project discussed in chapter 2. Air quality monitoring devices provided by USC to THE Impact Project community groups, along with efforts to do "traffic counts" of the diesel trucks leaving or entering the ports, railyards, or warehouses, reframed the technical research into an on-the-ground "citizen science." These included "neighborhood assessment teams" and other forms of community mobilization to create pressure points about specific policy changes and regulatory interventions. Other innovative evaluation tools, such as an electric vehicle mobile platform equipped with instruments able to track real-time emissions, further identified the extent of the air pollution exposures in neighborhoods such as Boyle Heights in Los Angeles (where the mobile platform study was conducted). Such studies reinforced the community's knowledge that they had become communities at risk.[37]

As knowledge about sources and impacts has deepened, grassroots campaigns have begun to advocate a "zero emissions" approach, as developed by one leading group, the Moving Forward Network. This campaign has targeted regulators such as the US EPA, the California Air Resources Board, and the South Coast Air Quality Management District to frame their regulatory powers and technology-forcing goals within a zero emissions framework.[38] Yet the effectiveness of such campaigns and the policy changes that have been instituted over more than half a century, while impressive in improving the quality of the air, has still left Los Angeles out of compliance on several key pollutants. "In the interim," the USC Children's Health Study researchers cautioned, "millions of people continue to breathe unhealthy air in an area where the word 'smog' has become a common household term."[39]

Hong Kong

Hong Kong's most successful policy intervention to reduce a particular source of air pollution took place in 1990, when a regulation limiting the sulfur content of industrial fuel to 0.5 percent was introduced. As a result of the regulation, sulfur dioxide levels at the heavily impacted industrial areas such as Kwai Chung and Kwun Tong dropped dramatically overnight. This example has often been cited as the most effective government intervention to address a major source of air pollution in Hong Kong. While industrial sources of air pollution declined after the 1990 regulation, they dropped even further toward the late 1990s as virtually the entire manufacturing base of Hong Kong had by then relocated to different parts of Guangdong Province. Subsequently, power generation emerged as the main local source of SO_2, NO_X, and PM emissions, followed by road transport. The situation changed once again during the late 2000s, when tighter emission caps were applied to power generation through the Scheme of Control agreement with the two utility companies. Both companies then switched to a fuel mix with more natural gas and less coal and installed advanced emission reduction equipment at the power plants. After emissions from power generation were slashed substantially, ships were then identified as a major source of local emissions.[40]

Moving Hong Kong's factories to Guangdong caused much of the air pollution associated with industrial production to move with them. Ironically, however, air pollution stemming from Hong Kong's factories in Guangdong impacted Hong Kong during the northerly and northeasterly winds of autumn and winter, since Hong Kong's location is downwind from Guangdong. With mainland China's lax environmental regulations and enforcement during the onset of industrial and economic liberalization in the Pearl River Delta from the 1980s through 2000s, the impact of cross-border industrial air pollution became significantly higher than the levels that had been generated by Hong Kong's local industrial base before the industrial relocation.[41]

With the growing impact from these regional sources, evidenced by the growing number of hazy days during both the winter and summer months, the Hong Kong government and residents began to blame the cross-border

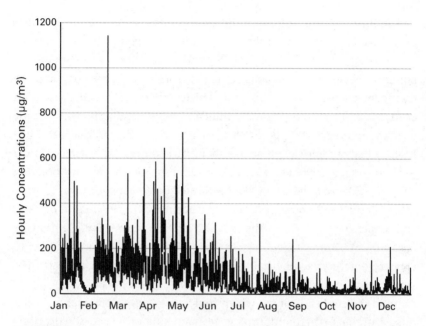

Figure 3.8
Sulfur dioxide levels dropped sharply in Kwai Chung (an industrial district) after fuel restriction regulation became effective starting July 1, 1990. Source: Simon Ng.

pollution from Guangdong for the city's air pollution problems. Yet the relocation of nearly 100,000 factories from Hong Kong to Guangdong had proceeded with few or no environmental considerations due to the cost of action as well as the recognition that enforcement would be lax. As a result, some of the air pollution from the Hong Kong-owned and -financed industrial establishments in Guangdong were externalized to Hong Kong and the wider region.

The Guangdong–Hong Kong air pollution connection also afforded opportunities to highlight the need for more effective intervention with stationary sources (as opposed to mobile sources, which were more local than regional in nature). One 2006 research report noted that "Hong Kong owners and managers of a large share of Guangdong's manufacturing have not faced up to their responsibility for much of the air pollution Hong Kong imports from across the border. ... For its part the Hong Kong Government has taken a laissez faire approach, despite the threat to the health of everyone in Hong Kong."[42]

The focus on Guangdong as the source for Hong Kong's air pollution provided an opportunity for the Hong Kong government to demonstrate its willingness to act. In 2002 a joint study was published on cross-boundary pollution issues by participants selected by both the Hong Kong and the Guangdong governments. The study, which received only limited attention at the time, set various pollution reduction targets and also identified the need to promote and support clean production practices in the Pearl River Delta. It had become clear that cross-boundary sources had emerged during the mid and late 1990s as an important factor in Hong Kong's deteriorating air quality. Data from the late 1990s indicated that Guangdong was contributing as much as 87 percent of SO_2, 80 percent of NO_x, 95 percent of PM_{10}, and 88 percent of VOCs in the Pearl River Delta region. But the growing tendency to blame Guangdong for Hong Kong's problems also tended to detract from the focus on locally created air pollution and the need to intervene around local sources.[43]

Amid the confusion about which sources were responsible for air pollution, a number of groundbreaking research studies helped situate Hong Kong's own role. A 2007 study by the Hong Kong University of Science and Technology with Civic Exchange, using a time-based apportionment method, identified local air pollution sources as the primary influence on Hong Kong's air quality about 53 percent of the time (192 days), whereas regional sources played a key role about 36 percent of the time (132 days). These results were seen as significant to policymakers and the general public because "by taking more environmental responsibility locally, Hong Kong can do much more to improve air quality and therefore public health," the study authors concluded.[44] As referenced in chapter 2, the research that led to Hong Kong's first-ever activity-based and AIS (automatic identification system)-assisted marine vessel emission inventory also identified that ships had been underestimated as a major local source of air pollution. With the new methodology and assumptions in this research, ships could be seen as a top emitter in Hong Kong that could be addressed through immediate mitigation and control.[45]

The medical profession and public health experts also weighed in on the need to control local emissions based on medical evidence. They viewed reducing emissions from sources such as road vehicles close to heavily populated areas as especially critical in order to reduce people's exposure to toxic air pollutants and the associated health risks. Health concerns were further magnified in 2013 when the WHO classified outdoor air pollution as carcinogenic to humans and, in particular, identified an increasing risk of lung cancer with increasing exposure to PM.[46]

Roadside air pollution had previously been identified as a serious problem by the government in the late 1990s. Former Chief Executive Tung Chee-hwa stated in his 1999 Policy Address that air pollution was serious and might get worse, and was affecting people's health. He pointed to deteriorating air quality in Hong Kong from the pollution at street level caused by vehicle emissions as well as to "regional air pollution caused by economic activities throughout the Pearl River Delta region including Hong Kong."[47]

At the street level, diesel vehicles accounted for 98 percent of PM_{10} and 85 percent of nitrogen oxides, and were responsible for 52 percent of PM_{10} and 60 percent of ambient nitrogen oxides in urban areas. A comprehensive emission control strategy for diesel vehicles was recommended in the 1999 Policy Address and has been implemented since then. These included tightening emission standards for new diesel vehicles, tightening the sulfur content of diesel fuel, promoting cleaner alternatives to diesel vehicles such as liquefied petroleum gas (LPG) taxis and minibuses, strengthening emission inspection and enforcement against smoky vehicles, and controlling idling engines. Plans were also identified to improve pedestrian facilities and railway services in order to encourage the use of low-emission transport modes. With these measures, Tung set targets to reduce total emissions of PM_{10} by 80 percent and NO_x from vehicles by 30 percent by the end of 2005.[48]

While the government's control measures since 2000 and subsequent initiatives have led to a gradual reduction of PM and SO_2 emissions from vehicles (but have been less successful for NO_x), roadside air quality still remains poor and unhealthy. In 2009, it was estimated that about 50 percent of the population lived and/or worked close to heavy traffic movements. "In practice," wrote Professor Hedley, "it is likely that a majority of the population is exposed to roadside pollution on most days of the week." In 2011, the last full calendar year under Donald Tsang's administration, annual average PM_{10} and NO_2 concentrations at the roadside were both about three times the respective recommended annual guideline values by WHO of $20\mu g/m^3$ and $40\mu g/m^3$. These numbers are likely to increase without greater intervention, as the number of motor vehicles on the road has continued to increase each year, despite Hong Kong's expansive and effective transit system, as will be described in chapter 6.[49]

The Hong Kong government's underachievement in air pollution control became subject to a review conducted by the Audit Commission, which had been established as an independent authority accountable to the Chief Executive through Article 58 of the Basic Law. Among other observations and recommendations, the Commission in its 2012 report highlighted the government's failure to reduce the number of high-polluting vehicles on the road. These vehicles are mostly old diesel commercial vehicles such as buses and goods vehicles with pre-Euro IV emission standards (i.e., pre-Euro, Euro I, Euro II, and Euro III standards). As an estimate, they represented less than 15 percent of the vehicular fleet in number but contributed about 88 percent of PM_{10} and 46 percent of NO_x emissions from all vehicles. To remove such vehicles from the road or to buy them out, as recommended by the Audit Commission (an approach that had been pursued in Los Angeles and California as a whole), was one readily available tool to reduce roadside emissions.[50]

The Audit Commission's report served as a timely reminder for the new administration under Chief Executive Leung Chun-ying, who took office in July 2012, as they were preparing their air quality management strategy and action plans. The sources and impacts from air pollution in Hong Kong due to local and regional factors were by then well studied and understood by academics and environmental officials. What had been lacking was a sense of urgency and commitment from top leadership to drive change and make a difference. With increasing attention by the public in both Hong Kong and China to the daily impacts of deteriorating air quality on their health and on the environment, the demand for clean air was not going to fade.

The Connection to China

The skies in Beijing are "APEC blue," China President Xi Jinping told the attendees for the opening session of the November 2014 Asia-Pacific Economic Cooperation meeting taking place in the Chinese capital. Speaking to the meeting delegates, including US President Barack Obama and Russian Federation President Vladimir Putin, China's president spoke of his daily routine of waking up in the morning to check Beijing's air quality. He had hoped that morning's readings would not be too bad "so that our distinguished guests will be more comfortable while you are here." The president also expressed his hope that "every day we will see a blue sky, green mountains and clear rivers, not just in Beijing, but all across China so that our children will live in an enjoyable environment."[51]

Clear skies over Beijing, however, were more episodic than not. They could often be found when an event like the APEC conference was taking place. To ensure reduced air pollution, the government instituted emergency measures, identified by China Vice Premier Zhang Gaoli as the "priority of priorities." For the APEC gathering, these measures included strict limits on car use, closing down factories, giving public employees days off, and even banning "the burning of clothes for deceased loved ones, a Chinese traditional practice to provide for them in afterlife," as the BBC noted. Air pollution did not entirely disappear during the APEC meeting, however; by the second day of the conference, pollution levels had increased again to more than ten times the safe levels designated by the WHO. Moreover, increasingly popular apps had begun to identify air pollution level readings, including those released by the US embassy, although they were only able to access the Chinese government's numbers, which were not considered fully accurate, for key pollutants such as $PM_{2.5}$. With China seeking to affirm its status as a global player, it was becoming apparent that it would also need to translate its "priority of priorities" around air pollution into more effective and ongoing action.[52]

Like other urban environmental issues, air pollution in China has been directly related to the country's ambitious industrial development and urbanization strategies after economic changes began to be introduced in the 1980s. Some air quality problems predated the 1980s, such as those caused by deforestation to clear land for the development of industries such as steel; deforestation also generated high particulate pollution by leading to dust storms that plagued cities like Beijing. But the major contributors to the more recent problems of air pollution were the development of new industries, including for export production; expanded energy development, particularly through the use of coal; and the rapid expansion of car use, which skyrocketed particularly after the 1990s. Those links between economic development, energy use, and the rise of the automobile in China also reflected some of the same shifts in Los Angeles and, to a certain extent Hong Kong, and their own histories of air pollution. What distinguished China was the speed and extent to which these types of developments and their related environmental consequences, including air pollution, occurred.

During the 1980s, China did not yet see air pollution as being as much of an environmental concern as resource depletion and stresses on the natural environment and rural communities. A 1979 proposed Environmental Protection Law was not formally adopted until 1989, while the focus on air pollution emerged a bit earlier with the 1987 adoption of the Prevention and Control of Atmospheric Pollution law that was subsequently amended in 1995, and revised in 2000 and 2015. Air pollution, seen as more of an urban issue, failed to generate any widespread grassroots action during the early and mid-1980s, as distinct from some other environmental concerns such as the construction of huge dams. Vaclav Smil, in his 1984 book on China's wide-ranging environmental problems, characterized the attitudes of China's rising urban middle class as not "overly concerned with environmental quality, and if some inconveniences are the price of higher city earnings, so be it."[53]

That situation began to change in the 1990s and into the new century, as air pollution became, along with water quality and food safety issues, the most visible and concerning of China's environmental problems. Air quality was particularly a rising concern in urban regions, where air pollutant levels spiked between the 1990s through 2007. Sources for such pollutants as NO_x, ozone, and ultrafine particulate matter ($PM_{2.5}$ and PM_{10}) were becoming clearly identified and included industry, power generation (particularly by coal), the large and ubiquitous construction sites and dust generation, and the rapid increase in car use. By way of example, in Shanghai, whose air quality was considered relatively good among China's megacities, WHO guidelines were met only 29 percent of the time for PM_{10}, 10 percent of the time for SO_2, and 50 percent of the time for NO_2. During the cool season when air pollution levels peaked, WHO guidelines for all three of the pollutants were met only 5 percent of the time.[54]

China's air pollution problems were substantially, though not exclusively, urban in nature and linked to regional sources. Nearly all the Tier 1 cities and many of the smaller (though still large by US standards) Tier 2 and Tier 3 cities experienced the air pollution and haze that enveloped urban and periurban places. The rapid increase in urban car use was increasingly identified as a major, if not the major source of a wide range of pollution problems. Industrial sources combined with emissions from coal-fired plants located outside but relatively proximate to urban areas combined with the urban-generated pollutants to create high pollution levels in the urban core and throughout the regions. This included the cross-regional (and cross-boundary) problems as well, such as those involving Hong Kong and the Pearl River Delta region.[55]

By 2007, as air pollutants climbed to their highest levels, the central government began to take steps to acknowledge the importance of the issue. New monitoring stations were established and the information about pollutant levels began to be made public. Air pollution began to be recognized as a more serious policy concern, heightened by increased international coverage of pollution problems in places like Beijing. The terms "crazy bad air days" and "airpocalypse" began to be used in this coverage, and although Chinese cities were not the most polluted in the world, China was nevertheless receiving the poorest media reputation for its global air pollution status, as one research study in the *Columbia Journalism Review* noted.[56]

The Chinese government's response, especially through its initial monitoring and public information efforts, was still seen as inadequate, if not misrepresenting the nature and extent of the pollution. In 2008, a young staffer at the US embassy in Beijing, concerned about her own and her colleagues' health due to air pollution, was given permission to set up air-monitoring equipment focusing on $PM_{2.5}$ (which was not yet being monitored by the Chinese government). The equipment was placed on top of the embassy roof, and its real-time information was shared with US personnel and residents through a Twitter account. Though Twitter had been banned in China, the information still went viral, as it was picked up through third-party apps and eventually posted by residents on Weibo, China's main social media outlet, with a huge audience. A handful of NGOs such as Green Beagle posted comparisons of the US embassy and Chinese monitoring numbers, whose discrepancies embarrassed the government. Humorous postings also proliferated, such as one by the well-known Chinese actress Yao Chen who joked that her son would not recognize a blue sky unless he saw one in a history book.[57]

Several weeks-long smog episodes in 2013 that affected almost the entire country had driven the central government to respond to air pollution with a flurry of new and major policy initiatives, acting with unprecedented swiftness, including the announcement by the State Council in September of the first national Action Plan on the Prevention and Control of Air Pollution, which is commonly known as The Ten Measures to Improve Air Quality, as well as a vastly improved air-monitoring system which included real-time data. The Ten Measures suggested "a new mechanism for air pollution control led by government, enforced by companies, driven by market and participated in by the public," set ambitious 2017 targets for air quality improvement (not air pollution reduction) in the Beijing-Tianjin-Hebei region, the Yangtze River Delta, and the Pearl River Delta, and defined ten measures to achieve the objectives.[58]

Research by Chinese scientists and economists on the extent of air pollution impacts had also become influential in this change of course. One 2010 study, for example, indicated that $PM_{2.5}$ had become the fourth biggest health threat to the Chinese population, while a 2012 Chinese Academy for Environmental Planning study, corresponding to equivalent studies by the WHO and the World Bank, concluded that between 350,000 and 500,000 people were dying prematurely due to air pollution. Studies by Chinese scientists further documented air pollution's health impact on outdoor activities, whether for school children or for senior citizens. In addition, concerns were raised about air pollution's role in the loss of economic productivity due to increased sick time as well as impacts on various economic sectors, such as agriculture, steel, energy production, and construction.[59]

At the same time, the Chinese government concluded that public concerns and protests about air pollution were not threats to the system. As an urban issue that impacted the overall population in the cities, individual reactions, such as the trade in gas masks or the installation of air filters in homes, were far more widespread than social actions. Some of that individual response was also class-based—wealthier parents would choose the more elite schools with better filtering systems or move to areas where pollution levels were lower or even travel abroad during the most polluted periods. This type of individual response was what motivated President Xi Jinping's remark about air pollution's impact on people's "perception of happiness." It also led the government to begin to be more transparent about air pollution levels, including during the most severe air pollution episodes when air pollution levels exceeded even the uppermost ranges of what was considered unhealthful.[60]

Even as the central government began to respond, one of its core challenges has been the question of implementation. By 2016, hundreds of environmental regulations by government entities were already in place at the provincial and local government level, where failure to implement was most acute. Against the punitive or monetary cost of failing to comply with regulations was the polluting industries' perception that, for them, the lower costs of production outweighed the costs from their lack of compliance. Moreover, local and provincial governments not only were wedded to the broad development and urbanization goals of the central government, which themselves constituted a barrier to effective implementation, but often had their own stake in the economic activities of those players who were the primary contributors to air pollution in their regions.[61]

Today, the idea that air pollution issues (and other key environmental issues) have equivalent status with economic development and urbanization is increasingly recognized through official and unofficial policies and documents. Smog, Chinese Premier Li Keqiang famously stated in 2014, represents a "red light warning against inefficient and blind development."[62] New punitive and monetary policies were developed to hold local government officials accountable for failure to implement environmental policies (i.e., through amendments to the Environmental Protection Law), and the environment (or "ecological civilization") was more directly elevated as a central policy objective. This latter goal was spelled out in a 2015 document of the Communist Party's Central Committee and State Council on the Promotion of Ecological Civilization by calling for "abandoning economic growth as the only criterion in government performance assessment [and establishing] a lifelong accountability system."[63]

Chinese government media coverage has also been changing, as witnessed by the November 2015 publication in the government news agency, Xinhua, of a series of images of air pollution haze in the northeast city of Shenyang under the title "Fairyland or Doomsday?"[64] Perhaps the most notable response by the government has been the identification of "urgent" triggers for temporary regulations to occur during severe air pollution episodes. Those triggers have included drastic restrictions in normal economic and daily-life activities, and represent a crisis management approach comparable to Los Angeles's response in earlier periods to its own air pollution crises. But whether a crisis management approach and the elevation of ecological civilization can be normalized into an actual paradigm shift has become perhaps China's greatest challenge.

Strategies for Change

In the Los Angeles region, on April 14, 2010, at the Four Seasons Hotel in Beverly Hills, German Chancellor Angela Merkel sat down for a meeting with then California Governor Arnold Schwarzenegger and the head of the California Air Resources Board (CARB), Mary Nichols. Merkel started the meeting by directly addressing Nichols to complain about California's nitrogen oxide air regulations that were "hurting German car makers." Nichols, the auto industry's bête noire during the early 1970s, was subsequently reappointed head of CARB by Governor Jerry Brown, who followed Schwarzenegger. In 2015, Nichols told the German business magazine *WirtschaftsWoche* that she had been surprised that Merkel had such specific knowledge about California's air regulations and their impact on German car manufacturers.[65]

It turned out that the timing of the 2010 meeting had been significant. German car manufacturer Volkswagen (VW) had just begun to install software known as "defeat devices" in its 2009 models. The software doctored the emissions levels for nitrogen oxides in VW's diesel passenger cars to be sold in the United States, including in California, in order to meet those regulations that had been of such concern to Chancellor Merkel. VW would go on to craft a media campaign (with the tag line, "Isn't it time for German engineering?") that promoted its diesel vehicles as environmentally preferable and efficient. Of the 11 million diesel vehicles VW sold worldwide between 2009 and 2015, more than 500,000 were sold in the US and were able to pass emissions tests. Yet, without the doctoring of the vehicles, the diesel cars would have emitted up to forty times more than California's upper limit for nitrous oxides. As the VW scandal intensified, it became, along with Chancellor Merkel's 2010 complaint, an important demonstration that California's air quality policy changes have had an impact, not just in Los Angeles (where the nitrous oxides rule has led to reduced emissions), but globally as well.[66]

In Hong Kong, in March 2013, a Clean Air Plan for Hong Kong was released by the government and was immediately recognized as an important blueprint for air quality management in Hong Kong. The plan emerged as a special document for a number of reasons, including that it explicitly identified protection of public health as a primary policy goal, and that it had overcome traditional agency divides by being a joint effort by the Environment Bureau, the Transport and Housing Bureau, the Development Bureau, and the Food and Health Bureau. Asserting that air quality should not be seen as just a passing concern, the plan further stated that strengthening the ability to address air pollution "saves billions of dollars in future public health costs, missed school days, [and] school absences, not to mention the discomfort and suffering from preventable illnesses and premature death."[67]

Soon after its 2013 rollout, a number of measures recommended by the plan were implemented. These included an incentive-cum-regulatory scheme to phase out all pre-Euro IV diesel commercial vehicles by the end of 2019; tightening the emission caps for SO_2, NO_x, and particulates by 63 percent, 40 percent, and 44 percent, respectively (also by 2019); expanding the regional air-monitoring program with the Guangdong government; and passage of a regulation that required oceangoing vessels to switch to clean fuel with a sulfur content of 0.5 percent or less while at berth in Hong Kong. In the first year after implementation, the HKSAR government identified encouraging readings from its air quality monitoring network, with lower concentrations of SO_2, NO_2, and PM_{10} in 2014 compared to the 2013 levels. Persistently high ozone levels, however, remain a worrying sign. Despite a long-standing preference for market rather than policy interventions in Hong Kong, the Clean Air Plan directly pointed to the importance of policy change as perhaps the most direct source for improving Hong Kong's air quality.[68]

In China, during the first week in December 2015 and just as the Paris climate talks were getting under way, the central government, responding to increased levels of air pollution in the Beijing region, called its first ever "red alert," a monitoring trigger it had established in 2013 to address the most severe episodes of air pollution. The red alert led to a number of temporary restrictions, not dissimilar to those imposed during the APEC conference the previous year. Schools were closed, cars were allowed to drive only on alternate days based on their license plate numbers, and fireworks and outdoor barbecuing were banned. Government agencies were also required to keep 30 percent of their automobiles off the streets. The official announcement of the red alert came via the state news agency's Twitter account and even included a photograph of the Bird's Nest, the stadium built for the 2008 Summer Olympics, shrouded in charcoal-gray smog and barely visible.[69]

While the timing of the red alert was criticized initially by environmental groups like Greenpeace East Asia because high air pollution levels had been registered the previous week without such a response, it soon became clear that the government's action was nevertheless unprecedented. Just two weeks later, a second red alert was issued, with some of the same restrictions, when high pollution levels returned to the Chinese capital. Though residents might well have continued to experience what one writer called "pollution fatigue," the issuance of the red alerts nevertheless indicated a different kind of approach by the government in its willingness to make more transparent the seriousness of the air pollution problem.[70]

For Los Angeles, Hong Kong, and China, policy changes and a change in the discourse had come about in part due to pressures from below, whether in the form of community mobilization, expanded research, or individual discontent. In Los Angeles, for example, the groundbreaking 2006 Clean Air Action Plan was a direct product of litigation, grassroots organizing, and government officials' beginning to recognize that a "greener" port and reduced emissions from goods movement traffic was potentially in the port's (and the region's) best interest. In Hong Kong, the elevation of the air pollution issue to at least equal status with the long-standing priorities of economic development and market-based solutions had been strongly influenced by the research and advocacy of groups like Civic Exchange, whose founder and former Chief Executive Officer had pulled together the Clean Air Plan after she was appointed to a leading environmental policy position in the government. And in China, new policy approaches coincided with the government's fears of widespread discontent about air pollution, especially among its urban middle-class constituency, considered central to the country's economic growth engine and its achievement of the "China Dream."

Air pollution, as noted earlier, is not just a local or metropolitan or even national concern. A key element of Hong Kong's Clean Air Plan, for example, is its connection to Shenzhen and the Pearl River Delta and the sources and levels of pollution experienced in both places. It has become clear to Hong Kong that a China connection is essential in developing an effective air pollution strategy for policy and implementation *at the regional level*. This is true not only for port issues but for the cross-border nature of air pollution overall, whatever the source of the emissions. A key element of Hong Kong's Clean Air Plan, then, is how such a connection could involve effective implementation and increased monitoring at both sides of the border.[71]

As described earlier, Hong Kong's connection to China with respect to air pollution has had its own complex history. While there have been complaints in Hong Kong that its air pollution problems have been due to the air pollution coming from factories in Guangdong, Hong Kong capital had played a key role in the development of Guangdong's dirty industries, particularly its manufacturing base for exports. Already by 2005, these Hong Kong-related Chinese factories employed nearly twice as many workers as Hong Kong's entire population. But in the early 2000s when Guangdong's government began to explore shifting some of its export manufacturing base, particularly to convert Shenzhen into a high-tech center and (presumably) a less polluting economy, responses in Hong Kong, including from Hong Kong's then Chief Executive Donald Tsang, were more ambiguous. It was only after air pollution became a more prominent problem in both China and Hong Kong that a shift took place at the policy level that recognized a transboundary approach would be necessary. This included Hong Kong's "Clean Production Partnership Program," which gave government grants to Hong Kong-owned factories in Guangdong so they could upgrade their technology for greater energy and water efficiency and pollution control.[72]

The connection to China has become a factor for Los Angeles as well. In 2014, a study by Chinese and US scholars indicated that China was exporting its air pollution to Los Angeles and the United States. The study noted that Los Angeles experienced at least a 2 percent increase in ozone emissions or one extra day of noncompliance with LA's ozone standard due to China's air pollution. This included emission sources related to the global trade and goods movement system, as global wind patterns carried the pollutants across the ocean along with the ships that carried goods that had been made in China. Yet the community groups advocating for clean air raised concerns that the focus on China took attention away from the local sources of pollution, which continued to be seen as the most important arena for action and change.[73]

The fear that China could create a global environmental threat was not new. In 1967, a *Time* magazine cover story entitled "The Polluted Air" showed a Los Angeles skyline obscured by a hazy smog. Fifty years later, that same type of image could be seen in pictures from Beijing or the view from the Peak in Hong Kong during its air pollution episodes. The 1967 *Time* article quoted UCLA professor and meteorologist Morris Neiburger speculating that if 800 million Chinese (its 1967 population) owned cars, then "a globe-encircling smog girdle could develop, and civilization would perish, not in an instantaneous cataclysm but in a prolonged suffocation, like dying in a caved-in coalmine."[74]

By 2016, China had failed to reach the 800 million car ownership level, but it had surpassed the number of automobile drivers in the United States and was approaching 300 million cars and trucks on the road. Los Angeles, however, still remained the representative of the car-dominant city, having paved the way for a development model that had led to a serious and debilitating air quality problem.

Air pollution in Los Angeles, Hong Kong, and China is in fact a global issue. Air pollution is linked to climate change, which stems from various air pollutant sources such as black carbon. But these issues are at once local *and* global. Local sources, whether car and truck use in Los Angeles, ships in Hong Kong, or coal-fired plants and the greater number of vehicles on the road in China, create both local and global impacts. Local air pollution represents the primary target for locally based community action and policy change. But, as one of the Chinese authors of the 2014 study pointed out, the transport of air pollution across oceans and borders is not just a question of where the pollution is coming from—including its role as a production-related issue—but where the goods end up—a consumption issue, as reflected in the global trade system's impact on goods consumption. This interplay between production and consumption is also a core contributor to climate change. The emissions from the production of traded goods, for example, grew from composing 20 percent of global greenhouse emissions in 1990 to 26 percent in 2008. This was especially true in the growth of those goods earmarked for export, such as textiles, electronics, furniture, and cars. According to the Intergovernmental Panel on Climate Change's fifth assessment report (2014), "a growing share of total anthropogenic CO_2 emissions is released in the manufacture of products that are traded across international borders." Those products in turn have formed the basis for consumption's contribution to climate change—and to what has come to be seen as the continuing problem of air pollution.[75]

This cycle of production, distribution, and consumption is also tied to the global movement of capital and labor, whether in the form of foreign direct investment or the perennial search for the cheapest labor sources. It is tied to a series of choices about how cities have developed and how the systems of production, distribution, and consumption have evolved. Yet, while air pollution is perhaps the most discussed environmental problem in the global city, it is only one of several urban environmental impacts that preoccupy community members, researchers, and policymakers alike.

4 Water for the City

Along the River

In 2011, a group of researchers, photographers, and student interns led by Civic Exchange undertook a fifteen-day journey along the Dongjiang (also known as the East River) in Guangdong Province. Their goal was to document conditions along the river, a source of water for Hong Kong for more than fifty years. A few years prior to the Dongjiang expedition, in 2008, a group of Los Angeles kayakers and environmental advocates had made a three-day expedition along the Los Angeles River with the goal of demonstrating that the channelized and much-maligned LA River was still a river. These two unusual river journeys were each designed in their own way to visually identify some of the key water issues facing Hong Kong and Los Angeles.

The Dongjiang journey began in Huizhou, one of the six cities dependent on the river for their water. (Aside from Hong Kong, they include Heyuan, Huizhou, Dongguan, Guangzhou, and Shenzhen in Guangdong Province.) Nearly forty million people live in those cities, in one of the densest regions in the world. The area also includes a range of industrial facilities, sewage and waste management systems, and agricultural activity.[1] Huizhou, with its nearly two million residents, is located at the upper end of the Dongjiang. Along with the smaller city of Heyuan,[2] the two cities' mix of petrochemical, electronics, paper, iron and steel, warehousing, and information technology industries and their rapidly expanding urban population (with greater sewage discharges and waste generation) provided an effective visual demonstration of the water quality and water supply problems the expedition sought to document. Both Huizhou and Heyuan are also sites for Guangdong's policy of industrial relocation to shift industries from one area to another, partly for development purposes and partly to relocate the pollution, including from areas with more labor-intensive industries.[3]

As the group made its way along the Dongjiang, they witnessed the results of the combination of factory relocation, industrial development, rapid urban expansion, declining agriculture, and the gap between rich and poor. At Huizhou, the team photographed the high-rises that had sprung up along the Dongjiang as it made its way through the city. They documented the hardening of the riverbanks and the river's increasing eutrophication and the growth of highly invasive and ecologically damaging water hyacinth along its sides. At Heyuan (whose name signifies "the origin of the river"),[4] the expedition photographed the industrial wastewater that had been discharged from one of the new industrial parks. Despite the construction of new wastewater facilities, the team learned that less than 60 percent of Heyuan's wastewater had been treated at the time of the expedition.[5] They also photographed a new urban development next to a reservoir in the city and the nearby mountainsides covered with fast-growth eucalyptus, another invasive. Further downstream, the group recorded an enormous open-air quartz quarry and processing plant also next to the river. Their most poignant photograph was of a father and daughter living in a garbage pile, representative of those displaced by the dramatic changes in the Dongjiang Basin area.[6]

The Los Angeles River kayak journey had its own core objective: to demonstrate that the LA River, which passed through the heart of the city, should not be considered just a desolate, dystopian urban landscape. Contained and straightened, the river, which has served as a backdrop in several Hollywood films such as *Terminator 2*, is now largely a bed of concrete. It was even once humorously characterized in 1985 by a *Los Angeles Times* reporter as a "threadbare coat of unspeakable slime."[7] After the US Army Corps of Engineers had completed cementing the river in the 1950s and had constructed the high walls along its banks (placing barbed wire along several of its sections), the river became a dry bed for most of the year. The exception was the relatively short rainy season from November through March. Then the river/flood-control channel was transfigured when larger storms took place, perhaps involving a few inches of rain, far milder than the "black rainstorms" and typhoon conditions that can descend on Hong Kong. During Los Angeles's largest storms, water cascades down from the

Figure 4.1
Garbage blocking water flow, Huidong County, Guangdong Province. Source: Civic Exchange.

San Gabriel Mountains, sending a torrential flow through the concrete bed until it reaches the river's mouth, where it empties out into the ocean at the Port of Long Beach. Since the river bed has been straightened and cemented, those occasional rain storms turn the LA River into a place of danger, sweeping everything in its path and reinforcing the river's reputation as a forbidding landscape, enclosed and separated from the rest of the city.

That pattern began to change in the late 1980s when a new sewage treatment plant near the source of the LA River began to discharge its tertiary treated water into the river bed, causing a small twelve-month flow of water along its fifty-one-mile path. In three "soft bottom" areas of the river where concrete had never been laid due to the high groundwater table, new life resurfaced—vegetation, trees, birds, even egrets and marine life. A very modest river began to be visualized, reinforcing the growing environmental advocacy about river restoration and a push for new environmental-oriented flood management strategies.[8]

These changes helped spark the idea of the kayak expeditions. This would be the first attempt to test the possibility—and politically promote the idea—that the river was "navigable" in order to place it under the jurisdiction of the Clean Water Act, with its various river protection mandates.[9] As described in his daily blog by LA River advocate and kayaker Joe Linton, the group took off about five miles downstream from the river's source, in eight bright-yellow kayaks and a couple of bright-green canoes. The launch area was one of the three places with earthen bottoms and tall trees, and became one of the more pleasurable parts of the trip, enabling the kayakers to see night herons, great blue herons, and mallards. But not all was so simple after that. During the next stretch of river, the kayakers had to walk their boats through the rocks and debris and other barriers, but they ultimately completed their journey—and made their point. "This is a River!" they proclaimed. Through that assertion they hoped to change the way Los Angeles residents could view this still bleak landscape and begin to reconstruct the way they thought about water in and for the city.[10]

Figure 4.2
Los Angeles River kayak expedition. Source: Joe Linton.

The two expeditions, as different as they were in what they encountered, shared the idea that water was precious—central to the landscapes they intersected and the places they served. Whether polluted, reconfigured, dried up, hemmed in, contested, or free flowing, the waters still had a "life of their own," as the Ute Indians in the southwestern United States said of all water sources, with the reverence they felt they deserved.

Water Supply

Los Angeles

Water has long played a defining role in the growth and development of the Los Angeles region, whether its water could be sourced locally or tapped from more distant locations. Local sources have included the Los Angeles River and San Gabriel River watersheds and related groundwater basins throughout the region. The push for a more distant or imported water supply was linked to the push for regional expansion, led by those promoting an extension of the city's (and later the region's) boundaries.

It was the development of those distant water sources that fundamentally reoriented Los Angeles's water supply and influenced its development patterns. Development began with the construction of the Los Angeles Aqueduct, completed in 1913, which brought water to the city from the eastern Sierra Nevada mountains (via the Owens River watershed), 230 miles northeast of the city's limits. As a result of this new imported water supply, the City of Los Angeles was able to expand four times between 1913 and 1928 by making the water available to adjacent areas contingent on their annexation to the city. The Owens Valley water also provided the basis for a frenzy of real estate speculation and new developments in Los Angeles far from the city's core.[11]

The next major imported water project involved the development of Hoover Dam in Nevada and the construction of the Colorado River Aqueduct, which diverted Colorado River water into California. After this imported water supply became available in 1941, it triggered new urban growth and territorial expansion. A pattern had been set: each new imported water source enabled new urban development, which in turn created pressure for yet additional imported water. With the Colorado River water came the extension of new urban growth well beyond the city's boundaries. The allocation of this supply was undertaken by a regional water wholesaler, the

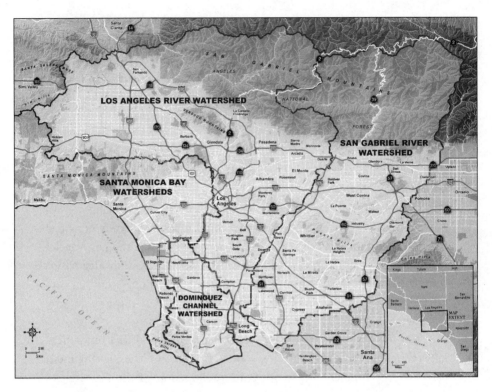

Figure 4.3
Los Angeles region watersheds. Source: Council for Watershed Health.

Metropolitan Water District of Southern California (MWD, or Metropolitan), which had been formed as a special district in 1928 in the wake of the Colorado River plans. Metropolitan subsequently divvied up this new imported supply to city and county water districts throughout the region (including but no longer limited to the City of Los Angeles) once they annexed to Metropolitan. Metropolitan's service area then stretched beyond Los Angeles as far as Orange and San Diego counties more than a hundred miles to the south and areas to the east (Riverside and San Bernardino counties) and west (Ventura County).[12]

The third major imported water system involved the development of a conveyance system from tributaries of the Sacramento River in Northern California through the Sacramento Bay Delta to the central and southern parts of California. The completion of the Southern California end of this California Aqueduct in the late 1960s reinforced the notion of continuing

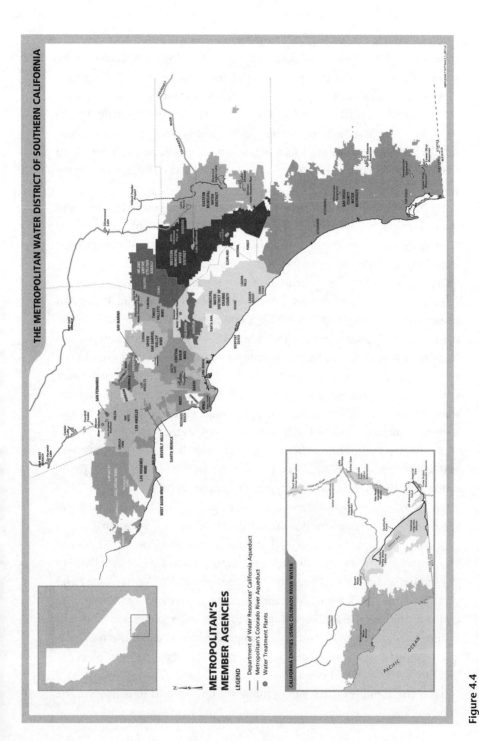

Figure 4.4

Metropolitan Water District of Southern California service area. Source: Courtesy of the Metropolitan Water District of Southern California.

future urban-edge expansion, accomplished in part through conversion of agricultural and undeveloped lands into urban development. Each of these annexations—first to the City of LA and then on an expanded regional basis to Metropolitan's system—disregarded any land use planning logic other than the ability to expand the outer edge of development. By the 1960s, Metropolitan had become a de facto regional land use planner, contributing to the horizontal pattern of growth and development in Southern California that became the prototype of post-World War II suburban expansion in the United States.[13]

From the late 1960s through the 1980s, the development interests and the water agencies assumed that this pattern of growth could continue by identifying yet more distant water supply projects. Some of the schemes were truly extravagant in scope—for example, a proposed transnational, interbasin water transfer program from water sources in British Columbia, Canada. This proposed project, known as the North American Water and Power Alliance, would have included a 400-mile-wide storage basin across the border in Montana before delivering water thousands of miles to Los Angeles as well as to southwestern states such as Arizona, and even to the Great Lakes area in the midwestern United States. There were even more fanciful water transfer ideas, such as creating an undersea aqueduct to transport Alaskan waters to the Port of Los Angeles that could then be sent to various locations in the Southwestern United States.[14]

But by the 1970s and 1980s, the era of the Big (imported) Water Project and a continuing cycle of urban expansion had reached its limits, partly due to prohibitive costs and partly due to recognition of the problems associated with horizontal/suburban expansion. In California, the culmination came with a 1982 referendum on construction of new water facilities (a Peripheral Canal, among others) that could better expedite the transfer of Northern California water through or around the Sacramento Bay Delta to the industrial agricultural lands on the west side of the Central Valley and then over the Tehachapi Mountains to urban Southern California. The defeat of the Peripheral Canal proposal (due to huge margins in Northern California, but a surprising number of opposition votes in Southern California) signaled that the old strategies of imported water expansion had conceivably reached its end point.[15]

Over the course of the 1980s, a subtle yet critical shift took place in the debate about future water policy, a change that became more noticeable in subsequent decades. During the 1982 Peripheral Canal campaign, dire predictions by advocates of imported water expansion ("people will have to drink water from the toilet bowl to meet mandatory cutbacks [if these facilities are not built]," one MWD board member argued)[16] brought a note of desperation to the debates about the future of water in Southern California. Yet, after the defeat of the Peripheral Canal and the failure two years later to get the California legislature to adopt yet another plan to transfer Northern California water south, the mood of crisis began to give way to recognition that the link between imported water and future urban-edge expansion had frayed. A series of dry years between 1987 and 1992 and consequent reduced demand due to restrictions on water use created a new kind of crisis—the need to raise water rates on an emergency basis, since lower sales of water to Southern California customers meant there was insufficient income to pay for the fixed costs of a system now dependent on imported water and its facilities.[17]

By the 1990s and into the new century, a combination of growing environmental advocacy coincided with an increased focus by the water agencies on water planning and water management strategies.[18] This new emphasis had multiple layers. Within the City of Los Angeles, the Department of Water and Power (or LA DWP, the public body that manages both the water and electricity systems for the city) finally resolved long-standing disputes with communities in the Owens Valley and Mono Lake watersheds where the DWP had been drawing water (and electricity from hydroelectric generation) since 1913. New water pricing strategies based on levels of use, indoor water and landscaping conservation measures, and even exploration of whether and how to once again use the LA River for storage and replenishment purposes began to modestly shift the city toward more of an efficiency-oriented, conservation, and demand management camp. The city's approach, however, was not without its critics, who felt it remained far too reliant on imported water and unwilling to fully address alternative supply and demand strategies.[19]

Beyond the City of Los Angeles, changes were uneven at the regional level at Metropolitan and among some of the cities and water districts in the urban-edge areas of the region. On the one hand, there was interest in developing new conveyance facilities for imported water. For example, San Diego's concern about limited local supplies—and its reliance on the Metropolitan system—translated into a desire to establish a pipeline from the Imperial Valley to pay for and access Colorado River water otherwise allocated to the agricultural water users in Imperial. Such a project could reduce the same supply source for other member agencies of Metropolitan, which led to sharp interagency conflicts. There was also interest in building large capital- and energy-intensive desalination plants to convert ocean water for use in Southern California; these proposals resurfaced periodically, particularly during drought periods. Some strategies oriented toward water management, efficiency, and conservation also began to be developed. These included incentives for less-water-intensive landscaping; the integration (or conjunctive use) of local, groundwater storage and imported water storage; and water reuse or recycling for places like golf courses.[20]

This shift at the regional level toward more of a water-planning and efficient use of resources approach became the focus of Metropolitan's "Integrated Resources Plan" (IRP). The IRP was first introduced in 1994, with a post-drought recognition that traditional supply strategies needed to be balanced with an efficiency and water management approach. With political support for the traditional supply strategies waning and climate change adding uncertainty to future supply sources, the new buzzword in the IRP and among water agencies became *reliability*. In an October 2010 and subsequent 2011 update of the IRP, this shift was made explicit. Metropolitan's IRP manager warned directly, in 2011, that reliability had become the central concern of the agency. "If no action to improve local and imported water resources occurs," he argued, "the region could experience significant water supply shortages once every other year (or 50 percent of the time). Of course this level of reliability would be unacceptable."[21]

Despite these efforts to address reliability, the water supply situation for Los Angeles and Southern California has remained challenging. It has become dependent not only on external factors (e.g., climate change) and problematic long-distance sources, whether from Northern California, the Colorado River, or the Owens River and Mono Lake watersheds, but also on whether and how changes in attitudes and practices regarding the *use* of water could be accomplished. Los Angeles could utilize local sources, especially its groundwater, but that supply has also been subject to challenges, not the least of which have to do with the *quality* of the water. Partly due to those water-quality-related problems, the city has been forced to shut down a number of groundwater wells and still receives as much as 89 percent of its water from imported sources. At the regional level, about 50 percent of the entire region's water supply is derived from local sources, with another 50 percent coming from imported sources, primarily from Metropolitan's imported water sources.[22]

The most recent drought, which began during the 2011–2012 water year (July 1–June 30) and reached crisis proportions by 2015, brought to the fore the unreliability and insecurity of the region's water supply and the need for a far more expansive water management focus. A mood of crisis—and a desire to act—culminated in 2015 when California Governor Jerry Brown announced 25 percent mandatory reductions in water use (and as much as 35 percent or more for some water agencies). These were the first mandatory restrictions ever put in place in California. With the heightened attention and the threat of major economic penalties if reductions failed to meet targets, Californians began to adopt major changes in how they used water, such as elimination of turf-based lawns and landscaping and, through regulations, installation of appliances and toilets with even greater water efficiency.[23]

The critique of the long-standing prevalence of what has been called the "super-green lawn" presents an interesting example of the efforts to mandate water use restrictions. For more than a half century, the highly manicured, continually irrigated, grassy lawn had been seen as an "American obsession," the quintessential American phenomenon. Associated with a suburban lifestyle and a manufactured form of nature in the city, its promoters happily characterized the lawn as a homeowner's "own little piece of the earth."[24]

In Los Angeles, the lawn had been prized as essential landscape, regarded as important for property values and for the look of the home. For many suburban developers and residents, it became central to the character and identity of the suburban home. In 1934, the City of LA passed a yard ordinance requiring that all residences have yard space, as the detached dwelling surrounded by landscaping became the region's dominant urban form.[25] This mutated into the super-green lawn look during the 1950s, reinforced by a kind of quasi-regulatory instrument to maintain that look and that uniformity. Almost from the outset of the development of the post-World War II suburbs, residents came to serve as gatekeepers, complaining about any lawns not well maintained by their neighbors and even pushing for local policies that mandated such upkeep.[26]

Table 4.1

Sources of Water Supply to the Metropolitan Water District Service Area, 1976–2015

Calendar Year	Local Supplies	Los Angeles Aqueduct	Colorado River Aqueduct	State Water Project	Total
1976	1,363,000	430,000	778,000	638,000	3,209,000
1977	1,370,000	275,000	1,277,000	209,000	3,131,000
1978	1,253,000	472,000	710,000	576,000	3,011,000
1979	1,419,000	493,000	784,000	532,000	3,227,000
1980	1,452,000	515,000	791,000	560,000	3,317,000
1981	1,500,000	465,000	791,000	827,000	3,583,000
1982	1,392,000	483,000	686,000	737,000	3,298,000
1983	1,385,000	519,000	850,000	410,000	3,163,000
1984	1,621,000	516,000	1,150,000	498,000	3,785,000
1985	1,535,000	496,000	1,018,000	728,000	3,776,000
1986	1,510,000	521,000	1,001,000	756,000	3,789,000
1987	1,465,000	428,000	1,175,000	763,000	3,831,000
1988	1,521,000	369,000	1,199,000	957,000	4,047,000
1989	1,542,000	288,000	1,189,000	1,215,000	4,234,000
1990	1,470,000	106,000	1,183,000	1,458,000	4,217,000
1991	1,426,000	186,000	1,252,000	625,000	3,490,000
1992	1,512,000	177,000	1,153,000	744,000	3,586,000
1993	1,408,000	289,000	1,144,000	663,000	3,505,000
1994	1,527,000	133,000	1,263,000	845,000	3,768,000
1995	1,590,000	464,000	933,000	451,000	3,438,000
1996	1,715,000	425,000	1,089,000	663,000	3,892,000
1997	1,759,000	436,000	1,125,000	724,000	4,044,000
1998	1,726,000	467,000	941,000	521,000	3,655,000
1999	1,887,000	309,000	1,072,000	792,000	4,060,000
2000	1,768,000	255,000	1,217,000	1,473,000	4,714,000
2001	1,708,000	267,000	1,245,000	1,119,000	4,340,000
2002	1,706,000	179,000	1,198,000	1,415,000	4,498,000
2003	1,659,000	252,000	676,000	1,561,000	4,148,000
2004	1,627,000	203,000	741,000	1,802,000	4,373,000
2005	1,590,000	369,000	707,000	1,525,000	4,190,000
2006	1,710,000	379,000	514,000	1,695,000	4,297,000
2007	1,852,000	129,000	696,000	1,648,000	4,326,000

Table 4.1 (continued)

Calendar Year	Local Supplies	Los Angeles Aqueduct	Colorado River Aqueduct	State Water Project	Total
2008	1,842,000	147,000	896,000	1,037,000	3,922,000
2009	1,857,000	137,000	1,044,000	908,000	3,946,000
2010	1,729,000	251,000	837,000	1,129,000	3,946,000
2011	1,664,000	370,000	445,000	1,379,000	3,859,000
2012	1,867,000	167,000	455,000	1,252,000	3,741,000
2013	1,866,000	65,000	984,000	974,000	3,889,000
2014	1,885,000	62,000	1,168,000	607,000	3,723,000
2015	1,676,000	27,000	1,180,000	550,000	3,442,000

Source: Courtesy of the Metropolitan Water District of Southern California. Figures are in acre-feet.

But water availability emerged as the super-green lawn's biggest problem. As early as 1980, a group of water activists began to talk about a "landscape ethic," focused on the 40 to 50 percent of water in Los Angeles used primarily for landscaping. Still, policy initiatives to address landscape use remained modest until the 2011–2015 drought led to a major shift in both consciousness and policymaking. Policy changes included the combination of mandated cutbacks and turf removal incentive programs that became hugely popular.[27] In 2014, California Governor Jerry Brown issued an executive order banning homeowners' associations from fining members for having brown lawns, a policy extended the following year by city and county regulations and fines. Social pressures shifted from residents with dry or unkempt yards to the "drought shaming" of households perceived to overwater, especially wealthy residents or celebrities who appeared to be buying their way out of conservation efforts. These policy changes and social pressures have helped transform the lawns (or at least front lawns) in many LA neighborhoods. Today, expanses of green, irrigated turf have given way to an assortment of succulents and cacti, crushed granite, bark or mulch, artificial turf, dead or dying lawns, raised bed gardens, and hardscapes like concrete, brick, or pavers.[28]

Turf removal can be seen as a type of low-hanging fruit in an evolving era of water use, particularly when implemented through incentives or rebates where the public response has exceeded expectations. The challenge has been how to extend what can be considered "dry year" conservation strategies, identified as one-time, drought-related changes, into "normal year" conservation based on changes that have already been undertaken or are about to be implemented. But as short-term ("dry year") conservation practices become institutionalized (becoming "normal" conservation practices), it reduces the margin for further reductions during any future dry-year events. It is quite possible, for example, that the turf removal programs will soon come to be seen as a "normal year" conservation practice, particularly after incentive programs are eliminated, as they were in 2016 when drought conditions eased. And while some of the discourse about water use has changed, it has underlined the need to extend a landscape ethic to deeper and more lasting changes in land use practices in the region. Climate change, in that context, becomes a critical, if not determining factor, reinforcing the need for even more dramatic shifts. While imported water had fueled Los Angeles development scenarios, the problem of future reliability poses the need for an entirely refocused water management approach.

Hong Kong

While Los Angeles became dependent on imported water, Hong Kong never had the luxury of fully relying on its local water, despite a history of efforts to capture whatever local sources could be developed. The interplay between local and more distant sources for Hong Kong was not so much a question of how to accommodate future growth, but how to meet existing and even immediate needs, especially in periods of insufficient supplies amid debilitating droughts.

In contrast to Los Angeles, Hong Kong gets plenty of rain. Its rainy season occurs during its warmest months, from May to September, when it receives 80 percent of its annual rainfall. The yearly rainfall may range between 40+ inches (1,000 millimeters) to more than 140+ inches (3,500 millimeters). Los Angeles annual rainfall, in contrast, may be as low as 3 inches (75 millimeters) or as high as 35+ inches (825 millimeters), less than even Hong Kong's driest period. But rainfall in both places is uneven and unpredictable, with major swings between dry and wet years. Even more than Los Angeles, Hong Kong can experience extreme droughts that can turn into crises. That possibility has shaped the city's approach toward developing its water supply.

Water for a Barren Rock is the title of a 2001 book by Chinese University of Hong Kong Professor Ho Pui Yin that was commissioned and published by Hong Kong's Water Supplies Department. It is an apt title for conveying Hong Kong's essential challenge—despite its massive storms and heavy rainfall, Hong Kong lacks sufficient *usable* freshwater resources and is largely devoid of any groundwater. Instead, it has a long history, dating back to the British takeover of the island in the 1840s, of constantly searching for ways to develop a water supply for a growing population. The early lack of any significant water infrastructure (both for water supply and sanitation purposes) also became a major contributor to health problems and poor living conditions during much of the nineteenth century and into the twentieth century. These problems were exacerbated by class and colony-related divisions. The colonial and wealthier residents who lived in private houses with their water containers, swimming pools, gardens, and fountains, for example, were not charged extra fees for their high water consumption, but paid the same amount as those with no running water. Water wastage was also a serious problem caused by the private home owners (called "Westerners") who would let their water cisterns run all night and would not automatically turn them off when they became filled. Similar tensions along a class and ethnic divide in water services took place in nineteenth-century Los Angeles, where the development of the city's water sewer services initially did not extend to the Mexican and Chinese neighborhoods, thus only reinforcing negative stereotypes about Mexicans and Chinese as "dirty and diseased."[29]

Hong Kong's lack of infrastructure and susceptibility to water quality problems, combined with periodic droughts, underlined the vulnerability of much of its population. The 1902 drought, for example, witnessed 460 cholera deaths traceable to contaminated water. The government was forced to shut down wells, even though this reduction of the available water created its own health-related impacts. An even more severe drought in 1929 forced the colonial government to ask mainland China to bring in water tankers to head off a wholesale calamity.[30] But it was the 1963–1964 drought that forever changed the dynamic of how Hong Kong would seek to meet its water supply needs.

Prior to the 1960s, the Hong Kong government had actively explored a range of local supply strategies, some of which were considered among the most innovative at the time. These included identifying locations for catchment basins, which led to the construction of two major reservoir systems to create some baseline water storage. One of these, Plover Cove, identified as a "reservoir in the sea," was recognized for its innovative design. The construction of Hong Kong's dual water system for seawater to be used for toilet flushing provided another important breakthrough that dated back to the 1950s. Not only has the seawater supply program been successful in reducing the need for freshwater resources for the city's residents, but it has been continually expanded to parts of Hong Kong not easily accessed. By 2015, the seawater toilet-flushing systems reached 85 percent of the city's residents and represented 27 percent of residential water use, with plans to extend it to 90 percent by the end of the decade.[31]

More than earlier periods, the 1963–1964 drought shifted Hong Kong toward far greater dependence on imported water, notably in the form of water transfers from the Dongjiang River and Pearl River Basin in mainland China. The 1963-64 drought also produced some of the most dramatic impacts in Hong Kong's history. Only forty-one inches of rain fell during that period (the contrast with Los Angeles is again striking). This shortfall was further exacerbated by increased water consumption in the two years before the drought. In September 1963, reservoirs were only 51 percent filled, just as drought conditions began to worsen. More draconian measures were instituted, including limiting public use to four hours of water every four days and even forbidding the customary use of wet towels in restaurants. Economic impacts were severe: a number of businesses were forced to shut down, and agricultural production suffered. The only businesses to flourish were the manufacturers of water buckets. Drought-related health impacts were also substantial, including incidences of cholera, dysentery, and typhoid fever.[32]

The immediate problems were relieved when Typhoon Viola hit Hong Kong on May 28, 1964, and brought eight inches of much-needed rain.[33] But the rain didn't eliminate the concern about Hong Kong's water supply. Plans were already under way to redirect Hong Kong's attention to the mainland—specifically, toward opportunities for water transfers from Guangdong. Beginning in the late 1950s, a series of agreements were developed between Hong Kong and the Chinese government to construct new storage and conveyance facilities that included the transfer of Dongjiang water to Hong Kong, as well as to expand the water supply areas in Guangdong. First conceived during the Great Leap Forward period in the late

1950s through the construction of the Shenzhen Reservoir, these agreements had the strong backing of the top Chinese government leadership, including Zhou Enlai, the official most directly focused on China–Hong Kong relationships. The first agreement, signed by the Hong Kong colonial authorities and the Chinese government in November 1960, provided for an annual supply of five billion gallons of water. This arrangement was interrupted by the 1963–1964 drought, but once drought conditions improved, it was reinitiated, expanded, and formally launched in March 1965 when the Shenzhen Reservoir connected to the Dongjiang. Hong Kong then began to receive a steady supply of imported water that not only helped meet emergency needs but presumably provided a reliable supply for future development and population growth. Meanwhile, China benefitted politically and economically from its expanding ties to Hong Kong, including Hong Kong's own role in fostering and financing Guangdong's economic development and subsequent rapid urbanization. While those benefits were less apparent with the first agreement in 1965, *political* benefits were notable, since Hong Kong offered an important foreign relations window for China in the midst of a turbulent period, at the dawn of the Cultural Revolution and amid the heightened Sino-Soviet conflict.[34]

In 1980, with the establishment of the Special Economic Zones (SEZ), including Shenzhen, Hong Kong's role was further magnified due to its roles in trade and goods movement, its investments in Guangdong's export industries, the shift of its own export-oriented manufacturing to the mainland, and its market-oriented system, which provided a link for China's later entry into the capitalist world economy. The handover by the British of Hong Kong in 1997 only reinforced the intricate relationship between Hong Kong and Guangdong, including their water supply agreements, which were periodically updated.

For more than three decades, Hong Kong not only had a reliable supply, but could even save some of the Dongjiang waters in its own reservoirs. That arrangement, however, also created a problem of overcapacity—that is, *too much* water storage, causing water to spill over the reservoir banks during particularly wet periods. Hong Kong's water use patterns also began to change—residential use increased but industrial use was reduced, as the low-wage export-oriented industries shifted from Hong Kong to Guangdong. The problem of having too much water was also compounded by the type of financing used for the expansion of various water projects, since

Hong Kong's share of the payments was in the form of an interest-free loan based on prepaid water purchase fees. On a number of occasions in the 1980s and 1990s when Hong Kong sought to *decrease* the amount of water it received from Guangdong, the Chinese provincial government refused. Even when a more flexible water supply agreement was established in 2006, the water deliveries to Hong Kong initially outstripped sales, resulting in the payment for seven years of water supply for only six years of its allotment.[35]

Still, the arrangement for Hong Kong has been viewed as enormously beneficial in addressing its historic lack of local water and unpredictable weather. How much water Hong Kong uses on a per capita basis remains a concern, given the fears about future reliability. That concern has been heightened by continuing population growth, including the large influx of tourists as well as overnight visitors from the mainland, which, in turn, adds to the increase in water use.[36] In addition, the explosion of industrial development in Guangdong, including some of the most polluting (and water-intensive) industries, and the rapid-fire urbanization in places like Shenzhen stimulated by the SEZ programs have raised a new set of concerns about the sufficiency of the water supply *for Guangdong*. This issue has been intensified by the water quality problems in the Dongjiang watershed and throughout the Pearl River Basin as well as the increasing industrial and residential demand for water despite the lower per capita use compared to Hong Kong. Hong Kong water officials remain concerned that any revised set of agreements could potentially reduce *their future supply* and increase uncertainty about water reliability.

Today, Hong Kong finds itself subject to a "one country, two water supply needs" system that underlines its complex relationship with China. The search for future reliability has become a political question about where and how Hong Kong might seek its water independence and find a new balance between supply and demand.

Water Quality

Los Angeles

Unlike its long-standing struggle with air quality, Los Angeles, up until the late 1970s, had not needed to confront the endemic problems with water quality that plagued other United States regions which primarily relied on their rivers and other surface waters for drinking water and other domestic, industrial, and agricultural uses. Polluted surface waters in the United States had long been a visible byproduct of industrial activity, agricultural practices, and urban development with its untreated runoff and wastes. From the mid-nineteenth century through the 1960s, numerous polluted waterways, from a dying Lake Erie to the burning Cuyahoga River in Cleveland, Ohio, became visual representations of the surface water contamination that impacted more than 90 percent of all US watersheds. The federal Clean Water Act of 1972, which focused on such surface water pollution, contained language about making waters drinkable, swimmable, and navigable. Its ambitious objective was to "restore and maintain the chemical, physical and biological integrity of the nation's waters."[37]

Water quality problems related to drinking water sources, however, were not considered a significant problem at the time of the passage of the Clean Water Act, including in Los Angeles. The Los Angeles River channel was certainly recognized as polluted—filled with debris and unregulated discharges. One memorable image from a science fiction film of the 1950s called *Them* had giant irradiated ants coming out of the storm drains that fed into the LA River. But by the 1970s the LA River had come to be seen as a flood-control channel, not a viable water source related to the city's water needs, as it once had been. Similarly, growing environmental concern about ocean pollution, heightened by a major oil spill in 1969 in the Santa Barbara Channel about ninety miles northwest of downtown Los Angeles, was not linked to concern about the management of the city's or region's water supply. Some water quality problems were associated with Los Angeles's imported water sources, including high salinity levels in the Colorado River water (producing a "hard" or more alkaline water) and taste and odor complaints about the water from Northern California stored in local reservoirs.[38] Still, the local groundwater sources in Los Angeles were considered exempt from the various water quality problems of surface waters. Even if agricultural chemicals or industrial effluent leached into the ground, it was assumed that groundwater basins provided a natural filtration capacity to eliminate any major water quality concern.

That assumption was undermined when new revelations of polluted groundwater began to appear in the mid and late 1970s across the United States. In Los Angeles, the first indication that parts of the Los Angeles and San Gabriel groundwater basins might be contaminated was the discovery in December 1979 that discharges by Aerojet had contaminated several wells in the San Gabriel Valley. The contamination had created a major pollution problem within a regional groundwater basin that provided as much as 85 percent of the water for more than a million people. Additional discoveries of contaminated wells in the early 1980s took more wells out of production, with fears that the contamination would spread as the groundwater plume moved from one area of the basin to the next. The connection between water quality and water supply also quickly became apparent, as several parts of the region, including the City of Los Angeles, found themselves needing to increase their imported water sources through the Metropolitan system, further reducing the regional reliance on local water sources.[39]

The second major water quality issue that emerged in Southern California during the 1970s and 1980s (as well as in other parts of the United States) was related to the treatment process used for drinking water. For much of the twentieth century, treatment of drinking water sources used chlorination to eliminate possible microbiological contaminants. Chlorination was seen as a major public health breakthrough, reducing or eliminating various diseases, such as cholera, caused by contaminated water. But in the early 1970s, researchers began to identify a new class of contaminants that resulted directly from the chlorination process itself when treating water with organic matter from both natural and human sources. The disinfection byproducts that formed as a result of the treatment process included potential carcinogens such as trihalomethanes (or THMs). Because chlorination was ubiquitous among water agencies in the United States, news of the research results led to explosive media coverage and a rapid legislative response. This included passage of the 1974 federal Safe Drinking Water Act which focused on drinking water quality.[40]

For Southern California, the problem of disinfection byproducts was particularly noteworthy since its imported water from Northern California contained a high degree of organic matter, the precursors for the formation of disinfection byproducts. For much of the 1970s and 1980s, Metropolitan sought to downplay this water quality concern, even as new water quality regulations at the federal level identified Southern California's drinking water quality as not always in compliance with its standard for THMs and other byproducts. Metropolitan's response only exacerbated growing public concern about water quality, including at the tap, and fueled a major jump in the sale of bottled water and water filter systems. It was even revealed that as many as two-thirds of the staff of the Los Angeles Department of Water and Power drank bottled water or used a filtered water system. With the subsequent strengthening of the federal regulations on disinfection byproducts, Metropolitan transitioned to new treatment strategies to comply with the federal regulations.[41]

The groundwater and drinking water treatment problems, and the defensive response of the water agencies regarding those issues, revealed a deeper problem with the mission of the water agencies. Both Metropolitan and the City of Los Angeles's Department of Water and Power had long considered their core focus to be providing an adequate water supply to accommodate existing and especially future demands, with the expectation that demand would expand, whether for residential, agricultural, or industrial uses, as well as for new urban-edge development where water availability was more limited. As water quality issues became increasingly prominent, they forced their way onto policy agendas, along with the recognition that the quality of the water was itself a supply issue.[42] Water quality problems in turn have increased the uncertainty and fears about reliability of the water for Southern California. Climate change has further increased that uncertainty, whether from reduced snowpack for the imported water, from increased algae blooms that have caused water quality problems, or from the unpredictability of future supplies.[43]

Nevertheless, attention to water quality problems has remained more episodic than mission-driven. US media pay more attention to water quality (and air quality) problems in China than to local, regional, or national US water quality issues. Yet for Southern California, groundwater contamination has reduced local water supplies in the region to only 50 percent, and, in some cases, such as the City of Los Angeles, to just barely over 10 percent. Policy concerns have shifted to other core problems with water quality, such as stormwater runoff, though water agencies like LA's Department of Water and Power and the Metropolitan Water District have only limited jurisdiction over how those issues can best be addressed. The problem of effective and integrated *water management* then looms large as a central policy consideration, with split jurisdictions and uncertainties about future supplies and water quality in an era of climate change underlying those considerations.

Hong Kong

In 2014, the Hong Kong government stated, in its online summary about drinking water quality, that "Hong Kong enjoys one of the safest water supplies in the world." Touting its program to monitor water quality—from water sources through treatment and distribution to consumers—the government compared its sampling results to World Health Organization water quality guidelines and concluded, "you can rest assured that the water in your home is safe for consumption."[44] The picture it described of water quality and drinking water safety, though, was incomplete. The Hong Kong government has not had the power either to improve the quality of its major supply of water, the Dongjiang, or to directly address the treatment of its Guangdong waters at their source. Where the government has had more of a direct role is with the water sources on Hong Kong itself, primarily in the New Territories where poor water quality had long been a factor.

In 1986, when the Hong Kong government initiated a water quality monitoring program, it identified local sources that "had become too polluted to support life and were black and foul-smelling." Conditions over the next two decades improved, due to restrictions placed on livestock farms (primarily poultry and pigs), the relocation of polluting industries from Hong Kong to Guangdong, and the development of more treatment facilities and sewer hookups. Nevertheless, some local quality problems still remained. These included, most notably, moderate to high levels of *E. coli* in the northwestern New Territories that stemmed from the continuing (albeit reduced) problem of discharges from livestock farms and nonsewered villages in the area, as well as from what the Environmental Protection Department called "the back flushing effect of the more polluted Shenzhen River."[45]

The issue of the polluted Shenzhen River is just one aspect of that more challenging issue for Hong Kong: water quality concerns about its imported water from Guangdong. Even more concerning is that despite some important improvements in the Dongjiang watershed, there is nevertheless the potential for future pollution from increased urbanization (and unregulated wastes and lack of treatment facilities) as well as from industrial and agricultural pollution. Existing water quality concerns include such sources as heavy metals, organochlorine pesticide residues, and various persistent organic pollutants such as PCBs that continue to be detected in the Dongjiang. The Hong Kong government has been able to counter some of those problems by contributing to the construction of new sewage treatment plants, by seeking to intercept wastewater flowing into the Shenzhen Reservoir which also supplies Hong Kong, and by diverting the discharge of polluted water away the Dongjiang. The Hong Kong government has also supported new initiatives by the Guangdong authorities to begin more aggressive monitoring and the development of new pollution control and treatment measures throughout the Pearl River Basin.[46]

One of the ironies of Hong Kong's dependence on the Dongjiang is that a substantial number—more than half—of the polluting Guangdong industries that discharge into the Dongjiang have been established and financed by Hong Kong investors, including those industries that relocated from Hong Kong to Guangdong. While such a shift in the industrial base from Hong Kong to Guangdong reduced some of the severe local water quality impacts in Hong Kong, it substantially increased the problems for its *imported* water supply. Hong Kong's connection to China thus included both the opportunity to expand Hong Kong's supply and the concern about the quality of that supply.[47]

The Connection to China

Similar to the experience of Los Angeles and Hong Kong, China's water issues are historically rooted, development-driven, and a continual concern. "In China," environmental researcher and advocate Ma Jun has written, "we have probably spent more time dealing with water problems than anywhere else on earth."[48]

China has about 20 percent of the world's population but only 6 percent of the world's total water resources, an oft-quoted statistic used to describe the country's water sources and the demands on them. According to a 2009 World Bank analysis, about 400 of China's 661 cities were short of water, with 180 of those cities experiencing serious shortages.[49]

Those numbers have only marginally improved since then. Moreover, some aspects of the decline in available water threaten to get worse. This is due to a number of variables: further increases in demand, especially due to industrial and urban growth; continuing major problems with water quality; poor control over industrial and domestic wastewater discharges; nonpoint source pollution, primarily from agricultural runoff; and multiple supply source problems, such as groundwater depletion and river and lake pollution. Climate change has only recently been identified as a major concern through higher evaporation levels, and it has the potential to reinforce the uneven distribution of rainfall and overall water supply.

The spatial distribution of China's water resources and the differences in the amounts and types of water uses between Southern and Northern China represent a particular predicament that has long plagued the country but has become even more pronounced since the 1980s. While the water basins in the north account for only 12 percent of the mean annual surface runoff and 20 percent of the groundwater, nearly half of China's population (45 percent) and almost 60 percent of its arable land are located within those watershed areas. This water divide has been a major policy focus among China's top leadership, several of whom had been trained as engineers and had experience with Soviet-style megaproject approaches. Far greater than Los Angeles's historical focus on big water projects, China's dream of moving water through massive transfer projects has remained a priority. Among other efforts, this approach has led to the development of China's most expansive and environmentally impactful massive water project, the South-North Water Diversion project, which is the largest water transfer project in the world. The expansion of just one of its facilities, the Danjiangkou reservoir, for example, has led to the relocation of 345,000 people, the single largest migration in Chinese history.[50]

While problems in the north have reached crisis proportions in some areas, problems are also widespread in the south. Throughout the country, lakes, rivers, streams, and essentially all surface waters are challenged, with many sources reaching levels that are categorized as unfit for human consumption or even unfit for any uses. As many as 270 million people have no access to potable water.[51]

While the safety and availability of water for drinking and other domestic uses represent a huge challenge for China's expanding urban population, it is a major concern as well for the industrial and agricultural sectors. China's ability to become the world's factory for such goods as paper, textiles, chemicals, and semiconductors has created its own challenges with water supply and quality. Several of these industries use water inefficiently, are themselves huge water consumers, and have contributed to major pollution problems. In some cases, an industry that relies on a particular water source will send its discharges back into that same source and contribute to its pollution. In recognition of those problems, Guangdong's

Environmental Protection Bureau has targeted electroplating, textile dyeing, chemical production, tanneries, and poultry farms as the province's highest polluters, even as they have become a core part of its regional economy.[52]

Groundwater sources in China are also under stress. As much as 70 percent of China's overall population and more than 60 percent of urban residents rely on groundwater as a primary drinking water source. And yet in the densely populated North China Plain, a government Land Ministry report identified as much as 70 percent of the water as unfit for human consumption. A combination of a shrinking supply (more groundwater taken out than replenished) and deeper (and more expensive) drilling of China's groundwater basins has further led to reduced capacity and availability. When combined with the severe problems of groundwater pollution from agricultural, industrial, and wastewater discharges, China's groundwater becomes as challenged as its polluted lakes and rivers, a less visible but just as serious indicator of the eroding conditions of China's water supply. By way of example, according to China's Land Ministry, among 4,778 sites in 203 cities tested for groundwater quality, 43.9 percent had "relatively poor" quality, while another 15.7 percent of the sites tested as "very poor" (that is, unfit for human consumption even after treatment).[53]

It is not surprising, then, that Chinese Premier Li Keqiang's 2014 report to the National People's Congress, which proposed a "war on pollution," identified groundwater as well as such surface water sources as the Yellow and Yangtze rivers as needing action to protect drinking water sources. "As China is witnessing rapid development in new industrialization, informatization, urbanization and agricultural modernization, the prevention and treatment of water pollution remain a demanding task for us," commented a government-issued 2015 Action Plan for Water Pollution Prevention.[54]

The link between polluted river basins and groundwater resources and possible water-supply shortfalls has been described by the Hong Kong-based NGO China Water Risk as "pollution driving scarcity."[55] As a water supply question, water quality emerges as a key determinant of future as well as present water supply due to what one study calls a "'quality-adjusted' supply-demand gap [that in China] is therefore larger than the quantity-only gap because some water is of such low quality that it can no longer be considered supply."[56]

For the Chinese authorities, water quality stressors, which also impact water supply, involve the demands for water for domestic and industrial uses for an expanding region (and continuing pressures from rapid urban growth, expanding industrial output, and a large, albeit modestly declining, agricultural base). While environmental regulations have been established and some infrastructure, such as new treatment facilities, has been built, a gap remains between full-scale implementation of those environmental efforts and the continuing and persistent problems with water quality and supply that plague China's waters. While the central and provincial governments have elevated the response to water problems as a major goal, there are counterpressures from local governments that continue to limit or even ignore enforcement of environmental standards. Lax enforcement is particularly the case with industrial discharges or urban development impacts, due to the role those same industries play in local economic development and the power of urban development interests.[57]

These pressures for economic growth and urban expansion have also turned water supply and particularly water quality into central environmental and public health concerns. Water quality and soil contamination present enormous challenges to human health, environmental degradation, and economic development, whether for drinking water quality, agricultural production, or industrial activities. A number of water quality episodes in the past several years, such as dead pigs floating down the Huangpu River in Shanghai or the emergence of what have been called "cancer villages," have been directly associated with contaminated water. Water pollution impacts have even crossed national boundaries, such as the 2005 chemical plant explosions and resulting chemical spill into the Songhua River in northeastern China that caused more than four million people in the city of Harbin to be without water for a week, and threatened dozens of cities and villages in Russia.[58] Such episodes have highlighted and intensified the debates about where and how to intervene at the water's source. A number of local protests have also been held, including in the "cancer villages" across China, which have pitted residents against industrial and urban development interests, with local government officials often allied with those interests.

For each of the causes contributing to the problems of water supply and water quality, tension has emerged between the fundamental structure, goals, and practices associated with urban growth, industrial development, and agricultural practices and the increased recognition of water problems and the growing public pressure (and government policies that have been established) to address it. The challenge is illustrated, for example, by the efforts to create a modernized industrial agriculture, with its huge production and use (or overuse) of chemical fertilizers and pesticides, and the parallel problem of unregulated discharges. The location of highly polluting industries adjacent to farm land and freshwater sources is another challenge, causing them to then face a double burden of pollution. The rice crop in several major farming areas, for example, has become contaminated from heavy metals such as cadmium from industrial discharges that enter the soil and water.[59] The rapid expansion of megacities such as Guangzhou, Tianjin, and Shanghai and the explosion in population in the middle-tier cities has also led to major infrastructure challenges for sewage and wastewater as well as for drinking water treatment. As soon as one set of facilities is built, the system gets stretched as population and urban boundaries are extended, leading to yet more wastewater discharges and pollution problems. China is in fact the top-ranked country in the world in the amount of wastewater discharged, its volume exceeding 68 billion tons, greater than the annual flow of the Yellow River.

While the magnitude of the problem has begun to be recognized and has even been highlighted by the mainland government, the continuing development that extends the water problems and the efforts to address them has led to an impasse on water supply and water quality issues. That tension has created what China's former Minister of Water Resources, Wang Shucheng, characterized as China's preeminent challenge: "To fight for every drop of water or die."[60]

Strategies for Change

In 1982, in the midst of the referendum campaign over the Peripheral Canal in California, Evan Griffith, then general manager of the Metropolitan Water District, contemplated writing a book about the huge engineering challenges involved in bringing Northern California water to Southern California. Griffith had played an important role in the construction of the California Aqueduct, including tunneling through the Tehachapi Mountains north of urban Southern California. Water agency officials like Griffith had once been lauded for their engineering skills and vision for big water projects, but now large, engineering-driven imported-water projects were increasingly seen as part of the problem, not the solution in addressing Southern California's water future. Criticisms included the environmental and cost concerns that would play a role in the defeat of the Peripheral Canal project. "I would love to tell my story about tunneling when I retire," Griffith lamented at the time, "but I don't know who would be interested in it."[61]

Yet Griffin's lament was not entirely accurate, at least at that moment of time. Despite a shifting terrain, the large transfer and interbasin projects the water engineers had helped to design and construct in Los Angeles, as well as in Hong Kong and China, characterized the approach to water development in those places for much of the twentieth and early twenty-first century. The huge water projects in China, whether the South-North Water Diversion Project or the massive Three Gorges Dam on the Yangtze River, have not only been implemented by water engineers in China but conceived, overseen, and nurtured by several of its top leaders, such as Zhu Ronji, Li Peng, Jiang Zemin, and Hu Jintao. In Hong Kong, with its limited options, the role of the Water Supplies Department has primarily focused on securing its own water supply by facilitating projects that import water from China. One of its most significant *local* engineering feats, the construction of a dual piping system for seawater for toilet flushing, identified a different type of management focus that sought to stretch the notion of available supply. Still, Hong Kong's reliance on the imported Dongjiang water supply, which would be shared with cities in Guangdong that are experiencing rapid growth and increased water demand, has remained far more pivotal to Hong Kong's water management framework.[62]

The limits of the water agency focus on engineering and imported water to best manage the problems of both the supply and quality of the water is increasingly apparent, given the integrated nature of the supply-quality connection and the range of other water issues outside the engineering purview. *In Los Angeles*, rainwater and stormwater runoff, for example, has been seen as a Public Works or Flood Control District concern and less of a water agency matter, since the paved, urbanized landscape creates problems of runoff and flooded streets, while treatment facilities have been primarily focused on sewage disposal problems. In Hong Kong, responsibility for those issues, though identified by the Water Supplies Department as part of its "total water management" approach, still primarily resides with other agencies, such as the Drainage Services Department, which is focused on flood control and sewerage issues. And in mainland China, rapid urban growth and the parallel need to build new wastewater treatment facilities for urban and nearby agricultural and industrial discharges have created a dual problem of lack of facilities and the local governments' unwillingness to enforce existing regulations to control wastewater discharges or storm-water runoff.[63]

Yet an issue like rainwater capture provides a twin benefit of reducing runoff and providing a modest supply source. Similarly, sewage and waste-water, which have been addressed as a treatment concern for sanitation districts and the Public Works agencies, can present an opportunity for water agencies to recycle and thereby stretch existing supplies.[64]

What the water agencies and other governmental jurisdictions have also been missing is a way to broaden their agenda to consider managing water in its many forms (including the water at its source rather than at its end point). At the same time, water agencies confront the question of their own mission and whether to reposition their approach, such as through an equity or environmental or public health framework. Access to drinking water, for example, involves core equity and public health issues, yet is seen more as a private rather than a public concern. This is symbolized by the rise of the bottled water industry and the growing interest in water privatization in Southern California as well as in Hong Kong and China. While drinking water itself represents just a fraction of the supply demands on a water agency, the cost differential between a private source (a small water bottle) and a public source (at the kitchen tap or drinking water fountains) can be as much as a thousand times, invoking considerations of equity.

When sales of bottled water by both smaller companies and large beverage manufacturers such as Nestle and Coca Cola increased geometrically in Southern California during the 1980s and 1990s, the public water agencies such as the LA DWP and the MWD reacted defensively and minimized the equity issue involved in bottled water sales. Much of the rapid increase in those sales was due to concerns about tap water quality as well as taste considerations. But even as the water agencies began to address those concerns, particularly with new treatment strategies, they still distinguished between their primary public role as water supplier and the *private* role of other companies—their erstwhile competitors—in the business of selling water. "We can't—and don't want to—compete with the private sector," water agency officials argued about their inability to tackle this equity concern, given the cost of bottled water.

In Hong Kong, the Water Supplies Department has sought to emphasize the safety of tap water, while also seeking to respond to public concerns about the quality of the water imported from Guangdong. Those concerns have produced substantial growth in sales of bottled water in Hong Kong, led by companies like Nestle, Danone, and Coca Cola. One market study estimated that more than 418 million liters of bottled water (including flavored water drinks) were sold in Hong Kong in 2013, or approximately 58 liters per person. The shift toward bottled water coincided with the lack of access to tap water, similar to problems in Los Angeles. "It is easier to find a 7-11 than a water fountain in Hong Kong," quipped Debra Tan of the China Water Risk research group.[65]

In China, bottled water sales have paralleled the dramatic population growth of urban centers. From just a single, domestic bottled water plant in 1980 (when concern about water quality was far less pronounced), bottled water sales in China, led by both private companies and local government entities that have entered the bottled water market, has soared in the past two decades. As a result, China catapulted into the second-largest bottled water market, after the United States. The biggest bottled water companies such as Danone and Nestle have gained a foothold in China through purchasing or merging with existing local companies and/or by acquiring a local brand name. In some cases, such as the over-100-year-old state-owned Guangzhou Water Supply Company, the water has been sourced from one of its existing treatment plants, with additional treatment applied to it, and then been marketed as a special "safe and healthy" bottled water product.[66]

The issue of water privatization further highlights the question of water agency mission, including, as with the bottled water issue, whether water should be treated as a commodity or as a public good. Most water utilities and water agencies in the United States are publicly owned and managed (as opposed to investor-owned electric utilities). Similarly, regional water wholesalers, such as Metropolitan, are also publicly owned and managed. However, some of the activities of the agencies are sourced to private entities. More importantly, the web of relationships involved in water development that include, along with the public water agencies, construction interests, engineering firms, water-oriented law firms, and real estate development interests that are dependent on accessing water supplies, have established what has come to be called a "water industry." This public-private framework for water management translates into a mission of meeting the needs of particular interests, whether industry, agriculture, or urban development.[67]

In Hong Kong, the main water agency, the Water Supplies Department, has remained an important government outpost, even as some public services have become outsourced or privatized or are threatened with such changes. Debates about the public versus private role for water services erupted soon after the 1997 integration of Hong Kong into China. The Asian financial crisis of 1998 added to the pressures on the newly formed Hong Kong SAR government to reduce the investment and operational costs of its water supply system by partially or completely privatizing it, whether through outsourcing (contracting for services) or a transition to full privatization of the system. Those efforts were suspended, due in part to the strong civil service–public agency ethic that had developed within the Water Services Department. However, partial contracting, through various public-private partnerships, has taken place, an approach bolstered by the prevailing market orientation and the lauding of the private sector in Hong Kong—an orientation also strongly supported by the mainland.

In China, the advent of market reforms, especially after 1990–1991, led to a shift in water system control from the central government to municipal governments. A modest, though increasing, role has been played by global water companies, such as Suez Environnement, Siemens, Veolia, and Thames Water, which have invested in China's water infrastructure, including new treatment facilities. In 2002, the Chinese government embarked on a full-fledged shift to privatization and marketization for its water sector, but without any clearly defined regulatory mechanisms. By 2008, private and foreign interests had obtained stakes in 20 percent of China's public water utilities and 70 percent of the country's wastewater utilities. Like many other aspects of the economy, water systems, whether owned or managed by the government, or by local or global private companies, or through joint public-private ownership, have come to be defined as commodity-based or profit-making enterprises, identifying water as a commodity rather than a public good.[68]

The battle over water as a commodity or a public good, the role of private interests (often in conjunction with public bodies) influencing the uses (and abuses) of water for various purposes, and the need for better water management integration have all become important issues of the urban environment, whether in Los Angeles, Hong Kong, or mainland China. For change to occur, a new water ethic needs to be developed based on such principles as fair and equitable pricing, greater accountability in decision making, and elevating water quality as a priority for action. Such a water ethic also points to the importance from an urban environment context of establishing a green infrastructure approach and a public or community role in promoting such goals for health, the environment, accountability, and equity. A green infrastructure approach needs to focus on demand as much as on supply. It needs to view the management of the watershed in an integrated manner as opposed to how best to transfer water out of the watershed. It requires a fundamental shift in the mission and modus operandi of the water agencies and policymakers, who need to assume a role of promoting health and the environment as well as contributing to public needs such as greater equity. Such a shift would become part of a broader agenda of "livability" in the urban environment—how we protect, manage, and use our resources and understand that they too have "a life of their own."[69]

5   The Food Environment

Tyson, Chicken Paws, and School Food

In the fall of 2011, the Food Services Department of the Los Angeles Unified School District (LAUSD) rolled out a new menu for its school lunch program. The menu had been designed to provide healthier options for the 750,000 meals that are served each day. These included more fruits and vegetables, baked goods, and meat and poultry sourced locally. The school district, however, had yet to implement specific environmental, workplace, health, and animal welfare criteria that might have guided their selection of vendors. Such an approach to establish a "good food" purchasing template for institutions that sourced food, like LAUSD, was then at the discussion stage at the Los Angeles Food Policy Council on whose board the LA School Food Services Director served. Implementing such an approach would not be easy. Just the year before these discussions, a contentious debate at the LA School Board had resulted in a five-year contract being awarded to Tyson Foods, the huge global food company, to supply the district's chicken products.[1]

Tyson had been a controversial choice due to its history of environmental, labor, and animal welfare abuses. These abuses included the treatment of its contract farmers, the poor workplace conditions at its various plants, and the treatment of its chickens housed in its concentrated animal feeding operations (CAFOs). Several of its facilities also had major environmental hazards due to water and land contamination.[2] These types of problems extended to Tyson's global operations, which had become an increasingly important component of its holdings. Tyson exported its products to 130 countries, including Hong Kong and China, with China the largest and most rapidly growing component of Tyson's global reach.[3]

As part of its global expansion, Tyson had discovered a valuable marketing opportunity for its export business: chicken paws (the area below the feet), which had been a waste product in the US domestic market but were widely used in soups or stews, as snacks, and as a flavor enhancer in Hong Kong and China. Hong Kong also played an important role for Tyson in its reexport of Tyson products such as chicken paws into China. This three-way relationship became particularly noteworthy when China and the United States engaged in a trade battle in 2010, and China established a high tariff on its chicken imports from the US (including a 43.15 percent penalty for Tyson's products). Chicken sales from the United States to China fluctuated significantly while the trade dispute was taking place, but Tyson also set records in 2011 for the quantity and value of its exports to Hong Kong. Of its chicken paw exports, 90 percent were shipped that year to Hong Kong, up 28 percent from 2010, with some of that total reexported to China.[4]

Aside from its growing global presence (which made up 13 percent of Tyson's overall sales), food service sales in the United States involving customers like LAUSD were a major market for the company (31 percent of its sales).[5] Despite its claims of efficiency and its ability to meet large-market demands, Tyson had experienced a major bottleneck in its supply chain at the very moment when LAUSD was set to roll out its new menu. Chicken items were an important component of the new menu, particularly with LAUSD's desire to create a more ethnic-focused approach. But Tyson's chicken deliveries failed to arrive the first day of the school year and for another six weeks. Forced to scramble and come up with primarily an all-vegetarian menu—a different experience for students used to fast food— LAUSD got mixed results during its rollout from some students, parents, and principals who clamored for them to bring back the pizza, hamburgers, and chicken nuggets, a Tyson core product. Conservative, right-wing commentators had a field day attacking the "Michelle Obama-inspired" LAUSD menu. LAUSD, however, was able to weather the criticism, and some of the new menu survived.[6]

In 2012, LAUSD signed on to the Los Angeles Food Policy Council's Good Food Purchasing Policy. The policy provided a ranking system based on a robust evaluation of any supplier's labor, environmental, animal welfare, and health records as well as whether it was a local or long-distance supplier. The first major request for proposals for suppliers was issued in early 2015 and included all the meat and poultry suppliers to LAUSD. Tyson, despite its labor and environmental record, wanted its contract renewed, and it plied the district with contributions, including funds it had donated earlier to promote the new menu. The language of the request for proposals indicated that Tyson might no longer qualify to be an LAUSD supplier—an "extraordinary outcome that could reorient the school food system," according to one school board member.[7]

Once again, a contentious process unfolded, related to whether and how a groundbreaking policy would be implemented. A compromise was proposed, with Tyson obtaining the contract for some though not all of the chicken products (which Tyson subsequently withdrew), while a local distributor that had already worked with the district and had met some of the Good Food Purchasing criteria obtained a contract for some of the other chicken parts. As the Tyson/Good Food Purchasing debate unfolded, it raised larger questions about food systems, including where and how the food is produced, what inputs are utilized, where and how it is sold and consumed, and whether a good food framework is possible in an age of global food.

Growing Food

Los Angeles
Between 1910 and 1950, Los Angeles became "the farming capital of the nation." From citrus to cattle, alfalfa to the production of grapes, oranges, asparagus, lettuce, lima beans, melons, cabbages, and many more crops, agriculture in Los Angeles laid the foundations for the massive citrus, wine, ranching, and dairy industries in California.[8]

The image of Los Angeles as a fertile and fruit-bearing land was part of the narrative of the Spanish explorers who arrived in Los Angeles in the eighteenth century. In Father Juan Crespí's famous comments from his diary, the Los Angeles environment represented a food and garden oasis, a "very lush pleasing spot" that would be capable of producing every kind of grain and fruit imaginable.[9] The agricultural activities of the Spanish colonists, who set up the San Gabriel Mission on the banks of the Los Angeles River, became the testing ground and incubator for introducing a European-style crop system in their newly colonized land. In the process, their agricultural interventions undermined the foraging system and food security of the native population (the Tongva) who consumed wild nuts and berries and wildlife. By the early nineteenth century, the Spanish colonists had established the huge ranchos that stretched 1.5 million acres from the coast to the eastern edge of the region at San Bernardino. Grazing activities caused erosion, had an impact on the groundwater, and reconfigured the Los Angeles basin's prairie environment, altering its native grasses, clovers, and other plants.[10]

When the transcontinental railroads were completed in the late 1860s, a number of Chinese immigrant rail workers settled in Los Angeles and became the primary workforce for the newly planted vineyards and citrus groves. The immigrants, mostly from Guangdong Province, had a wealth of farming experience handed down for generations. They soon became the earliest vegetable farmers, growing leafy greens, cabbage, green beans, and other crops on unclaimed land, using water from the LA River and establishing a form of truck farming to sell the produce door to door. Of the 234 market gardeners in Los Angeles County, 89 percent were born in China, according to the 1880 census.[11]

The exploitation of labor in the fields—the citrus industry prominently relied on thousands of Native American, Chinese, Sikh, Japanese, Filipino, and Mexican workers—masked the nature of development. Citrus became an iconic image of a lush Southern California while real estate speculators created "citrus towns" and other new developments throughout an expanding Los Angeles. Citrus growers were themselves more investors than farmers, as they hired managers and large crews to work the groves. Farm labor was often hazardous. Tent fumigation with cyanide to control pests was introduced in the late 1880s, anticipating the use of various poisons as essential inputs, such as metals (e.g., lead arsenate) and chemical pesticides and fertilizers. The size and scale of the citrus industry, its requirements for investment capital, its use of farm labor, its developing model of the corporate farm, and the corporate consolidation of the citrus enterprise as a whole, established Southern California as a leading region in transforming agricultural development into a form of "modern corporate capitalism," as historians Ronald Tobey and Charles Wetherell argue. This model, which extended beyond citrus into other areas of agricultural production, also had a global component with the Pacific trade for fertilizer inputs, such as Peruvian guano, and a labor supply that extended to China in the form of Chinese coolie labor.[12]

This evolving industrial agriculture system changed the dynamic between the laborers in the fields and the food growers, who hired the managers who then oversaw an increasing labor pool of farmworkers. The conditions in the fields led to a series of strikes during the 1930s, from pickers in the berry fields in El Monte to the east, to Mexican vegetable workers to the west of downtown Los Angeles, to workers at dairy farms. In some cases, the land was leased by the grower/owners to Japanese tenants, with the growers fudging the law by extending to the Japanese farmers the leases allowed under the Alien Land Acts, one of several anti-immigrant laws established in the United States in the first several decades of the twentieth century. Violence was common, as the growers turned to groups like the Associated Farmers which served as a paramilitary vigilante group, and the "Red Squad" of the Los Angeles police department, which targeted the foreign immigrants as well as the Depression-era domestic workers from places like Arkansas and Oklahoma.[13]

By the end of World War II, Los Angeles farm production had begun its decline. This was due in part to the relentless expansion of real estate development and suburbanization that stretched the urban boundary lines by converting agricultural land to residential and commercial development. Air pollution, already a problem in the mid to late 1940s, also contributed to LA's loss of agricultural lands.[14] At the same time, with the expansion of the industrial agriculture model in the Central Valley, California was able to emerge after the war as the leading agricultural state in the US. The California industrial model came to include huge land concentrations, continuing reliance on immigrant labor, a major reliance on energy and chemical inputs, imported water, a reliance on finance capital, and a robust export, or out-of-state, market (primarily to other parts of the United States but increasingly to markets outside of the country).

Today, even as California agriculture consolidates its leading position and global presence, only a very modest local farm economy can be found within the city and county limits of Los Angeles. Its leading crops include ornamentals (shrubs, trees, and flowers), alfalfa, and carrots, much of which is grown in the less developed areas of the region.[15] Yet while local farm production has become more limited and reliance on food grown outside Los Angeles has become the norm for the large food retail outlets (supermarkets and big-box stores), a strong interest in local food has also emerged. The popularity of buying and eating locally produced food is now widespread in Los Angeles and across the United States, along with new strategies and initiatives seeking to draw a stronger connection to food grown as well as food consumed. This search for local food has paralleled the advocacy for the ability to access fresh, affordable, and healthy food in all communities and schools, yet it is occurring at a time when the industrial model of food growing, prevalent for nearly a century in Los Angeles and California, has become increasingly challenged.

Hong Kong

When the British took control of Hong Kong in the 1840s, agricultural activity had been primarily related to fishing and salt production. The British colonialists established two initial goals: the creation of a revenue farming system, especially including opium production, and an expansion of Hong Kong's role as a trading center for various goods. The revenue farm was part of a franchise system designed to collect state revenue. Each farmer was granted a monopoly right for a limited time period and at a particular location, based on what was paid to the colonial government. The opium farm was the leading income generator for the British, while other revenue farms included gambling farms; farms for the sale of liquor, salt, and pork as well as for pawnbroking or fishing; and farms that imposed duties on goods or transit along roadways and seaports.[16]

Though agricultural production was limited by the area's hillsides and rocky terrain, Hong Kong was able to produce much of its own food, including rice, livestock, and vegetables, until the 1960s. Peasant smallholder farmers who emigrated from South China began to farm in otherwise barren land in the New Territories in the 1940s and 1950s, helping to maintain and even to expand Hong Kong's local farming economy. But the rapidly expanding industrial base, the huge expansion of export (and import) shipping, and subsequent urban development in the New Territories eroded local agriculture. The shift toward imported food and away from local food production finally reached a symbolic end point in 2003 when the Hong Kong government abolished all import quotas on rice.[17]

Today, local agricultural production represents barely 5 percent of Hong Kong's food supply.[18] Several factors have come into play in the shift toward imported food. The rapid expansion of Hong Kong's local industries, which restarted in the 1950s and peaked by the late 1980s to early 1990s, turned government policy away from agricultural supports. In addition, policies that favored new residential construction helped foster urban developments on lands in the New Territories that had been dedicated to farming. As much as one-half to two-thirds of the arable land otherwise available for farming was instead made available for future land development and residential construction. Today, a large part of the New Territories has become "land for apartment blocks, container yards, car parks and garbage dumps ... [where] land can be turned into cash [and] the local market is flooded with imported food," writes Lingnan University Professor Chan Shun-hing.[19]

Changes in Hong Kong's food environment have also been linked to changes in China after 1979, particularly in the Pearl River Delta region. Both agricultural production and industrial development expanded substantially in Guangdong, creating an even stronger synergy between Hong Kong and China. As agricultural production increased in that region, Hong Kong investors worked with local government officials to facilitate numerous land purchases that displaced the small peasant farmers from their lands in China and established new industrial operations, including the relocated industries from Hong Kong. The displaced farmers, in turn, became a source of labor for those industries. In the 1990s, industrial production in Hong Kong began its transition to a service and logistics-based economy, which further influenced where and what kinds of food was sourced.

By 2015, only about 2,400 small farms remained in Hong Kong, with 4,300 farmers and their workers. The average size of a farm was only 0.2 hectare, or less than half an acre. Six percent of the land in Hong Kong remained designated for agriculture, even as half or more of that land continued to be slated for further development. Similar to Los Angeles, ornamentals have emerged as one of the primary products, though a small volume of vegetables, fruits, pigs, poultry, field crops, cattle, fresh milk, and eggs continue to be produced. Even as late as 2015, local poultry production, particularly live poultry, accounted for the lion's share (95 percent) of the poultry consumed in Hong Kong, whereas only 1.8 percent of fresh vegetables and 6.1 percent of live pigs came from local sources.[20] However, poultry production in both Hong Kong and Guangdong have periodically experienced major setbacks due to scandals over avian influenza, the SARS epidemic, the H7N9 virus, and other poultry-related diseases. In 2004 and again in 2008, the Hong Kong government, in response to those scandals, introduced a "voluntary surrender and buyout scheme" for poultry farmers.[21]

The various poultry-related food safety crises have highlighted two key factors of the Hong Kong food system. On the one hand, the long-standing connections to China for access to fresh food have expanded, with the mainland representing a major source of food imports, including fresh fruits and vegetables and some animal imports. China's fresh vegetable imports are far cheaper than what is locally produced; as a result, several Hong Kong farmers have moved their operations to China and now sell their products back to Hong Kong. Aside from China, Hong Kong food imports have come from several other countries, including the leading exporters, Brazil and the United States. The United States continues to be Hong Kong's leading source of fresh fruit, tree nuts, and fruit imports as well as vegetable juices and organic foods, with much of that shipped from California through the Los Angeles and Long Beach ports. At the same time, a significant portion (29 percent in 2015) of the food imported into Hong Kong is then reexported, including to China.[22]

Food safety issues have remained an important driver of the growth of the food import markets, although such issues have also been a problem for the importers as well. The United States, for example, lost its number-one ranking as a poultry exporter to Brazil when avian influenza cases were reported in the US in 2004. Brazil continues to remain the leading poultry exporter, with the most popular US chicken products including chicken paws, legs, and wings which are popular in both retail and restaurant establishments, such as cafés.[23]

The growth of the organic market is perhaps most notable in relation to imports. Hong Kong has recently experienced a fledgling but increasingly visible expansion of local organic growers, and more consumer interest in greater transparency and direct connections to where and how the food is grown. This is expressed as a desire to reconnect with the soil, associated with the concept of *nong*, as distinct from the use of the term "land," more often associated with "property" and real estate development.[24] This interest has translated into direct farmer-to-consumer programs, such as Community Supported Agriculture, as well as an emerging urban agriculture that includes community gardens and rooftop gardens. According to the Hong Kong Agriculture, Fisheries, and Conservation Department, there were 514 organic farms in Hong Kong as of August 2014, with overall estimates of the organic market (locally grown and imports) at $500 million and an annual growth rate of 10 to 15 percent. Responding to the growing interest in organic and local food, the Hong Kong government announced a new "Sustainable Agricultural Policy" in December 2014, which included the establishment of an Agricultural Park and Sustainable Development Fund and the creation of a 70-to-80-hectare AgriPark in the New Territories.[25]

The AgriPark proposal, however, has been criticized as a stalking horse for reducing farm land by setting aside one small "demonstration" area for farming, while the inexorable spread of urban development continues around it. This parallels previous developments in the New Territories where farmers have been displaced, such as the 2010 demolition of Choi Yuen Village and neighboring villages in Pak Heung to make way for the Hong Kong–Guangzhou High Speed Rail Project. The Choi Yuen Village displacement led to major protests and crystallized interest in a renewal of local farming and sustainable and organic agriculture advocacy. But urban land development continues to trump local farm production, thus reinforcing the role of imported food as the dominant food supply for Hong Kong.[26]

As in Los Angeles, a distinction—and occasional tension—between "local" food and "organic" food (much of which is also imported, including from China) remains a factor in sorting out future changes in the growing of food that is made available in Hong Kong. It suggests, among other factors, that a local Hong Kong food system (food grown within Hong Kong) will likely remain limited for the foreseeable future, despite the growing interest in sustainable agriculture. Its most significant opportunity for further development resides in what has become more of a niche market for both farmers and consumers, such as community-supported agriculture programs. The future of Hong Kong's relationship to its neighboring Pearl River Delta foodshed may also become a factor in how and what type of food is grown. Such a foodshed approach could facilitate the further development of a regional (and potentially more sustainable) Hong Kong/South China food system while extending the notion of *nong* to improve the environment and rebuild community.

Food Retail

Los Angeles

From its more humble origins as the neighborhood general store and the "self-service" market, the rise of the supermarket in Los Angeles and throughout the United States became central to the restructuring of the food system. The first stand-alone food and general store was established in 1871 in a brick building in what is now downtown Los Angeles by three partners who became major civic and economic leaders in the city, including Jacob Heilman who founded one of the city's first banks. A few decades later, a new form of food retail in LA, the "groceteria," was introduced, which subsequently evolved into a supermarket-style format. During the 1920s and 1930s, supermarket chains emerged through the consolidation of some of the largest of the food retail operations. In Los Angeles, Charles Merrill (of the brokerage firm Merrill and Lynch) put together the Safeway food retail chain through a merger of a major LA operation with large store operations in Northern California and other western states. Some of the largest of these stores also sought to attract a customer base by promoting their free parking.[27]

As supermarkets expanded food retail market share, they increased their floor space and overall land requirements. By the 1950s and early 1960s, supermarkets in Los Angeles had grown to 20,000 to 30,000 square feet, creating an increasing need for land dedicated to parking, as the car became a more dominant mode of transport for shopping. The largest supermarkets emphasized brand products (linked to the growth of major food manufacturers such as Kraft) and increased product proliferation, as represented by the number of food items on store shelves. Supermarket chains shifted many of their operations to suburban locations while reducing the number of stores in denser low-income urban core areas such as South Central Los Angeles. Suburbs offered a larger land base and convenient access for the ubiquitous automobile (many of the supermarkets were located adjacent to freeways or high-traffic roads), enabling longer-distance shopping so that customers might come less frequently for perhaps a week's worth of items to purchase. Given the location of the large suburban stores, shoppers had almost none of the attachment to place or community that the urban core neighborhood stores had provided. Land in suburban locations was more available and cheaper (per square foot) in contrast with the denser urban core where land was more expensive and adjacent lots were harder to come by. Parking regulations that stipulated the number of parking spots required also did not take into account communities where residents were more transit-dependent and where stores were located within walking distance of denser neighborhoods.[28]

Supermarkets' shift from urban core to suburban areas created a "grocery gap" in low-income communities in Los Angeles and throughout the United States. Inner-city neighborhoods now depended on small stores that focused on cigarette, candy, and liquor sales but had only limited or no fresh fruits, vegetables, and meats. These "food deserts," as they were called (though a more appropriate name would be "food swamps," due to the proliferation of fast food restaurants and liquor stores), became the norm in South, Central, and East Los Angeles, the poorest and densest areas in the city.[29]

In 1987, a major new player entered food retail—Walmart, the largest retailer, employer, and global behemoth with its vast supply chain systems. Walmart quickly captured the largest share of the US food retail market and extended its emphasis on large store size and huge parking lots by introducing its "Supercenter" format, which combined food products with Walmart's wide array of consumer items.[30] Ideally suited for the company's already ubiquitous presence in rural and suburban areas, these massive (200,000+ square feet) supercenters impacted the grocery business in multiple ways: they furthered supermarket chain consolidation, put downward pressure on store employee wage levels thanks to Walmart's low-wage strategy, and further increased the trend toward supermarket influence over supply chain relationships. Brand items became even more ubiquitous on shelves, displacing products from smaller food production enterprises. The trend toward "private labels" also increased (food items that had the supermarket's own label, even if based on a similar set of supply chain sources). Walmart's supply-side strategies that squeezed prices from its suppliers and utilized its own vast logistics operation caused even huge food companies such as Tyson to become dependent on Walmart's decisions about which products to source, through what locations (globally or otherwise), and at what price points. In this manner, the largest of the food retailers became the centerpiece of what UK researchers Tim Lang and Michael Heasman call "retailing industrialization," with its "new ways of packaging, distributing, selling, trading and cooking food ... all to entice the consumer to purchase."[31]

Walmart also established food retail operations in other countries, including China. Unlike in the United States where it was often able to overwhelm its competition,[32] Walmart faced tougher rivals at the international level with such global competitors as France-based Carrefour and

UK-based Tesco. Walmart had sought to enter the UK market by taking over one of Tesco's key competitors but ultimately failed to displace Tesco as that country's leading food retail chain. Tesco, in turn, decided to enter the US food retail market during 2006–2007, selecting Southern California along with Las Vegas and Arizona as its entry point and initial challenge to Walmart. Tesco was aware that Walmart's supercenter strategy had been unable to expand into the largest urban markets, including Los Angeles. In 2004, Walmart had experienced a rare defeat when local community groups in the city of Inglewood (in the southwestern part of the greater Los Angeles area) successfully defeated Walmart's plan to build one of its trademark 200,000-square-foot supercenters. The land use requirements for its store format, along with its low-wage labor and supply chain operations, generated strong opposition while also representing a major constraint, particularly in those dense urban areas.[33]

Tesco, in contrast, adopted a far smaller store size format (10,000–15,000 square feet), expanded its use of private labels (up to 50 percent of all items in the store), reduced its full-time (and nonunion) workforce, and utilized an automated checkout system. It also sought to emphasize a "fresh and healthy" product selection to fit what it considered the Southern California lifestyle (its US stores were in fact called Fresh & Easy). But Tesco misjudged what consumers wanted when it came to "fresh" (not plastic-wrapped food items that included non-US imports, with markers that were required for the checkout system); it was disconnected from a changing Los Angeles and its diverse populations with multiple food cuisines; and its problematic checkout system was not perceived as convenient. By 2014, Tesco had given up and sold its Fresh & Easy stores to a US buyer, after its US affiliate declared bankruptcy to facilitate the sale.[34]

Walmart, meanwhile, also reconfigured its own strategies for breaking into the urban market, including Los Angeles, by deciding to go smaller, and stating that it would source more organic items and secure locations in food desert areas. But Walmart's stated commitment was challenged as misleading, and internal Walmart communications to its employees were later revealed in which the company encouraged employees to donate to food banks that could be used by other Walmart employees to achieve their own food security. After being blocked from entering the City of Los Angeles for ten years, Walmart did succeed after a protracted struggle in locating one of its small stores (its "Neighborhood Market," at 33,000 square feet) in the

Chinatown district just north of downtown Los Angeles. But its success was short lived, as Walmart closed the Chinatown market (along with 269 other US stores) in 2016 as part of the company's downsizing in what had turned out to be "a difficult retail climate," as the *Los Angeles Times* put it.[35]

Walmart's strategies, as well as those of Tesco and several other major chains, reflected an effort to capitalize on the growing interest in local and organic food. This included a new emphasis in Los Angeles and elsewhere on supporting truly local neighborhood markets, such as the ubiquitous corner store and the wide number of ethnic markets. A "Healthy Corner Store Campaign" was launched in Los Angeles, initially developed by the Los Angeles Community Redevelopment Agency (CRA) and subsequently shepherded by the Los Angeles Food Policy Council after CRAs in California were eliminated.[36] Many of these smaller stores also catered to the enormously diverse Southern California population and its strong orientation toward ethnic and immigrant food. Other trends included the expansion of farmers' markets that became a favored source of local (and organic) food and, in some neighborhoods, also emphasized a wide range of products grown by immigrant farmers to serve immigrant populations. Farmers' markets, as well as the increase in community gardens and a modest "urban agriculture," further established the farmer-to-consumer connection that several of the Los Angeles supermarket chains began to recognize as a growing consumer interest; the chains responded by highlighting local or California-grown fruits and vegetables in their produce section. In an age of global food retailers and retail industrialization, these new interests began to influence the trends in how, where, and what food was sold, whether in Los Angeles or elsewhere.

Hong Kong

In contrast to Los Angeles, supermarket chains developed much more recently in Hong Kong. Due to numerous barriers, including land requirements, the largest of the global food retailers such as Walmart, Tesco, and Carrefour have not successfully entered the Hong Kong market.[37] Selling food in Hong Kong has historically and traditionally been associated with the small open-air stands where fresh produce and live animals brought by local farmers are sold and then used for that day's food preparation. This practice was particularly the case in the period prior to refrigeration, due to Hong Kong's climate. Shopping was a daily experience, and the brick-and-mortar food stores that appeared after World War II were limited in size and product mix.[38]

In the 1980s, growing concerns about food safety and food spoilage led to the creation of "wet markets" housed in municipal buildings which were perceived to be more sanitary due to the hosing of the building's floor areas (thus the name). The wet markets serve the same function as the open-air markets: a wide variety of fresh produce, rice, poultry, and meat is available to be consumed that day from the same-day purchase. Even after refrigeration became more available, the size of homes in such a high-density region often precluded large appliances and storage space. Nevertheless, as a Consumer Council report in 2003 noted, supermarket chains were able to take root in Hong Kong during the 1990s, due in part to their ability to secure prime store sites close to new residential housing estates that provided "an almost captive customer base of residents."[39]

By 2016, two major chains—ParknShop and Wellcome—had secured the largest market share of supermarket revenues, upward of 75 percent. ParknShop is part of the Hutchison Whampoa conglomerate. Aside from its port holdings discussed in chapter 2, Hutchison Whampoa owns or controls real estate developments in Hong Kong and China, one of the two electric utilities in Hong Kong, a Canadian oil company involved in oil sands development, and numerous other interests. Hutchinson Whampoa, along with Cheung Kong Holdings, is part of business magnate Li Ka-shing's empire, whose ParknShop group also operates thirty hypermarkets (the equivalent of a supercenter) in southern China. Around the time of the 2014 Occupy Central protests for more democratic elections in Hong Kong, Li Ka-shing restructured his holdings to shift away from some mainland China investments, spun off the company's vast Hong Kong real estate businesses into a new entity, and continued to expand his global presence in more than fifty countries around the world, including in telecoms, infrastructure, energy, retail, and port-related services.[40]

Hutchison Whampoa's main competitor, the Wellcome Company, is part of the Dairy Farm chain of retail outlets, which includes supermarkets based in Hong Kong, Taiwan, and the Philippines. Dairy Farm is controlled by the Jardine Matheson group, a name that dates back to the colonial origins of Hong Kong. Dairy Farm also controls other types of outlets in Hong Kong, including ThreeSixty (organic and natural foods stores) and Oliver's

The Delicatessen (fine food and wine stores). Together, Wellcome and ParknShop represent a food retail duopoly.[41]

Hong Kong supermarkets are quite different from their Los Angeles counterparts. They are far smaller in store size and format due to limited land availability and reduced shopping loads, as there is far less car use in Hong Kong. They tend to carry a smaller array of products, and those are primarily processed and frozen food, some imports, and a limited selection of fresh produce. For new food products, including imports, the Hong Kong supermarkets charge listing fees to suppliers to enable those products to be put on shelves; these fees allow the stores to maintain leverage over their supply sources (similar to the "slotting fees" for new products that the Los Angeles and US supermarkets charge for products to secure shelf space). In addition to listing fees, the Hong Kong supermarkets exert considerable control over their suppliers, including the use of private labels.[42]

Despite the increased presence of supermarkets and the role of the duopoly chains, Hong Kong shoppers still rely on the wet markets for a significant percentage of their purchases, particularly fresh food items. The wet markets are complemented by the very small stores not linked to the chains which employ fewer than ten people but also contribute to the concept of daily shopping by their ubiquitous presence across Hong Kong. Street vendors or hawkers are another significant, although endangered, part of the Hong Kong food scene. Overall, daily shopping and the strong preference for fresh foods that can be prepared the same day remain central to the Hong Kong experience, keeping street food and the wet markets going, and also requiring the supermarket chains to adapt their strategies accordingly.[43] Thus, while change continues to occur in the Hong Kong food retail sector, the supermarket chains, even as more people come to shop in their outlets, have nevertheless remained a somewhat limited force in the food system due to the complexity of Hong Kong's built environment and its deeply embedded food culture.

Eating Food

Los Angeles
In 1993, a UCLA report entitled "Seeds of Change" profiled food issues in Los Angeles while highlighting conditions in one low-income neighborhood in South Los Angeles. The report documented a number of changes in

the food environment, including lack of access to healthy and fresh food (e.g., the abandonment of full-service food markets in low-income neighborhoods), the protracted problems of food insecurity, and the need for new programs and policies to better address a changing food system. The report found that as many as 27 percent of the residents in this neighborhood experienced hunger an average of five days each month, a phenomenon characterized as "continually dropping in and out of hunger." Hunger was defined as an issue of *community* food security, identifying a community's food environment—namely, what residents in a community eat, where they eat, and their access to food.[44]

Another team of UCLA researchers, following up on the "Seeds of Change" report and the media coverage about hunger in Los Angeles, undertook a study in fourteen low-income schools in the Los Angeles Unified School District of what high school students ate on a daily basis. Meals eaten in school were included in addition to what was eaten outside of school. Utilizing a daily food intake analysis, the researchers evaluated whether there was a connection between what students (and their families) ate and the "dropping in and out of hunger" phenomenon. The study confirmed that a high percentage of the students experienced hunger at particular times during the week but also found that many of those same students were overweight or obese. "We were astounded by the numbers," one of the researchers told one of us at the time of the study, and the findings caused the researchers to rethink their research focus.[45]

As subsequent studies have documented, in Los Angeles and elsewhere, there continues to be a coincidence of hunger and food insecurity and a huge increase in overweight and obesity in the same population. The increase in weight gain in the last several decades has led to an "obesity crisis" that crosses all income levels, racial, ethnic, and age groups, as well as neighborhoods, rich or poor, in Los Angeles and elsewhere, although it's most pronounced in low-income neighborhoods like South LA. As a consequence, there has been a dramatic rise in diet-related health outcomes, such as diabetes, and the diet-health connection has in turn helped elevate food environment issues into a major policy focus in Los Angeles, in California, at the national level, and globally.[46]

Central to the discussion of the food environment and what people eat has been the rapid penetration of a fast food culture that has influenced food choices and diets and the parallel changes in how much people eat (that is, food portions). Los Angeles in particular has long been associated with fast food; it was the home of the first McDonald's restaurant and the early fast food drive-ins that further indicated how the car was shaping urban life. Los Angeles has been at the forefront of the shift from cooking food at home to eating out (including at fast food chains), which has come to average more than 42 percent of residents' food dollars spent outside the home.[47] The increase in highly processed foods that require limited time and few cooking skills has at the same time influenced food preparation and food choice, as has the increase in the consumption and size of sugar-sweetened beverages, snack foods, and foods high in sugar, fat, and salt content.[48] These issues are particularly pronounced for those under eighteen years old. A number of studies have noted in this age group a direct correlation between high percentages of increased weight gain and eating frequently at fast food restaurants as well as consuming limited amounts of fresh fruits and vegetables. Moreover, the correlation between diet and weight gain for younger people has the potential to create a greater long-term risk of diet-related health problems.[49]

Aside from its homogenous fast food offerings, Los Angeles has become the home of a rich and diverse food culture, as has been noted earlier, thanks to the influx of a huge immigrant population, especially from Asia and Latin America, in the past thirty years. Building on earlier histories of immigrant street vendors, LA's street food scene now includes a thriving food truck and sidewalk vendor presence that dot both low-income and middle-class neighborhoods. It also entails the small restaurants and small stores specializing in regional cuisine and food items as well as the restaurants in such self-characterized neighborhoods as "Oaxacalifornia," "Little Bangladesh," "Little Ethiopia," or "Koreatown." Los Angeles's reputation as a food place of "insane diversity," as one of its chefs put it, has been enhanced by its vast array of food options, cataloged by a new breed of food writers, such as Pulitzer Prize-winning journalist Jonathan Gold, and made available by street food chefs such as Roy Choi, who made LA's "Korean taco" famous.[50] Ethnic cuisine has even extended to the lunch menu items of the Los Angeles Unified School District.

With all of these changes taking place, Los Angeles has emerged as a central battleground for the future of food and food environments, including whether and how these could be changed. A campaign in 2001–2002 to ban the sale of sodas in vending machines in Los Angeles schools was successful despite opposition from the soda companies like Pepsico. As the second school district to ban the sodas (Oakland, California, was the first), the Los Angeles school district inspired campaigns at schools around the United States. As early as 1997, one of the Los Angeles area schools was the first to embrace "farm to school" programs, in which school districts source from local farmers for their school lunch program, encourage school gardens, and shift toward a healthier menu. These initiatives have been part of a larger focus on changing the Los Angeles food environment to emphasize greater access to healthier and fresher food choices and a food system that reinforces rather than undermines such choices. Once considered the headquarters of a fast food culture, Los Angeles has emerged as a place central to the debates in the United States about what to eat.[51]

Hong Kong

In April 2010, the Centre for Food Safety at the Chinese University of Hong Kong released a comprehensive food consumption survey that had been commissioned by the Hong Kong government's Food and Environmental Hygiene Department. The government was interested in the food consumption patterns of the Hong Kong population to assess concerns about food safety, including but not limited to food imported from China. The study documented that food safety was considered very important for more than half of those surveyed (57.7 percent among males, and 68.7 percent among females), far more than other factors such as nutrition, taste, price, or ease of preparation. Dietary considerations (for example, fruit and vegetable consumption) were also considered significant, particularly as they revealed what the study identified as "health-driven behaviors." At the same time, the survey reported that a high percentage of the population (47.1 percent) was overweight or obese.[52]

The concern about food safety and the rise in obesity and overweight is occurring at the same time that hunger and food insecurity have been prominent concerns, exacerbated by the huge income divide among the poorest and richest in Hong Kong.[53] A 2011 Oxfam Hong Kong study on how the escalating costs of food prices and housing have impacted Hong Kong's low-income population identified widespread problems with food insecurity and hunger, and how much low-income residents need to spend on food compared to wealthier individuals and households. Nearly half of the low-income residents surveyed by Oxfam (45.9 percent) experienced food insecurity, and nearly one-sixth experienced "high food insecurity." As in Los Angeles, high food insecurity meant that a significant number of Hong Kong residents did not have enough food to meet their needs at all times—the "dropping in and out of hunger" phenomenon. The report also indicated that 40 percent of low-income residents' total household budget went to food purchases, whereas the highest-income groups spent 21 percent of their budget on food purchases. In order to cope with these issues of food insecurity, the Oxfam study noted that households purchased the lowest-cost foods and eliminated at least one meal a day when unable to purchase enough food. These numbers underline the income and poverty gap that other studies have documented for Hong Kong, including the statistics that 20 percent of the population is living in poverty, the homeless population is growing, and as many as one in three seniors is unable to meet basic nutritional needs. Hong Kong's income inequality has been called its "badge of shame" because it is the highest income gap between the rich and the poor of any developed economy.[54]

What Hong Kong residents eat and where they eat has become a factor in diet and food choices, particularly among youth. Although a fast food culture has not penetrated as deeply as it has in Los Angeles, there is a greater trend toward eating out in Hong Kong. One study indicated that as many as 30 percent of the population ate out for breakfast five or more times a week, with more than 50 percent eating out for lunch and more than 10 percent for dinner. Foods high in fat, sugar, and salt, as well as sugar-sweetened beverages, have become more prevalent, including sugary drinks, deep-fried food, burgers and fries, ice cream, and hot dogs.[55]

Yet, like that in Los Angeles, the food culture in Hong Kong is diverse, combining traditional foods with the multicultural environment that characterizes both cities. Street vendors with traditional offerings vie with cafés, restaurants, bakeries, and other places that provide a variety of cuisine and ways in which the food is prepared. At the same time, food safety scares have led to an increased interest in healthy and organic foods. Moreover, a renewed emphasis on growing food that one consumes as well as having a direct connection to where and how the food is grown and produced have become part of the Hong Kong food environment.

Hong Kong still remains dependent on food imports, both fresh and manufactured food. The city's residents continue to be sensitive to fluctuations in food prices that are often globally dictated, further exacerbating the income (and food) divide. Changing diets and limited access to and participation in physical activity have further contributed to the increased trends toward weight gain and related health diseases, even as food insecurity continues to be a concern. Hong Kong remains a complex food environment, as different food choices vie with each other, even as food remains a central part of residents' lives and culture.

The Connection to China

For millennia, China relied on a food economy based on rural, small peasant farms with a goal of food self-sufficiency. In recent years, Chinese academic researchers and sustainable agriculture advocates have characterized those six thousand years of agricultural practice as "always ecological" with their emphasis on resource conservation and recycling.[56] In the last half century, such rural farming traditions have undergone a series of cataclysmic changes, influenced by shifts in political regimes, the invasion of the Japanese, and the huge changes in rural and urban policies after 1949. During the Maoist period (1949–1976), the supply and marketing of food for urban areas remained a state responsibility, with the goal of minimizing imports or long-distance supply sources. Urban areas were able to obtain more than 85 percent of their vegetable supply from adjacent rural or periurban districts. The food supply for rural areas, on the other hand, remained the responsibility of the communes which contributed to a national goal of self-sufficiency in food. In 1978 the household responsibility system began to shift land use rights to rural households and allowed individual farmers to select crops and to sell whatever they produced beyond their quota of grain sold to the state.[57]

During the 1980s, while economic conditions in the rural farm communities improved with the development of village and township enterprises and private household-based food production, the uncertainties about land tenure placed limits on rural development strategies. The contracting of land holdings created incentives for short-term profit maximization but not longer-term sustainability. Local governments facilitated the conversion of agricultural land to rural industrial development, increased urbanization, road construction, and house building. The process of rapid urbanization and the continuing distinction between an urban and a rural residency permit (the *hukou*), combined with the push for development of export-oriented industrial production, led to an enormous expansion of a rural-to-urban migrant labor force of either displaced or migrating peasant farm members. This "floating population" of internal migrants who "float and move" without permanent registration become highly vulnerable and exploited, not dissimilar to undocumented migrant labor in the United States and around the world. In the nearly forty years since the introduction of market reforms in 1978, almost two-thirds of this Chinese rural labor force has ended up working in off-farm employment.[58]

As earning a sustainable livelihood from farm activities grew more difficult, the number of people directly engaged in farming, which had peaked at 340 million in the 1980s, began to decline at a rate of about 6 million each year. Moreover, the conversion of agricultural land to industrial or residential development transformed rural farming villages into new urban agglomerations, sometimes at a phenomenal rate. When agricultural land is seized, rural smallholder farmers are often not adequately compensated, and options for younger people working the land become even more limited. This lack of work, in turn, has led to a demographic crisis, similar to that in parts of the United States, with young people increasingly leaving rural areas to find employment in urban places.[59]

In Guangdong, the rural migrant labor force includes workers from distant provinces, speaking different languages from their Cantonese-speaking bosses, and experiencing the kind of migrant worker abuses found in California. This has created a type of semiproletarianized peasantry, with peasant family households still trying to cling to their small landholdings while family members join the rural migrant labor force with its indeterminate status, neither urban nor rural.[60]

The changes in the rural population and related loss of farmland have been exacerbated by farmland lost to pollution, due to both the use of chemical pesticides and fertilizers and industrial discharges of heavy metals such as cadmium that pollute the soil. A shift to higher-value products has further contributed to soil decline. China has become the largest user—and producer—of pesticides in the world, and its loss of farmland due to contamination has reached as high as 19.4 percent, according to a 2014 report by the Ministry of Environmental Protection and the Ministry of Land Resources. Even more dramatically, eight million acres of land had become so polluted "that farming should not be allowed on it," its vice minister warned in a December 2013 news conference.[61]

Strategies to achieve agricultural modernization have included the development of vertically integrated organizations that link the production, processing, and marketing of food. These include "dragon-head enterprises" as well as various brokerage organizations and new types of cooperatives that focus on specialty markets. The dragon-head enterprises, a mixed public-private type of Chinese agribusiness, have become a critical component of the development in China of a capitalist framework for agricultural modernization, at the level of both the central government and (especially) the local government. Large-scale dragon-head enterprises have been subsidized by the central government, while provincial, municipal, and county governments have aggressively sought to subsidize them at the local level through tax and credit benefits, advantageous loans, and the availability of land and electricity.[62]

Agricultural modernization has further led to a major influx of foreign direct investment, increased food imports, and an increasing number of global food players involved in integrated global food operations. This is particularly significant in the area of pork, beef, and poultry production, where changing diets and new consumer demand has had ripple effects throughout the Chinese food system. China became a net importer of soy and grain for use as feed, undercutting its long-standing desire to be self-sufficient in grain. The transformation in the pig industry from backyard and small household farms to an industrial model led to a quantum jump in global company investments and imports for feed, especially from Brazil. With this change, as Emelie Peine has written, there emerged a "Brazil-China-soy-pork commodity complex" that provided "a lens on global agro-food restructuring." In this case, four primary global soybean brokers and processors (including the US-headquartered Cargill and Archer Daniels Midland) have played a central role in controlling the supply chain and how this new global commodity complex is governed.[63]

The poultry industry provides another illustration of how large global companies, including Tyson and its subsidiaries, have entered the Chinese market. Tyson had originally depended on small farm suppliers for its various integrated operations, which included breeder production, hatcheries, broilers, and feed production. However, the company decided to reorient toward greater control of its China supply chain by beginning to establish its own farms and related infrastructure. By utilizing its own "brand" in China, Tyson hoped to gain market advantage by asserting that, as a global,

US-headquartered food company, it would have a better track record on food safety.[64]

A number of other global players have carved out a place in the poultry business in China, such as Cargill and even Goldman Sachs. At the same time, China has coveted the US market for its own exports of poultry by its China-based producers, which, until 2013, had been restricted. Then, in a convoluted ruling in July 2013, the United States Department of Agriculture determined that four China-based processors met equivalent US food safety standards for exporting processed chicken parts to the United States *as long as the chickens were sourced from the United States* and a couple of other countries. While the USDA decision generated strong opposition in the United States concerning China's food safety history, the irony of sending the whole chicken from the United States to China for processing and then exporting that processed product back to the United States was not lost on some observers.[65]

China's own integrated global food companies have also been active in entering other markets, whether through exports or through investment and in some cases by purchasing major foreign food companies or farmland. One such notable acquisition occurred in October 2013 when the Chinese food producer Shuanghui International (now the WH Group) purchased the largest US pork producer, Smithfield Foods (itself a major conglomerate which had consolidated its position through purchases of other US operations).[66]

These changes in China's food system have also been reflected in the rise of its supermarkets. A relatively new phenomenon for China, supermarkets have rapidly increased in the number and size of their outlets since the 1990s, particularly in the larger cities, gaining upward of 30 percent of all food retail sales. China's still vast rural and periurban populations, however, continue to shop at small single-store outlets and buy fresh meat, fish, fruit, and vegetables from traditional street markets on a daily basis.[67]

After the introduction of market economy changes, the food retail sector was one of the first sectors to be privatized. The shift toward greater consolidation and larger stores became more pronounced after 2001 when China entered the WTO and several of the major global food retailers, such as Walmart, Carrefour, and Tesco, began to enter the China retail market.[68] These global food retailers contributed to major changes in China's food retail environment, such as the rise of the hypermarkets. Hypermarkets combined the features of large supermarkets with some department-store-type features, and offered a variety of prepared foods, fresh and frozen meat and seafood, food service counters, imported foods, and highly processed food products.[69]

Despite their increased presence, the global food retailers have had their challenges. Growth rates have been slower than anticipated, and they have faced competition from domestic or Asia-based food retailers, including local ones that have played a greater role outside the largest cities and most developed regions. Global retailers have also experienced difficulties in managing a more complex supply chain and varied customer preferences. Walmart, for example, which opened its first store in China in 1996, waited until China joined the WTO before making a big push to expand its number of stores. Seeking to use its long-standing supply chain management strategies along with a high-volume business based on multiple large stores located throughout the country (432 stores in 174 cities in 2016), Walmart found that it had overreached and began to close some stores, even as it also pursued new store openings in lower-tier cities and a focus on its "safe foods" and online shopping platform.[70]

Tesco, which entered China in 2004, had perhaps the most difficulty of the global retailers. With ambitious plans to expand to as many as 200 stores by 2016 and a reliance on its well-developed hypermarket format, Tesco experienced multiple challenges in its China operations. As a result, in 2013, Tesco accepted the offer from China's largest retailer, China Resources Enterprises (CRE), to establish a joint venture, with Tesco the minority partner at 20 percent. The joint venture linked the 134 hypermarkets and 19 shopping malls that Tesco still managed (although not under the Tesco brand name). Similar to its Los Angeles experience, though, Tesco in China was unable to fully adapt to specific Chinese consumer interests and preferences.[71]

The changing pattern of food consumption in China, linked to rapid urbanization and China's growing middle class, has also been a factor in the expansion of the food retail sector and where food is consumed. These changes in diet have occurred in just a few decades. After the market economy reforms were introduced, small food stores, kiosks, and restaurants sprang up, and, as soon as the late 1980s, department stores were offering large food sections, anticipating the rise of the hypermarkets. Food processing expanded, creating more manufactured and processed product selections, while away-from-home food spending, almost nonexistent in the early 1980s, increased to 22.8 percent by 2010. While percentages for Chinese cities are smaller than those for Los Angeles and Hong Kong, they represent a more dramatic shift in where food is consumed.[72]

The changes in food consumed include the trend toward a diet heavier in meat, poultry, and other protein sources and a reduction in the consumption of the traditional sources of grains and starchy roots. One study estimated that in the space of a few decades, the ratio of consumption of grains, vegetables, and meat items shifted from an 8-1-1 to a 4-3-3 ratio, with rapidly growing consumption of meat, pork, and poultry creating a more protein-heavy diet.[73] The shift toward more meat and poultry improved important dietary needs, but also became part of a food consumption pattern that favored more processed foods as well as snack foods higher in sugar, salt, and fat content. Global fast food chains such as KFC and McDonald's have played a role in this transition since they entered the Chinese market in the late 1980s. KFC, which with its 4,600 outlets has the largest market share among the global fast food companies, became increasingly dependent on sales in China (upward of 50 percent) for its overall corporate performance. An unexpected drop in KFC sales in 2015 led to the decision to pursue a sale of its China franchise, making it a separate business, but without initial success due to disagreements with bidders over price.[74]

China's nutrition transition, fueled by increases in fast food and processed foods and combined with other factors such as decreased physical activity, has led to increases in overweight and obesity. A study in *The Lancet* identified 62 million obese people in China in 2013, which represented nearly 10 percent of the world's obese population (the United States had the highest numbers, at 86.9 million). Books and articles with such titles as "Fat China" or "Obesity Rate on the Increase" have identified another outcome of China's modernization strategies.[75]

The changes in diet and reduced physical activity in urban areas have further led to a sharp increase in the prevalence of diabetes, similar to those in Los Angeles and Hong Kong. According to national surveys, diabetes prevalence increased from less than 1 percent of China's population in 1980 to 11.6 percent (or fourteen million adults eighteen or older) in 2010, with more than half the population having possible risk factors for the development of diabetes or cardiovascular disease. With China's marketization in mind, some of the global drug companies have perceived advantages in these numbers, with their promise of "double-digit pharmaceutical sales growth and significant opportunity for multinational pharmaceutical companies," as one research report for such companies characterized the opportunity.[76]

Because both diet and the food system have changed, fears about food safety have been heightened. The term "food safety" first became prominent during the 1990s as concerns shifted from historical problems with "poisoned foods," due to unsanitary conditions, spoiled foods, and unsafe storage, to contemporary concerns about hazardous food additives, toxic food preservatives, fake foods, chemical inputs, and cost-cutting abuses along the supply chain up to and including markets and restaurants. Such products as adulterated baby formula and powdered milk, fake soy sauce, cooking oil adulterated with sewage oil, and adulterated meats have led to nationwide scandals and heightened food safety scares.[77]

Those scares have stimulated a search not just for "safe" food but for a connection to where and how food is grown and prepared, especially for urban consumers. As in Hong Kong, new "farm to table"-type initiatives such as rooftop gardens, CSAs, agricultural ecotourism, and a revival of urban farmers' markets have sprung up in the past five to ten years. China's growing urban population, especially its emerging middle class and upper class, has developed an almost nostalgic desire to "get closer to the food" that has also generated a rise in ecotourism at periurban farm sites.[78] Similarly, the desire to relearn the process of "growing food" (or, in the case of the rural migrant workers, to maintain some food-growing opportunities in their villages in the city) has generated interest in urban agriculture. An organics market, originally conceived as a form of export production, has also mushroomed among China's urban food consumers.

The central government, recognizing the impact of food safety concerns, diet and nutrition changes linked to greater meat consumption, the urban focus on healthier food, and declining arable land and increased water pollution, among other factors, established in February 2014 new guidelines on food and nutrition development. These included such goals as "inheriting a healthy dietary tradition of consuming mainly vegetable food and less animal food" and "protecting food with local characteristics." Yet abuses have continued, and mistrust has been even further exacerbated by the development of organic and alternative farming-based "special supply farms" that serve only a selected constituency, including government officials.[79]

All of these trends further emphasize that China's food system is in flux, with an ongoing debate about the impacts of modernization and marketization on the food system, on the urban environment, and on the people who prepare and consume the food.

Strategies for Change

In October 2011, New York State organic farmer and food justice advocate Elizabeth Henderson visited Taiwan and mainland China to talk about the CSA movement in the United States and to learn about CSAs and other farm-to-table initiatives in Taiwan and China. In Beijing, she was hosted by Shi Yan, a young CSA advocate who was completing her PhD at Renmin University under Dr. Wen Tiejun, a leading academic in China who has written extensively about a rural reconstruction approach to income disparities, food system change, and rural economic and community development. Shi Yan, who had previously spent time on a CSA farm in the United States, brought Henderson to Little Donkey Farm, one of China's first CSA farms, located just outside Beijing. A year after Henderson's trip, a conference in Hong Kong brought together CSA advocates, rural reconstruction researchers, and other environmental and food justice participants to explore opportunities for changing the food system. Similarly, other global food system critics and alternative food advocates in Hong Kong and China have begun to explore establishing a food sovereignty network that would parallel food sovereignty movements throughout the world.[80]

These gatherings symbolize a growing body of research, project-based initiatives, and related policy and community-based efforts that has emerged in Los Angeles, Hong Kong, and mainland China to change the culture and practices of the dominant food system, including how food is grown, where it is sold, and what is consumed. This effort is embedded in the concepts of food justice and food sovereignty which extend to changes for the entire population, including those most vulnerable, such as the urban poor or farm laborers. The movement includes establishing farm-to-consumer connections (such as CSAs or community gardens and urban agriculture), expanding new kinds of rural-to-urban relationships, and celebrating and sustaining local and long-standing ethnic and place-based food cultures. It further identifies the need to reduce or eliminate the wide range of environmental, health, and land use food-system hazards and impacts (such as efforts to scale up local and organic food production and the challenges associated with identifying what constitutes "organic" food).

In Los Angeles, a relatively robust and innovative food justice and farm-to-consumer approach began to take root in the 1990s and has significantly expanded in the last decade. This has included the development in Los Angeles of the farm-to-school program mentioned earlier (and its sister program, farm to preschool), both of which were among the first to be established in the United States. Farm to school is a derivative of the farm-to-table concept but with a direct food justice orientation. Instead of a set of individual subscribers purchasing food from a single farmer most often located in the same region, as in the CSA model, farm to school involves an institution such as a school district purchasing directly from several local farmers, with the goal of providing healthier items for the school lunch and breakfast programs. The program also benefits local farmers, many of whom rely on direct farm sales, either through farmers' markets, farm stands, or CSAs. Farm to school also increases school-aged children's awareness of where food comes from, which is reflected in parallel efforts to establish school gardens and their related learning opportunities. The food justice dimension directly comes into play since many public school students who have benefitted from the program are low income (for example, over 80 percent of the 640,000 students in the Los Angeles Unified School District) and otherwise do not have access to "farm direct" or organic food, given its expense. The Urban & Environmental Policy Institute (UEPI), which helped initiate one of the first farm-to-school programs, also established the first "farm-to-preschool" program in Los Angeles (involving children under five and focused on growing and learning about foods), and that program has now also spread throughout the United States.[81]

Los Angeles, with its twelve-month growing season due to a favorable climate and its history of a once flourishing agriculture, has witnessed a modest revival in new forms of growing food in the city and the surrounding region. Community gardens have been established on vacant or underutilized land, or where private developers have allowed temporary use of their land for food growing until they are ready to develop the land. More innovative uses of land have been pursued, such as planting in the median strip on roadways, growing food in alleyways, or harvesting food

Figure 5.1
Farm to preschool program: preschoolers from Taylor Family Childcare in South Los Angeles. Source: Emily Hart.

from fruit trees or backyard gardens that are made available to low-income residents.[82]

These urban agriculture initiatives face several challenges, including land use barriers. Available land for an urban farm or even a smaller community garden is very limited, and the land title/private land ownership constraints are a constant factor. An effective policy mechanism is needed to overcome or at least address these and other types of barriers as well as to help facilitate or implement the variety of new food initiatives that have mushroomed in the region.[83]

In 2009, several community-based food organizations, researchers, and food policy experts teamed up with officials in the City of Los Angeles mayor's office to help develop such a new policy-related entity. This group, initially organized as a task force and subsequently as the Los Angeles Food Policy Council (LAFPC), is essentially a hybrid of civil society groups and government officials. It has secured an unofficial or advisory relationship

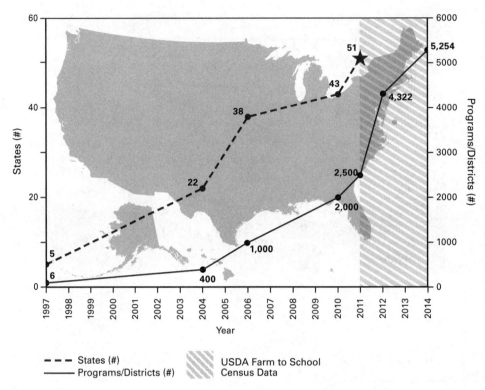

Figure 5.2
Growth of Farm to School in the United States (1997–2014). Source: National Farm to School Network.

with the mayor's office while maintaining its own independent status. The council has focused on program development, community advocacy, and policy development, and has established several working groups on such issues as urban agriculture, street food, food access and food retail, school food, sustainable sea food, food waste and resource recovery, hunger and food security, and procurement. Its detailed and groundbreaking Good Food Procurement Policy program identified five key criteria for procurement that it uses to assess supplier practices; these criteria, which have become the heart of its orientation toward food justice, include local sourcing, environmental impacts, animal welfare, health and nutrition, and fair labor practices. Seeking to identify where and how to pursue changes within the food system while also influencing the discourse about food in the urban environment, the Food Policy Council has become a unifying force in Los Angeles to help stimulate and convene ideas and constituencies dealing with such food system change. It has also helped formulate important food justice policy innovation and helped frame a new relationship between community groups and government, where much of the policy development has been driven by the nongovernment players.[84]

In Hong Kong, the momentum toward food system change has focused less on policy than on innovation and new food alternatives. The biggest influence on the interest in change and the opportunities for innovation has been the concern about food safety, particularly as Hong Kong has become more intertwined with China's (and especially Guangdong's) food-growing and production systems. Survey after survey has pointed to food safety as the most prominent issue for Hong Kong's consumers. This concern has been intensified by heightened media coverage of each new food scandal, including ongoing daily coverage in the *South China Morning Post* under the heading "China Food Scandals."[85] Whether related to the trade in live poultry or contaminated ingredients in products, such as melamine in baby formula which led to serious illnesses and even deaths, each scandal has only magnified such fears and begun to shift certain government policies, food product selections, and individual choices of what to eat. For example, after the baby formula scandal in China led to a huge increase in purchases of baby milk formula by mainland visitors and by black market traders, the Hong Kong government imposed a limit of two cans of formula that could be carried across the border. As a result of the food safety scares, food imports from countries other than China have been increased, and consumers have been purchasing larger amounts of processed foods. In addition, organic food products have experienced double-digit growth rates for the past several years.[86]

As in Los Angeles and the rest of the United States, debates have begun to take place in Hong Kong over what constitutes "organic" food and whether organic represents a different food system ethic or simply a new economic market opportunity. In Los Angeles and the United States, the debate has centered on the rise of what has been called "industrial organic"—corporate takeovers by large firms of smaller organics producers, or the expansion and consolidation of individual producers who capture an increasing share of the growth and production of particular food products, such as leafy vegetables or dairy. Some global behemoths, such as Walmart,

Coca Cola, and Con Agra, have entered the organics business and have had a huge impact on organics consolidation through their supply chain and purchasing arrangements. Industrial organics has become the symbol of how a market-driven approach can undermine or simply ignore key organic growing and producing practices. Workplace conditions and detrimental environmental practices have been substantial among industrial organic companies, causing many Los Angeles and US food justice groups to link "local" with "organic" as well as environmental and labor conditions in evaluating "good food" criteria.[87]

In Hong Kong, the development of an organics sector is relatively recent. The first organic farm was established in 1988 by Produce Green (later Produce Green Foundation), an environmental education group of self-styled "local enthusiasts concerned about modern farming and the environment." Several other educational and research groups have been formed to promote organic farming methods to help conventional Hong Kong farmers make a transition to organic growing methods. New organic farms were also established, and by December 2002 the Hong Kong Organic Resource Center was created to develop organic production and processing certification standards in response to this budding sector.[88]

In some cases, the concept of "organic farm" has been extended to the development of new community gardens and demonstration farm initiatives that have come to represent Hong Kong's emerging urban agriculture. Given Hong Kong's vertical development (as opposed to Los Angeles's horizontal sprawl), rooftop gardens have emerged as an additional type of urban agriculture. By December 2013, according to government figures, there were 480 organic farms in Hong Kong, including traditional family-operated farms, enterprise-operated farms, educational hobby farms, and noncertified farms that identified themselves as organic.[89]

Some groups also operated their own farmers' market and served as training and demonstration sites. The most notable of these is the Kadoorie Farm and Botanic Garden (KFBG), originally established in 1956 as a support service for Hong Kong farmers and for refugee farmers from mainland China. Support for KFBG was provided by two brothers, Sir Horace and Lord Lawrence Kadoorie (Lawrence was the first Hong Kong person to be named a British Lord and serve in the UK House of Lords). The Kadoorie brothers owned China Light and Power, the utility that played a leading role in Hong Kong's postwar industrial boom.[90]

Through the Kadoorie holdings, land for KFBG was donated and resources were provided to establish a new education and training center on 148 hectares on the northern slopes and foothills of Hong Kong's highest mountain—Tai Mo Shan. The KFBG land is an impressive landscape, with its streams, woodlands, and vegetable terraces that have enabled the group to establish its conservation-based and environmental education programs. Its mission was expanded further in 1995 when the Hong Kong Legislative Council formally established KFBG as a corporation, helping to consolidate its transition as a center for promoting sustainable and organic farming. In recent years, KFBG staff members have become linked to a loose-knit set of advocates promoting new initiatives such as CSAs and organic and sustainable agriculture in Hong Kong as well as China.[91]

Although groups like KFBG and Green Produce have emphasized the environmental side of organic farming, the rapid growth of the organic market has been seen as a major business opportunity by food producers, retailers, and farming enterprises in both Hong Kong and China. The debate over what constitutes "organic" food has emerged as a consequence, as seen in the complex and at times misleading certification process that has arisen, especially in China where marketization has become so pronounced.

In China, the transition to a market economy in agriculture improved crop yields and expanded food choices, but it also led to soil erosion and other harmful environmental practices that could undermine China's strategic goal of self-sufficiency, particularly in grain production. Those concerns led the Chinese government, already by the early 1980s, to support the development of a Chinese Ecological Agriculture (CEA) initiative. CEA highlighted such practices as traditional crop rotation, organic fertilizers, and a "virtuous cycle of production" involving the recycling and utilization of organic wastes. But CEA didn't succeed in attracting the newly privatized small farmers, who focused primarily on where they could create demand for their products. While the CEA began to fade as a government-supported approach by the late 1980s and early 1990s, the growing fears about food safety, already evident during that period, caused the government to introduce a "green food" concept, complete with a certification strategy and a Green Food Logo to help stimulate demand and reduce the food safety fears.[92]

Eventually, three labels emerged: hazard-free (the weakest), Green Food (still essentially transitional), and an organic standard, which is also the most expensive. With higher costs of production and certification, the organic standard helped establish a niche market for wealthier consumers.[93] Yet, even as the demand for "hazard-free," "green," or "organic" products and the amount of land dedicated to them has increased substantially, concerns about safety as well as abuses related to the certification process have remained widespread. The problem of trust continues to be an issue for green and even organic food. In a 2009 interview with Tony Guo, the sales director of Shanghai's largest grocery chain, City Shop, which specializes in imports and also sells many organic items from its own affiliated organic farm, Guo stated that "not many people, including myself, believe the organic label." Guo estimated that only 30 percent of the organic labels were authentic.[94]

The continuing food safety scandals in China and high degree of mistrust regarding such products as baby formula have led the mainland government to extend its regulatory approach, including the most recent 2015 Food Safety Law which updated and expanded the previous 2009 food safety regulations. Specific targets ("special food regulations") included infant formula milk powder, health food labels and menus, food for special medical use, small food manufacturers and processing mills, and online shopping (itself an increasing trend in food retail).[95]

The issue of trust has also become a rallying cry for the emerging CSA and alternative food groups that constitute the beginnings of a civil-society-led food movement. One key figure in this emerging movement has been Shi Yan, who has placed the issue of trust between farmers and consumers/eaters as central to the mission of a CSA-type initiative. However, as the Little Donkey Farm and some other CSAs began to grow, Shi Yan felt some of the farms were turning into more of a market-driven model where plots of land are rented and local farm labor is employed. Food safety scares have tended to influence the motivation of those consumer/subscribers who are less concerned about environmental issues or support for local farmers than about obtaining food that would have zero food safety risk, and would be willing to pay a premium price in order to obtain it. Shi Yan and her allies have sought to create a farm-to-consumer model that identifies a new type of food and environmental ethic that also emphasizes the importance of elevating the role and value of rural farmers and some of the traditional environmental-oriented growing strategies. The program that Shi Yan established in 2012, Sharing the Harvest farm, recast the CSA as a "value-based" approach in which consumers and farmers share risks and environmental farming practices are renewed and elaborated, but where consumers, instead of just seeking hazard-free food, could also become "ethical eaters," as one California CSA leader characterized the approach.[96]

One example of this new type of ethical arrangement is a small CSA operation founded in 2004 in the Tai Po section of the New Territories in Hong Kong by Li Kai Kuen, a local housewife. Gathering other women residents in the area to serve as volunteers and linking up with Kadoorie Farm staff, the indefatigable Kuen has worked closely with small farmers from the area to create a highly motivated group of farmers and consumers strongly committed to the support of the farmers and the local community. Emblematic of those relationships, one of the farmers, an eighty-year-old woman, insists that she teach the CSA participants and volunteers how to cook and prepare her special crop of noodles.[97]

Whether a new ecological-based farming strategy or an ethical farmer-eater relationship can be part of a larger sustainable development approach

Figure 5.3
Tai Po CSA volunteers; Li Kai Kuen is in front. Source: Marjorie Pearson.

impacting both rural agriculture and urban consumption has also preoccupied Shi Yan's academic mentor Wen Tiejun and other academics and
researchers who have sought to identify a renewed rural development or
rural reconstruction framework for China as a whole. This includes, as Wen
Tiejun wrote in 2003, "the reduction of the enormous disparity in the levels
of development between urban and rural China, as well as between Western
and Eastern China." As Wen Tiejun subsequently wrote in 2007, China's
focus on wealth creation (Deng Xiaoping's famous comment "to get rich is
glorious") based on a market-based export-oriented production and a fast
rate of urbanization has failed to address what he calls "the three big disparities"—between incomes (China's huge gap between the wealthiest and
the poorest), between urban and rural areas, and between regions. What is
it that we really want, Wen Tiejun asks—a continuing fast rate of GDP
growth or "the sustainable development of human beings, resources, and
society"? Issues related to food reside at the center of that debate about
sustainability, justice, and livability, as do the multiple issues of the urban
environment in the global city, whether in Los Angeles, Hong Kong, or
in China.[98]

6   Transportation in the City

Crossing Streets

In 2006, eighty-two-year-old Mayvis Coyle stepped from the sidewalk to cross the street at a wide intersection in Sunland, an older suburban community in the foothills of the San Gabriel Valley, on the northeastern edge of the City of Los Angeles. Cars rule in Sunland. Its streets are organized to best facilitate their flow, with the shortest amount of time allotted for a person to cross an intersection before a "Don't Walk" sign starts flashing. To pedestrians racing to cross such an intersection, the car (whether in Los Angeles or any car-dominant city) can seem like an invader, harmful not just to the pedestrian but to the very nature of "urban and social life," as French philosopher Henri Lefebvre has written.[1]

In Hong Kong, the street is a continuing transportation battleground. Hong Kong's density and its many narrow streets preoccupy transportation officials, who favor the free flow of vehicle traffic, whether for cars, buses, or trams. Take the case of Tsim Sha Tsui, a popular tourist area in Hong Kong. Salisbury Road in Tsim Sha Tsui is the main roadway, which a constant flow of people must cross to get to the harbor-front promenade and other shopping and leisure facilities. In 2004, the Transport Department decided to remove the main crossing at Salisbury and Nathan roads, arguing that because the number of pedestrians crossing the intersection was growing, it would be safer for them to cross Salisbury Road by going *underground* via the entrance and exit from the subways below this intersection. Then, with pedestrians out of the way, traffic flow could be optimized. With this change, local residents and tourists would now be obliged to use the subway system to cross Salisbury Road below street level, provided that they could find the entrance to the subway, pick the right turns along the way, and navigate unerringly in a maze of underground pathways. A straightforward thirty-second walk suddenly became a five-minute expedition into the "tunnels of agony," as depicted in a YouTube video filmed by the nonprofit organization Designing Hong Kong.[2]

In September 2014, thanks to the growing advocacy of groups like Designing Hong Kong and other NGOs, as well as politicians, business groups, the district council, and government departments, an at-grade pedestrian crossing was reinstated at the junction of Salisbury and Nathan roads. The reopening represented a victory for a people-based planning approach, especially in a city like Hong Kong where vehicles are nearly always given priority over pedestrians, although it took almost ten years to put back the pedestrian crossing and reconnect one of the many missing links in the city.

In China, the rapid increase in the number of cars on the road in its largest cities has been a challenge not just for pedestrians but increasingly for bicyclists, whose former command of the streets has continued to decline.

Figure 6.1
Tunnels of Agony. Source: Simon Ng.

The large, widened avenues and boulevards can sometimes become a chaotic mix of cars, bicycles, motor bikes, and pedestrians. Pedestrian-based
street life is better found in the alleyways and narrow streets where people
strolling on sidewalks and in the streets can encounter street peddlers, bike
riders, mahjong players, and the occasional cars. On the boulevards and the
wider streets, cars take control.

In Los Angeles, Mayvis Coyle's crossing a street in Sunland appears to
reinforce the long-standing adage that nobody walks in Los Angeles. To
quickly walk across a seventy-five-foot-wide intersection, where speeding
cars would often go through flashing and even red lights was a challenge,
let alone for an eighty-two-year-old. Coyle had just come from the grocery
store and needed to carry bags of groceries to her trailer home. Traffic sig-
nals were timed for an average pedestrian speed of four feet per second in
order to not impede traffic flow. Transportation officials calculated that
such an average would be met by only 85 percent of the population.[3]
Seniors were an obvious vulnerable population by this standard. And for
Mayvis Coyle, when the "Don't Walk" sign at the corner of Foothill Boulevard and Woodward Avenue began flashing, she was not yet able to get to
the other side. When she got there, she was met by a policeman on a motor-
cycle who had been waiting for her. He greeted her with a $114 ticket and
stated that she was obstructing the flow of traffic. Ms. Coyle, who told the
press that she felt she was being treated like a six-year-old, was not going
to let it rest. She fought the ticket—and lost, although the judge waived
the fine.

At a subsequent pedestrian safety forum, residents of the trailer park and
other residents in the area expressed their anger and frustration at the
"short signal timing, lack of safe pedestrian crossings, excessive [car] speeding, exaggerated crown to the roadway and high curbs, and some confusing
intersections." Complaints were relayed to the City Council representative
and to the General Manager of the LA City Transportation Department, but
to no avail.[4]

But Mayvis Coyle continued her fight. As a result of her battles, she
became a Los Angeles media figure, with coverage extending beyond the
United States to places like Scotland and Taiwan. Coyle was even contacted
by television host Ellen DeGeneres to tell her story to her millions of view-
ers. She received so much coverage that the Los Angeles Police Department
told her to stop talking to the media.[5]

But it was too late: at eighty-two, living in the Mar Vista Estates trailer park and as feisty as ever, Coyle had become the newest symbol for the struggle over the right to the city's streets, even in suburban Los Angeles. Foothill Boulevard in Sunland, Tsim Sha Tsui's Salisbury Road, and the boulevards and alleyways of urban China are today the places where the outcome of those struggles can help define the future of the urban environment in the global city.

Evolving Systems

Hong Kong

In 2008, the Hong Kong Transport Department published a brochure commemorating its fortieth anniversary. The cover provides a sleek picture of heavy car traffic along the Gloucester Road in Wan Chai, with its four car lanes in the middle, three lanes in a separated roadway on one side, and two lanes in a separated road on the other side. No pedestrians are visible in the photo, even on the sidewalk, which itself is barely visible. Speeding cars and a handful of buses dominate the picture.

In his introductory comments for the brochure, Commissioner of Transport Alan Wong highlighted Hong Kong's extensive system of roadways, touting its length of 2,000 kilometers that carry 560,000 vehicles. "The territory," Wong proudly stated, "is now one of the cities with the highest road traffic density in the world." The brochure also discussed Hong Kong's public transit system, whose 11 million passenger trips make up 90 percent of Hong Kong's daily commutes. These are numbers that Los Angeles transportation planners would obviously envy. Yet the highlighted cover image of cars and roadways rather than of Hong Kong's widely recognized and admired rail system, as well as the absence of any discussion of pedestrian and walkability issues, underlined the Transport Department's emphasis on roadways for vehicles.[6]

Hong Kong's policymakers have long prioritized the movement of vehicles over the social functions of streets. While the city's high densities might lead one to expect a highly walkable city, hyper-compactness produces its own challenges. Today, Hong Kong has low rates of vehicle ownership (94 licensed vehicles per 1,000 people) but among the highest traffic densities in the world (340 vehicles per kilometer of road). While it is generally possible to reach all destinations via public transportation, pedestrians

Figure 6.2
A view from Central, Hong Kong. Source: Simon Ng.

often encounter inconvenient and congested walking facilities designed to fit them around vehicle movements. Planning for pedestrians frequently takes an engineering approach, focusing on pedestrian flow rather than the quality of public space. Bike planning is nearly nonexistent, and bicycles are only considered for recreation in limited areas. Buses and trams have also become essential parts of Hong Kong's transportation infrastructure, although they also serve to further define streets as passageways for vehicles. Ferries still transfer people between destinations, but they are more favored by policymakers for their tourist appeal. Where Hong Kong's transportation system most excels is in its extensive subway and transit operations, the "gold standard for transit management world-wide," as a 2013 article in the *Atlantic* magazine put it.[7]

Before there were cars—or trains, or trams, or buses—there were ferries, the first major transportation system in Hong Kong. The first ferry docked in Victoria Harbor in 1874, but it was nearly twenty-five years later, in 1898, that the Star Ferry Company established the first ferry service between Central in Hong Kong Island and Tsim Sha Tsui in the Kowloon Peninsula and a few years later from Central to the Yaumati area in Kowloon. The colonial government, eyeing the increasing popularity of ferry service, established a "Ferries Ordinance" in 1917 to regulate the service and capture some of the profits from the licenses it granted.[8]

The most expansive period for passenger ferry service took place between 1950 and the early 1970s as Hong Kong's population grew and development expanded on Hong Kong Island and Kowloon as well as in the New Territories, where ferry service had also been established. During this period, the Hong Kong and Yaumati Ferry (HYF), the largest ferry service, alone created thirty-one new passenger ferry lines with double- and triple-decker ferries, expanded the number of the passengers it served, and redesigned several of the ferry piers. While still remaining very much part of Hong Kong's identity, ferry service began to lose its status as a primary transport system with the opening of the Cross Harbour Tunnel in 1972 and the opening up of roadway linkages between Hong Kong Island and Kowloon. Today, about 135,000 passengers still ride the ferries each day, including commuters and tourists.[9]

Trams and buses were also an important part of Hong Kong's early transportation history and continue to be a feature of today's transportation mix. The first trams began operation in 1904 with service from Wan Chai to Shau Kei Wan. The fleet, owned by a British company, Hong Kong Tramways Limited, consisted of twenty-six single-decker trams imported from England. From the outset, the "ding dings," as they affectionately came to be known, are considered along with the ferries and other boat traffic to be a part of Hong Kong's cultural heritage. During much of its rule, the colonial government continued to play a role in managing the tram system, including having the tickets printed in England, except for a short period during World War II when the Japanese took over the island. But after the anticolonial, pro-China riots of 1967, the government reversed course and had the tickets printed in Hong Kong, and stamped "Hong Kong" instead of "England" on the tickets.[10]

The trams today are celebrated for their role in short-distance travel up to four kilometers. While attracting tourists, they remain an inexpensive mode of transport for lower-income residents. In 2009, a subsidiary of the French-based global player, Veolia Environnement, purchased a 50 percent share (and the remaining 50 percent in 2010) from property developer Wharf Holdings (Wharf was controlled by Hong Kong billionaire Peter K.C. Woo, whose company, Wheelock & Co., also owned the Star Ferries). Veolia saw the purchase as an entryway to the Chinese market where it already had water and sewer treatment holdings, and wished to expand into China's transportation sector.[11]

Buses also have had a long history in Hong Kong's transportation mix. The two primary bus companies, Kowloon Motor Bus (KMB) and China Motor Bus (CMB) Company, were founded in the early 1920s and competed with a number of other small bus companies that crisscrossed Hong Kong, Kowloon, and the New Territories. In 1933, the colonial government set up a bus franchise system for different routes and facilitated the takeover of several of the small bus companies to help consolidate the reach of the two dominant companies, dividing service between KMB in the northern part of Hong Kong and CMB on Hong Kong Island. After a series of mergers and corporate takeovers, KMB emerged as the largest operator. Citybus Limited, originally involved in shuttle and residential bus services, came into the picture in the 1990s. They took over twenty-six franchised routes from CMB in 1993, and in subsequent years added new routes and buses. After the expiration of CMB's franchise in 1998, more routes were transferred to Citybus, while the remaining ones went to a new operator called New World First Bus, owned by NWS Holdings Limited. In 2003, Chow Tai Fook Enterprises, parent company of NWS, acquired Citybus, putting both Citybus and New World First Bus under the same roof.[12]

Both KMB and NWS Holdings are themselves subsidiaries of huge conglomerates controlled by two of the wealthiest families in Hong Kong. KMB is owned by Transport International Holdings Limited, a subsidiary of Sun Hung Kai Properties, which is controlled by the Kwok family. Sun Hung Kai has built some of the tallest towers to shape the Hong Kong skyline and is considered the world's second-largest property developer by market capitalization. NWS Holdings is a subsidiary of Chow Tai Fook Enterprises, which is controlled by the Cheng Yu-tung family and is the largest jeweler in the world. Cheng Yu-tung himself is considered to be the third or fourth richest individual in Hong Kong, but has turned over operations to others in his family, including his son and granddaughter, who run the real estate and hotel components of the Chow Tai Fook empire. Among its vast holdings are roads, ports, water, energy, ferries, and logistics operations; it also manages the Hong Kong Convention and Exhibition Centre and the Qianhai Mall in Shenzhen, which opened in December 2015 and was designed to bring the Hong Kong mall experience (brands and lower prices) to China.[13]

While the bus subsidiaries of these two family empires are dwarfed by their other holdings, they are nevertheless large bus operations that are an important part of the transportation system in Hong Kong. KMB has the largest fleet, with 3,826 mostly double-decker buses that carry 2.6 million passengers a day. Bus prices are significantly higher than the trams' fares, with fares higher the longer the distance. New World First Bus Company has a fleet of 708 buses that carry 466,000 passengers daily, while Citybus operates two bus networks under two franchises, with 751 and 182 buses respectively, and with daily passenger rides of 555,000 and 76,000. There are also mini buses that operate without fixed routes and another bus service that is primarily focused on routes on Lantau Island.[14]

While Hong Kong's major bus systems are privately owned and managed by some of the largest companies and wealthiest individuals who also have investments in China, its rail systems have a long public history and role. The first rail line in Hong Kong was developed in 1910, with an additional line put in the next year to Canton (Guangzhou). During the next sixty years, rail services were relatively limited compared to other transit modes, and they deteriorated during the Japanese occupation. But as rail regained currency with the combination of increased population and new housing and economic development, the government funded a consultancy study, published in 1967 (the "Hong Kong Mass Transport Study"), which identified the need to develop a new, separate underground rapid rail transit system. In 1975, the colonial government established the Mass Transit Railway Corporation (MTRC), with the government as sole stockholder, to develop and operate this new mass transit system for Hong Kong (which in turn was called the MTR) under prudent commercial principles.[15]

The first of the new rail lines, from Kwun Tong to Shek Kip Mei, was completed in October 1979. Four months later, the first cross-harbor rail link connecting Hong Kong Island and the Kowloon Peninsula began operations, resulting in "profound changes in the way Hong Kong people commuted," as one government document put it. With strong government support and funding, the railway network expanded rapidly during the 1980s and 1990s as MTRC established itself as one of the premier new transit operators in the world. In 2000, MTRC's status was changed from a wholly government-owned statutory body to a company listed on the Hong Kong Stock Exchange, with 23 percent of its share capital offered to private investors. MTRC further consolidated its role in Hong Kong's rail system when in 2007 it took over the multiple rail lines of the publicly owned Kowloon-Canton Railway Corporation (KCRC), the oldest rail system in Hong Kong, which had served locations in the New Territories and other areas throughout Hong Kong.[16]

This new public-private or quasi-public entity, publicly funded, independently managed, and connected to government policy goals, became perhaps most noted for its "integrated rail-property development model." This model linked MTRC's rail operations with high-density property development above or adjacent to its various rail stations. Initially, most of MTRC's rail lines were aligned along the densest urban corridors and population parcels, which guaranteed high volumes of patronage. The robust transit-oriented development framework that placed new residences and businesses next to rail stations further reinforced that advantage. In the process, MTRC became central to expansive real estate development (including construction of some of the highest towers in Hong Kong's vertical land form) that was integrated into its extended transit system. MTRC also extended its reputation as a unique international model that especially attracted the interest of mainland Chinese cities that were just beginning to explore building their own mass transit systems; for these cities, a rail-property and transit-oriented development approach met both their transportation and urban development goals.[17]

With this reputation, MTRC sought to become a global operator, particularly in mainland China. From 2004 to 2011, as China moved with breakneck speed to develop its own metro system, MTRC was able to establish full or partial ownership and management relationships with metro systems in Beijing, Hangzhou, and neighboring Shenzhen. Partly in response to the Hong Kong government's interest in its own connections to China, MTRC also began to construct a 142-kilometer high-speed rail link connecting Hong Kong to Shenzhen and Guangzhou. Construction of this new intercity rail network, which commenced in 2010, became controversial due to land takeovers of agricultural villages (as described in chapter 5) as well as continuous delays and cost overruns. Critics argued that the initial $67 billion price tag, which had mushroomed by 2015 to $85 billion to be paid by Hong Kong taxpayers, was not worth the justification that the high-speed system would reduce the time to travel between Hong Kong and Guangzhou from one hundred minutes to fifty minutes.[18]

Where the MTR system continues to maintain its strong reputation is in its efficient, on-time rail operations within Hong Kong. As of 2016, the system includes a total of nine lines stopping at eighty-seven stations and carrying an average of 4.71 million passengers a day. Its thirty-five-kilometer airport express service from Hong Kong Station in Central to Hong Kong International Airport has become a model rail-to-airport system that has reduced the number of cars seeking to access the airport. In 2014, a government transportation planning report identified the MTR system as the "backbone" of the overall public transit system and concluded that its further development would "reduce the reliance on road-based transport (and the need to build more roads), alleviate road congestion and lessen vehicle-induced air pollution."[19] When all the recommended new rail projects are completed by 2026, about 75 percent of Hong Kong's total population and 85 percent of job opportunities will be within reach of rail services.

But while cars cannot compete with the MTR and Hong Kong's overall public transit systems, they have contributed to the overall problem of street congestion due to the competition of cars with buses and trams in the central corridors. Unlike Los Angeles, Hong Kong, despite the Transport Department's picture of fast-moving cars and roadways, has a far more limited infrastructure and role for automobiles. Due to the steep terrain on Hong Kong Island and the narrowness of its streets, Hong Kong was a fairly late adopter of the private car. In 1910, there were only 20 cars in the city, mostly owned by the British, and in the aftermath of World War II, there were still only 3,986. Car use increased incrementally over the next several decades, dipping slightly during economic downturns. By the 1990s, however, private car registrations continued to climb, reaching 388,000 in 2005, and expanding to 567,000 in 2015. Rates of growth have been close to 5 percent annually. While not necessarily considered a luxury, car ownership has come to be seen as a status symbol, similar to owning a car among the wealthier urban residents in China.[20]

Today, public transit in Hong Kong, with its expansive rail and bus systems, remains the dominant transport mode. This continues to be the case despite concerns about problems of sufficient coordination between transport systems and the steady increase in car use. In contrast to Los Angeles, public transit in Hong Kong, particularly rail service, has secured its role without significant competition from the automobile, and in keeping with Hong Kong's compact, vertical development and land use constraints. Cars nevertheless have become a visible part of the street scene, impacting the environment and people's health and quality of life.

Los Angeles

Before Hong Kong's MTR system had been built and before Los Angeles had become auto-dependent, LA had had the most expansive rail and streetcar system in the United States. And even as the rail lines began to cross multiple jurisdictions at the turn of the twentieth century, the bicycle had achieved popularity and prominence. The sequence of transportation systems—from bike to rail and then to car—helped lay the groundwork for the next system that replaced it, undermining or at least narrowing the options of each system displaced.

During the 1890s, the bicycle was widely used for both recreation and transport in Los Angeles, due to the city's mild year-round weather and largely flat land. In contrast to the horse-drawn carriages that led to manure in the streets, the bicycle was recognized as a modern, clean, and efficient way to transport individuals from place to place. Bicycles helped establish the first paved roadways and provided the first transportation system widely used by women. Forty years before the first freeway was built, an elevated cycleway was proposed and partially built to connect an eight-mile stretch of Pasadena to downtown Los Angeles—the very same route that the first freeway in Los Angeles would follow forty years later.[21]

While Los Angeles during the 1890s was considered the bicycle capital of the United States, the bicycle had already begun to be eclipsed by the extensive rail systems that were being built. The major railroad companies—the Union Pacific, Southern Pacific, and Santa Fe, among others—quickly established themselves as the most powerful players in the development of Los Angeles. They negotiated the terms of whether and how the city would grow, including its ability to be connected to other parts of California and the rest of the country. In exchange, they received huge land holdings, often as a condition for laying the tracks to connect the city.

The railroad magnates, led by the ambitious Henry E. Huntington, became a powerful force within the city and the region through construction of a vast electric railway system. This included Huntington's Pacific Electric Railway's "interurbans," named for their larger size and faster speed than rail lines operating just within the city limits. Railroad magnates like Huntington sought to influence and control every aspect of urban development—from the location of the harbor to the new subdivisions that sprang up at the edges of the central part of the city. Although not always successful in achieving their goal—the harbor battle in the 1890s represented one of the few losses—they were most successful when they were able to control or make common cause with parallel interests, such as the electric utilities and the real estate syndicates organized in the first decade of the twentieth century.[22]

The story has often been told about how Los Angeles extended outward thanks to the land speculators and syndicates that subdivided undeveloped or agricultural land for new urban subdivisions. The electric railway was central to that process, often taking the lead along with the electric utility to establish a new town plat, also made possible with the availability of imported water. Huntington liked to boast that laying out the electric rail lines to the new lands *preceded* the actual development of new townships and home construction. This was a process not dissimilar to what the railroad interests had achieved in other parts of the United States through their "streetcar suburbs." Los Angeles took it a step further in the ability of the interurbans to shape the city, particularly through the sheer scale of the system. By the 1920s, the interurbans had more riders in more areas and more track laid than any other city in the United States. Each day 6,000 street cars utilized 115 routes, with over 1,000 miles of track laid in a complex set of grids. The rail lines crisscrossing the region corresponded to the crazy quilt pattern of development in Los Angeles, which in turn complemented how the imported water system had influenced the patterns of growth. It was the largest urban electric rail system in the United States. And for the railroad companies, Los Angeles had become nothing less than "an electric railway paradise."[23]

Even at the height of their reach, the privately owned rail companies were already facing challenges. Passenger service began to deteriorate during the 1920s when the railroads started to shift resources toward their freight operations, which offered them greater profit potential. By 1925, three-quarters of the rail system was dedicated to freight rather than passenger service. Rider discontent in the 1920s and 1930s also led to a revival of calls for public ownership, though the rail companies were able to defeat those efforts. But the discontent with rail service also fed into the increased receptivity for the automobile.[24]

Both the bicycle and rail systems enabled the car to achieve its prominence in Los Angeles. The paved roadways for the bicycle facilitated the expansion of paved roadways for the early rise of the automobile. The extensive rail systems that had facilitated the geographic expansion of the city and established LA's streetcar suburbs also provided an ideal environment for the automobile to become the main form of transportation—as long as the roadway infrastructure was set in place.[25]

Some of the earliest challenges for the automobile in Los Angeles in the 1910s and the 1920s were related to inadequate road systems. As automobile ownership increased and the car had to compete with the interurbans, the car was not initially the most efficient or quickest mode of transport. A short-lived parking ban in 1920 in downtown Los Angeles was one of the first battles over roadway and street primacy. Similarly, various proposals to separate the interurbans from car traffic, such as a proposed subway system and an elevated rail line in downtown Los Angeles, failed to gain support. More effective roadway design and efficiency, particularly for the automobile, preoccupied Los Angeles planners and business interests alike throughout the 1920s, even as Los Angeles achieved the distinction of becoming the number one city in the nation for car ownership.[26]

It was in the 1930s, thanks in part to Depression-era public works spending, that Los Angeles began to put in place its extensive system of dedicated roadways. A 1930 report by nationally renowned planners Frederick Olmsted and Harlan Bartholomew, prepared for but never supported by the LA Chamber of Commerce, laid out a bold vision for Los Angeles that linked parkways for auto transport with parks and open space. This included parklands adjacent to the Los Angeles River that could also serve as floodplains.[27]

Later in the decade, infused with funds from the federal government designed for job creation, two parallel, ambitious infrastructure projects were proposed and successfully implemented. One involved the channelization of the Los Angeles River, including its tributaries such as the Arroyo Seco. The second involved building a roadway along the same route of the original cycleway proposal. The proposed roadway was originally conceived as a scenic parkway with graceful curves adjacent to the Arroyo Seco stream while that waterway would simultaneously be laid with concrete as part of the ambitious LA River flood control project. But even before construction was completed in 1941, the parkway had come to be defined as the first *freeway* in the western United States. This would be a roadway that emphasized speed and destination rather than a connection to the places it passed through, as intended in the original parkway design.[28]

The debate over how to plan for the Hollywood Freeway, the second freeway project to be built in the mid-1940s, became emblematic of the shift from rail and bus to the automobile as the primary if near exclusive focus of transportation planning for the Los Angeles region. Instead of establishing complementary systems (for example, a busway along the median strip of the freeway, or feeder systems between rail, bus, and car), the freeway instead became the centerpiece of an expanding transportation infrastructure to an ever-expanding geography of urban subdivision and real estate development.[29]

Freeway expansion got an enormous boost when the 1956 national highway bill was passed, providing as much as 90 percent federal funding for highway construction. Thanks to this enormous funding windfall (complemented by gasoline taxes that had been introduced as early as 1919 and were dedicated to roadway construction and maintenance), Los Angeles launched a massive construction program during the late 1950s and 1960s. New freeways were built throughout the region, disrupting the built environments of whole neighborhoods and creating further class and racial divides in the process. Freeways were also built to extend outward to the new suburbs, at distances between suburb and urban core far greater than the longest electric rail lines. These freeway suburbs became almost entirely auto-dependent, as the distances between job and home or home and shopping and other daily-life needs required travel by car.[30]

At the same time, the rail system began its own rapid decline. Entire rail lines began to be dismantled during the mid and late 1940s, with the railroad companies ready to abandon passenger service entirely. The oft-cited popular notion that the rail system was destroyed by a conspiracy of automobile, rubber, and oil companies—while partly true (there was court action taken against those companies for their takeover of the main transit systems in the 1940s that helped further the shift away from rail)—misses the point that there were lost opportunities in the 1930s and 1940s to design a multimodal transportation system for rail, bus, and car, even as the car began its ascent in Los Angeles. By that time, the private rail operators had already helped facilitate the end of rail passenger service, as they continued to expand their freight operations. The railroad companies even joined with the Automobile Club of Southern California and suburban developers to defeat a 1948 rapid transit bond measure that might have been able to improve passenger options.[31] By the 1950s, public ownership of passenger rail lines, the earlier dream of advocates, took place not as an opportunity but as a bailout. The shift from rail to bus ("substituting rubber for rail," as one cement company owner and Auto Club executive put it) that also took place in the 1940s and 1950s only further reinforced the dominance of the auto-related paved roadway systems. The massive electric rail system of the 1920s and 1930s became literally nonexistent in Los Angeles, a historic artifact subject to nostalgia and museum exhibits, with the very last interurban service from downtown Los Angeles to Long Beach shut down in 1961.[32]

Los Angeles assumed its place as the auto and freeway capital of the country. Car ownership far outstripped that in any other city or region. Already by 1954, Los Angeles had achieved the distinction of having by far the highest number of vehicles but the lowest numbers of persons per vehicle—and also the lowest per capita rate of public transit use in the United States. Car-related institutions and places were springing up—the drive-in movie theater and drive-through fast food restaurant, the huge parking lots at giant suburban shopping malls, and the car-centered culture expressed in music, books, urban design, and architecture.[33]

But even at the height of their triumph in the 1960s and 1970s, the automobile and the freeway were becoming negative symbols and sources of conflict. Largely thanks to the automobile, Los Angeles achieved its notorious reputation for air pollution. When the connection between cars and air quality was identified in the 1950s, it eventually led to increasing calls for regulation and a search for alternatives. At first, a few freeway projects had to be shelved, particularly when they had been planned to pass through wealthier neighborhoods, such as Beverly Hills. Some of the residents of the poorer neighborhoods, such as the ethnically diverse and culturally rich enclave of Boyle Heights on the Eastside, put up a fight to stop freeway construction that would split their neighborhoods, raze some of the classic Victorian homes that were the pride of the community, lead to the loss of scarce parkland, and create enormous noise and air pollution impacts. But through the 1960s, most of these neighborhoods were unable to stop the freeway construction juggernaut. Boyle Heights, as one example, found itself subject to three massive freeway interchanges, while as many as seven freeways intersected and passed through adjacent neighborhoods on the Eastside.[34]

By the 1970s and 1980s, the combination of the air pollution, increasing freeway (and street) gridlock, episodes of road rage, and the two or three hour commutes from the freeway suburbs had begun to shift the nature of the debates about transportation, albeit modestly.[35] Nostalgia about the railway led to continuing efforts beginning in 1974 to raise funds (initially unsuccessful) and jump-start construction of new rail lines, an expensive proposition since nearly all the old electric rail lines had been torn up and new routes had to pass through settled neighborhoods. Bus service had expanded and was largely used by the poor and the transit-dependent, but bus riders often required transfers in order to complete their lengthy journeys from home to work or to basic services such as health care, child care, or shopping.[36]

As the push for transit alternatives gathered force, new conflicts between bus and rail advocates surfaced, leading to a landmark civil rights lawsuit in 1994 that resulted in a consent decree two years later forcing the regional Metropolitan Transportation Authority (the Metro) to shift some of its funding to improve bus service.[37] The lawsuit was based on the (accurate) assumption that primarily poor people and people of color utilized the buses, but it also assumed (inaccurately) that primarily middle-class and wealthier people would utilize the new rail lines. The assumption by the Metro was that rail service would be the preferred alternative for middle-class car users who would not be willing to give up their car commute by switching to buses. But as the new rail lines were put in place, the majority of their users were low-income, albeit a smaller percentage than the large majority of low-income bus riders.[38]

Despite the growing advocacy for alternatives, Los Angeles remains an auto-dependent region. There are 6.19 million cars registered in Los Angeles County and another 1 million trucks. Nearly 80 percent of all residents living in LA County own a vehicle.[39] Moreover, a significant amount of the funding and transportation planning is still earmarked for the automobile, whether at the street level, for parking, or for the freeways. For example, in 2011 and again in 2012, more than one billion dollars for freeway construction was earmarked for a single carpool lane over a ten-mile stretch on one of the most congested freeways. This nearly decade-long project culminated in closing the freeway for a single weekend in order to tear down a bridge overhang, with a repeat closure the following year to complete the bridge removal and replacement. Those weekend events were called "Carmageddon" (and the sequel was "Carmageddon 2") on the assumption that, without the freeway, streets would be overrun by desperate drivers frantically seeking ways to get to their destination. But Carmageddon never happened—car use on those weekends dropped significantly, while alternative ways of getting from place to place were also discovered. In the spirit of the event, some people even sought to surreptitiously enter the freeway, snap pictures, and, in one case, settle in for a gourmet meal they had brought with them to feast on what one writer called "the eerily empty freeway."[40]

Today, Los Angeles is seeking ways to redefine its transportation system and, by extension, its own urban landscape. Architecture critic and *Los Angeles Times* writer Christopher Hawthorne identifies this search in part as a transition to "the Third Los Angeles"—from the streetcar-based, pre-automobile period to the auto-dominated city and region with its freeway suburbs to a city that will potentially learn to use space, time, and the built environment without total dependence on the automobile.[41] The increased push for expanding bus and especially rail service, as well as growing bicycle advocacy and pro-pedestrian movements, have begun to identify where and how such a transition might occur. That transition will require not just a shift in resources and planning, as critical as that might be, but a cultural shift (and a far greater policy makeover) in how the city's and the region's residents identify with the places—and streets—they inhabit as well as pass through.

Impacts

Hong Kong

While Hong Kong's transit systems can be considered a major success story in their high-volume use and efficiencies (for example, the extensive and continuous on-time rail service), transportation has also had some negative effects on the daily lives of residents and commuters. These include problems of congestion, air quality, noise, and safety, as well as constraints for pedestrians and bike riders.

Congestion in the core areas of Hong Kong and Kowloon represents perhaps the most tangible impact. The high-volume use of the transit system due to its dense corridors and vertical commercial and residential development along transit stops has led to overcrowded subways and slower bus service during morning and evening commutes. As the number of rail users continues to grow incrementally, the high level of use has meant overcrowded trains.

The issue of congestion, however, is most notable in relation to *road congestion* which has become an increasing problem that has been significantly, though not exclusively, related to car use. Although the percentage of car use by population has held relatively stable at about 10 percent of all passenger trips, the number of cars on the road has increased at a higher percentage than other vehicles, with a nearly 5 percent annual growth from 2010 to 2015. More than 567,000 passenger cars are in circulation compared to the other 200,000 vehicles on the road, such as buses and goods vehicles. Cars account for as much as 40 to 70 percent of the total traffic flows on most major roads, according to a 2013 survey, but only accounted for 16 percent of the total passenger journeys. Buses and light buses, on the other hand, carried about 71 percent of the road-based passenger journeys but took up only about 2 to 5 percent of the total traffic flow. Road congestion problems worsen when cars search for one of the 17,900 on-street parking spaces, whose costs have remained flat for more than two decades.[42] Double parking and curbside loading/unloading activities further exacerbate the congestion.

The rise in the number of passenger cars also reflects a class distinction, as car owners are wealthier than transit riders. As the number of cars on the road has increased, the speed of the average car journey has dropped by more than 10 percent to 22.7 km/hr. One study noted that car journey speeds on major traffic corridors has declined even further. During peak travel times such as weekday mornings, car travel time on Des Voeux Road West declined to about or even lower than 10 km/hr—not much faster than one might walk a similar stretch. Commute times from the more distant New Towns in the New Territories to urban core areas have also increased continually.[43]

The efficiency of the bus systems have also been impacted by road congestion. Buses have experienced greater delays and increased travel time through the most congested parts of the city due to the competition for road space. These delays have caused a public call for Hong Kong to address the congestion issue, though the public has been far more focused on the need to control the growth of car use (about 70 percent, according to one survey).[44]

Traffic congestion has become a key quality of life issue. Congested roads and increased car use have generated significant environmental, noise, and safety impacts. Motor vehicles have become the major source at street level for such air pollutants as nitrogen dioxide and particulates. Motor vehicles also represent the second-largest source of greenhouse gas emissions. The increased NO_2 and PM levels have led to a jump in the number of hours that the roadside air pollution has been considered unhealthy (with high, very high, and serious health risk). From a public health standpoint, road vehicles pose the biggest threat to the general public, given Hong Kong's compact urban morphology and street canyons, intense pedestrian movements and street activities, and hence the direct and close exposure of large numbers of people to toxic air pollutants on a daily basis.

Noise and safety have become issues as well, particularly in the most congested areas of the city. Noise sources include road transport as well as construction, commercial, and industrial activities. Road transport is recognized by the government as the biggest contributor, and the most congested areas have had the highest noise levels. As many as six hundred roads generate sound levels greater than 70 decibels, above which impacts are intensified. The development of road infrastructure has also caused noise pollution, with some highways built just outside of people's living space. Impacts from noise include stress, hearing loss, and disruption of sleep and other daily-life activities. As road congestion and road infrastructure development have increased since the 1980s, there has been widespread recognition that noise is a daily-life burden.[45]

Road traffic fatalities from motor vehicles are another concern, with about 20,000 fatalities a year. While the number of fatalities remained relatively constant between 2004 and 2014, Hong Kong's growing older population, particularly those over seventy years old, has experienced the largest percentage increase in fatalities among the overall 4,400 pedestrian deaths.

The number of those injured from traffic-related accidents has also witnessed an increase in the past ten years.[46]

These types of impacts from transportation, particularly through road congestion and increased car use, have led policymakers and business interests to worry that the image of Hong Kong as a world-class metropolis could be affected, despite the advantages of its widely admired rail transit system. Problems such as traffic-induced air pollution and noise have negatively affected quality of life and could reduce its global city aspirations, making it harder to attract overseas companies to establish branch operations or to maintain its role as a hub for transport and goods movement. Whether those impacts can be reduced or mitigated becomes a central question for Hong Kong's ability to achieve its global city goals.

Los Angeles

Even as efforts have been made to transition beyond an auto-dominated Los Angeles, the impacts from the automobile remain enormous. Although other transportation systems have had environmental and (in the case of rail) land use impacts, these are modest compared to the automobile's role in shaping the region.

The automobile's land use footprint—including freeways, streets, and parking—is vast. Forty-two different freeways and highways in Los Angeles County, used by cars and trucks, extend 915 miles. When combined with various other roadways, that constitutes 25,000 total maintained road miles in Los Angeles.[47]

Moreover, many communities throughout the region are dotted with heavily used, high-traffic streets that are designed for maximum traffic or vehicle flow. Street miles represent an overwhelming percentage of the urban built environment. For example, the City of Los Angeles has 86.5 square miles of its land area occupied by streets mainly filled with cars, or 28 percent of the city's entire developed land area. Sidewalks for pedestrians remain narrow and, in a number of suburban communities, are extremely rare. Streets are not made for walking or biking. Even though bike infrastructure—bike paths, bike lanes, and bike sharrows alerting drivers to share the road with bike riders—has rapidly expanded from 245 to 562 miles between 2005 to 2015 in the City of Los Angeles, this still represents only 7.5 percent of the streets used by automobiles and trucks. Similarly, even with a 56 percent increase in bike-to-work commuting between 2000 and 2010, only 1 percent of all commuters biked to work. To put it in perspective, up to 75.2 million miles are being driven in the city on an average day, with 53 percent of those miles being driven on the freeways.[48]

The bias toward car-related transportation and land use planning is reinforced by long-standing assessments of what is required to handle traffic flow (such as widening streets or building more freeway lanes) and even how such flow is determined in the first place. One study has identified "phantom trips" as a common form of traffic assessments that overestimate the number of car-related trips in order to expand road systems. "If trip generation estimates are inflated," writes transportation researcher Adam Millard-Ball, "then catering for 'phantom trips' may likely lead to roads with more and wider lanes and intersections with longer signal phases— and, in turn, urban places that are hostile to pedestrians and devote too much land to vehicle infrastructure."[49]

Car parking represents another massive auto-dominated land footprint, including but not limited to surface-street parking. Parking regulations play a major role in extending that footprint. Various parking codes and regulations such as minimum parking requirements require housing developments, restaurants, supermarkets, and other places to accommodate a specified number of parking spaces. These are often determined by set factors such as the building's square footage or number of residents or customers, regardless of context (for example, if a supermarket is located in a dense, transit-dependent neighborhood). An estimated four parking spots are available for each individual vehicle in Los Angeles.[50]

Historically, one of the first indications of the shift toward auto-dominant and pedestrian-unfriendly land uses in Los Angeles was the introduction of the concept of the "jaywalker," referring to someone crossing a street "illegally." As Peter Norton pointed out in his study of street use in Los Angeles, the concept of jaywalking was first introduced in the 1920s as a way to shift the meaning of how pedestrians could—and should—use the streets. Once jaywalking became a prohibited activity, the amount of land dedicated to the automobile on surface streets (or "floor space," as auto executives characterized it) shifted the perception—and the reality—of the purpose of streets. At the same time, the responsibility for safety and the question of limited access came to be assigned to the pedestrian to accommodate the ability of the automobile to "move with speed." Pedestrians, one auto industry advocate argued, needed to be educated that "automobiles have rights."[51]

Similar to Mayvis Coyle's situation in suburban Sunland, controversy erupted in 2015 when police in downtown Los Angeles began issuing a number of tickets to individuals crossing an intersection when the "Don't Walk" light began to flash. The tickets were issued despite a much touted gentrification-oriented change in the downtown area, with more housing, an expanded rail system with downtown hubs, and efforts to create a more pedestrian-friendly environment. Yet a *Los Angeles Times* assessment of car-bike collisions that resulted in pedestrian injuries or deaths, based on California Highway Patrol data, also identified downtown Los Angeles as the most dangerous area for pedestrian injuries and fatalities.[52]

Los Angeles also became known, as early as the late 1940s and 1950s, for road rage and ever-expanding traffic delays. In 1955, the term "Sig Alert" was introduced to identify major traffic problems in the transportation segment of radio newscasts, and the term itself became the dreaded indicator of huge delays for the car commuter. Time on the freeway turned into extended solitary confinement and an exercise in frustration. Commutes between home and work became longer and longer excursions from the freeway suburbs to workplaces scattered throughout the region. At the same time, low-income workers without cars had lengthier and longer rides as they took the bus (or multiple bus rides) from their homes to their final work destination. Los Angeles became a case study of the jobs-housing mismatch, amplified by a car-centered transportation system.[53]

The Los Angeles transportation system has a striking disjuncture between density and transit that poses a particular problem for the current efforts to rebuild a transit (and particularly rail) system for the region. Contrary to common wisdom, the Los Angeles region is the most densely populated *metropolitan region* in the United States. Despite the large number of freeway suburbs and their extensive land base, those same freeway suburbs are denser than the outlying areas in other regions in the United States. The problem for transportation planners is that the density is not concentrated in a few key residential locations, as it is in Hong Kong, or in a few key commercial and jobs-centered nodal points where jobs and housing and jobs and services could then be matched. Even as downtown Los Angeles becomes a hub for many of the new rail lines and efforts to repopulate downtown through the construction of new residential units, using the rail lines often requires getting transfers to second or even third lines to reach one's destination.[54]

The phenomenon of the dense freeway suburb, or rather the spread-out density that characterizes Los Angeles, combined with high-traffic roadways have led to serious health impacts, most directly for those living within 1,500 meters of a freeway, who are most often low-income residents. Air quality issues in Los Angeles, as described in chapter 3, have long been associated with the automobile. Besides the problems of air pollution, nearly every area experiences additional health and environmental impacts, from polluted runoff to increased noise and stress levels, to the urban heat-island effect caused by the vast roadway footprints. Many of those impacts are also concentrated in the denser low-income areas adjacent to numerous freeways and high-traffic roadways, where the per capita car ownership is lowest and where walking and biking are also most dangerous.[55] The tension between the increased desire and advocacy for a more walkable, bikeable, and transit-friendly Los Angeles in contrast to the overwhelming presence of the automobile and auto-dominated streets and freeways remains palpable, notwithstanding the hope for the emergence of a Third Los Angeles.

The Connection to China

Transportation modes and transportation strategies in urban areas, like so much else in China after the reform period, have changed rapidly, with speed a key objective. This has been true of its port developments, its transportation-related manufacturing objectives, whether for exports or for a domestic market, its development of metro subway and high-speed rail systems, and its expansion of car manufacturing, highway and expressway construction, and domestic car ownership.

But before there were cars and highways, subways, and high-speed rail trains, there were bicycles. For about four decades, from 1949 until the mid to late 1980s, China became appropriately known as "the Bicycle Kingdom." Traffic congestion was endemic in the largest cities, but it mostly involved bikes, not cars. Bikes represented the primary transport mode, with road competition between bicyclists, pedestrians, and the occasional cars that tried to weave their way past the bikes.

When the bicycle was king in Los Angeles in the 1890s, it was more a curiosity and relatively rare in China. Because the bicycle was perceived as more of a western import, bike production and use were relatively slow to develop. But the reorganization of both urban and rural life after 1949, including notably the development of the *danwei* or work unit in which the majority of the urban population was organized, greatly favored bike use. The danwei work unit connected the workplace with housing and other core urban places such as nurseries and eating places and thus facilitated the idea of short-distance travel, which was ideal for bike use. As a result, production of bikes for the domestic market and the use of bikes for daily trips increased dramatically, establishing China's reputation as the Bicycle Kingdom.[56]

Bike production in China took off even further during the 1980s with the advent of new centers of manufacturing in the cities. By the 1990s, production of bikes for the domestic market peaked but then began a decline into the new century, even though overall production, now primarily focused on export markets, continued to grow. As bike use declined, the streets increasingly became the province of the automobile, particularly in the largest cities, and bike traffic fatalities increased. In Beijing, for example, bike use decreased from 62.6 percent in 1986 to 17.9 percent in 2010, while car use increased from 5 percent to 33.5 percent. Due to the growing number of cars, cities like Guangzhou and Shanghai widened their streets and converted bike lanes to mixed-traffic service roads with fast-moving auto traffic, thus leading to the increase in bicycle fatalities even as overall bike use declined.[57]

In the 1990s, Guangzhou led the way among Chinese first-tier cities in deemphasizing bike commuting, to the point that bike use was banned in some areas of the city to make way for the automobile and its association with modernity. One 2002 study noted that areas where bike commuting had declined significantly were also areas where major auto manufacturers were located, such as the Guangzhou Motor Group Company's joint venture with Honda. Similarly, in Shanghai, which also had a major automotive group involved in a joint venture with General Motors and Volkswagen and had witnessed a major decline in bike use, transportation planners had become increasingly skeptical of the role of bike commuting. Another 2001 study related that planners described cycling as a "second-rate mode of transport," and complained that cyclists got in the way of motor traffic.

"Improved efficiency of junctions can only be achieved by taking the cyclists out of the equation," the planners argued, asserting that cycling's image was "inappropriate to a World City at the forefront of the technological revolution."[58]

That image continued to persist in China even as late as 2009. That year, in advance of the Copenhagen climate conference, the lead representative on the climate talks, China's ambassador to the United States, Yu Qingtai, was quoted as saying that "every Chinese citizen should have the right to all of the modern industrial and transportation options enjoyed by, say, Americans—including the right to own a car." "We should not be expected to stay forever as a kingdom of bicycles!" Qingtai told US National Public Radio's Robert Dreyfuss.[59]

However, the image of a "world city" as being amenable if not supportive of developing bicycle infrastructure has led to a renewed interest in bike use in several Chinese cities. Guangzhou, for example, reversed its approach on banning bikes by building new bike lanes and identifying the positive health and environmental benefits of bike riding. The 2008 Olympics in Beijing and 2010 Expo events in Shanghai helped stimulate the new interest in bike infrastructure and bike use, as those cities presented themselves to global audiences. New bike share programs were instituted and quickly surpassed other programs around the world, given China's speed in introducing new developments. Bike lanes that had been eliminated began to be restored, and even a Vice Minister in the Ministry of Construction, Qiu Baoxing, argued that it was important for China to regain its reputation as the "Kingdom of Bicycles."[60]

Despite the new interest in bike sharing and restored bike lanes and other bike-related initiatives, bike riding has continued to decline while car use has skyrocketed. Domestic bike production remains limited, while the manufacturing sector continues to emphasize exports. China has flooded the US market with cheaper bikes and even forced the major bike producer in the United States to move its production activities overseas. By 2015 nearly two-thirds of China's bike production revenues, which were growing at an annualized rate of 3.9 percent and had an overall value of $11.6 billion, came from exports, according to market research numbers. The numbers contrast with those in the United States, where 99 percent of the bikes in use were imports, with many of those (as many as 70 percent) coming from China, including nearly half arriving at the ports of Los Angeles and Long Beach.[61]

One of the striking visual discrepancies in places like Beijing or Nanjing today is the contrast of people riding older bikes along the edge of a wide major thoroughfare occupied by new cars that are most likely to have been recently purchased. While not yet the "Kingdom of Cars," in just two decades China's megacities such as Beijing, Shanghai, Guangzhou, and Shenzhen have witnessed an explosion of car buying, congested roads, a squeeze on available land, and air quality impacts. This transformation coincided with the development of the Chinese auto industry in the late 1980s and its elevation in 1994 into one of the "pillar industries" for economic development.[62]

With the twin problems of congested, often chaotic traffic in the central core and the dramatic declines in air quality, local governments began to impose restrictions on car driving and car purchases. One of the first policies restricting car use was introduced in response to the 2008 Olympics in Beijing. That one-time event failed to slow down the number of cars on the road, which continued to increase at a rate of 250,000 additional cars

Figure 6.3
Competition for road use, Nanjing, China. Source: Simon Ng.

purchased each year. In January 2011, the Beijing government introduced further restrictions, this time directed at car purchases, instituting a lottery system that stipulated no more than 20,000 car purchases a month by Beijing residents. Car purchase restriction policies were extended to seven other cities in the next few years, including Guangzhou, Shenzhen, and Hangzhou.[63]

The car purchase restrictions in Hangzhou and Shenzhen were notable in the way the policies were introduced. In May 2014 the Hangzhou government, responding to residents' concerns, suggested that it would not pursue a quota policy on car purchases. But then the very next day it announced that the policy was now in effect, causing an outcry. Shenzhen's government fared even worse. It had announced in early 2014 that economic strategies would be considered rather than restrictions on car purchases, and that it would give advance notice and would solicit public feedback if a lottery system were introduced. But six months later, on December 29, the Shenzhen government announced a quota policy limiting sales to 100,000 cars (including 20,000 electric vehicles). The announcement was made at 17:40 in the evening to go into effect *just twenty minutes later* at 18:00. That caused an immediate spike in Internet messaging, and people armed with cash rushed to car dealerships to beat the deadline. Several dealers took advantage of the chaos by raising their prices. "It seemed like buying a cabbage rather than a car," one of those who had rushed to beat the deadline said of the scene.[64]

Shenzhen's new policy of trying to slow down car ownership—and car use—was made necessary in part by its own development strategies, which had favored road building and the promotion of a car culture. At the time of the December 2014 car purchase restrictions, the number of cars in Shenzhen had reached 3.14 million, ranking it first in car density per population in China. By 2014, there were more than thirty-five cities in China with more than a million cars registered, and ten cities, including Shenzhen and Guangzhou, with more than 2 million. Already by 2009, China's own automobile manufacturing sector, almost all of which was domestic-oriented, had become the largest manufacturer (in terms of units produced) in the world. China's urban residents were also the largest car users (in terms of number of cars, though not per capita use). Even during the Great Recession of 2008–2009, unlike in the United States and Europe where car sales plunged, China's own economic stimulus program was able to counter the downturn in car sales. To put in perspective China's embrace of the car, the number of units produced annually had increased from 2 million at the time that China joined the WTO in 2001 to 13.6 million in 2008 and 18 million by 2010, while the number of cars on the road had jumped from 16 million in 2000 to 120 million in 2011. Similarly, highway construction to accommodate the new cars on the road had increased to 4.2 million kilometers in 2012, more than quadrupling the amount of highways since the advent of the Reform period in 1979. Moreover, the huge car-buying market in China also meant that US-based car manufacturers were themselves becoming dependent on the Chinese market. While 17 million cars from US car manufacturers were sold in the United States in 2014, 24 million were sold in China. China also represented 30 to 40+ percent of total profit for some of the largest global car makers (e.g., as much as 44 percent of GM's profits in 2014).[65]

When China's growth rate started to slow after 2014, car manufacturing numbers also slowed, though domestic car ownership and use continued their ascent. Problems of congestion, air pollution, parking limitations, road infrastructure, and land use remained as pressing as ever. In response to the air pollution problem, the central government sought to rapidly expand its electric and plug-in vehicle market, allotting more electric and plug-in cars in the car purchase restriction policies in Beijing, Shenzhen and other cities. Yet the overall number of alternative vehicles remained small: in 2014, of 1.8 million cars in China, there were 29,000 electric vehicles and 23,000 plug-in hybrids sold. While more impressive than numbers elsewhere, these totals still fell short of achieving China's goal of selling 500,000 such cars by 2015 and 5 million by 2020.[66]

Among its transportation strategies, the government's focus on rail (including high-speed rail) and metro development has been most impressive in the number of systems constructed and kilometers covered. The focus on rail and especially metro development for cities was relatively late in coming, but once the commitment was made, the level of spending and development of the required infrastructure rivalled the commitment to the automobile.

The first urban metro line was established in Beijing in 1969, but with the Cultural Revolution not yet fully abated, foreigners were not initially allowed to use the system. Thirteen years later in 1982, a second metro line was constructed in Tianjin. Although the turn to a market economy increased interest in infrastructure, additional new metro lines did not get built until the 1990s, beginning with Shanghai in 1993 and Guangzhou in 1997. Metro development also increased exponentially in Beijing in anticipation of the 2008 Olympics, and, nationally in the wake of the 2008–2009 recession and its emphasis on new infrastructure. In Beijing, five subway lines were constructed before the Olympics, with another five lines soon to come on line. The recession also gave new prominence to metro construction (as well as high-speed rail), with the State Council easing some policies governing construction and financing to further infrastructure spending as the key to stimulating the economy.[67]

By then, every major city in China began to put forth plans for the development of new subway lines in their cities, often without effective planning or financial considerations in the rush to construction. The underground subway, although far more costly, remained the preferred mode instead of cheaper elevated or at-grade rail lines, as land continued to remain at a premium. "In the eyes of city officials," the Chinese business and finance online newsletter *Caixin* commented about the subway choice, "land equals money." Those choices combined with the rush to build the subways led to financial difficulties fueled by inaccurate assessments. Some of the cities pushing for new metro lines manipulated their figures to meet the minimum requirements and obtain the approval to move forward.[68]

Shenzhen provides an illustration of the rush to construction and the financial issues involved. Shenzhen's first two subway lines came on line in 2004 during the first wave of the metro construction boom that extended to other cities besides Beijing, Shanghai, and Guangzhou. The decision to continue to go underground and keep fares low presented financing difficulties that had been based on 70 percent government funding and 30 percent from bank loans. To showcase its contemporary image in advance of the 2011 Universiade, an international sports event for which Shenzhen had successfully bid, new financing and construction strategies were required. That led to an agreement with Hong Kong's MTRC to undertake a thirty-year "build-develop-operate-transfer" contract, whereby the Shenzhen government would acquire land in and around the metro line and would sell the land at 60 percent of the estimated market value to MTRC. MTRC would then assume the financing, construction, and operating costs at agreed user fees while it maintained the right to develop the land it had now acquired (e.g., the commercial and residential developments), with profits to be earmarked for subway construction and operation. But after the contract was signed in 2004, continuing delays meant that MTRC was able to keep the initial profits from the land, which only intensified public concern about lack of accountability and whether the government was monitoring the program. In addition, as the 2011 Universiade approached and the MTRC-built and -operated subway was finally completed, the Shenzhen government created further controversy when it evicted eighty thousand people to make way for the metro development, including migrants who had been without employment or without an urban *hukou* (household registration). (Shenzhen at the time had a far greater number of migrants—6.5 million—than the locally registered population—2.5 million).[69]

As in Shenzhen, the great metro boom in China included a number of such private-public partnerships as well as "rail plus property" agreements that included the construction of apartment buildings and shopping malls. Some US transportation planners, such as Robert Cervero and Jennifer Day, saw such agreements as helping to promote transit-oriented development strategies that could offset some of the more negative features of urban (and suburban) expansion. This transit-oriented approach, they argued, could place "rapidly suburbanizing Chinese cities on a more sustainable pathway." Yet economics (including the deep focus on development) remained the prominent driver of policy, for both the metro lines and China's most ambitious rail projects, the development of its high-speed rail (HSR) network.[70]

A relative latecomer (after Japan and Europe) to the development of an HSR system, China's network was first launched in 2003, with passenger service becoming available in 2008. In 2012, the central government announced an additional $293 billion to connect all of China's major cities by 2020 with more than 16,000 kilometers of high-speed rail lines. But already by 2014 that target had been met, which secured China's ranking as the largest HSR system in the world.[71]

High-speed rail had the advantage of effectively shortening travel distances between cities and better facilitating three key regional network relationships around Shanghai, Beijing, and Guangzhou and the Pearl River Delta. The Guangzhou/Pearl River Delta network also included Hong Kong, which began to plan its own HSR system to connect with Guangzhou and beyond. The HSR network, according to Yang Yao, the dean of the National

School of Development at Peking University, had altered the idea of geography in China, transforming what had been separate cities into outer suburbs (e.g., between Hangzhou and Shanghai) where manufacturers and buyers of goods in places like the Yangtze River Delta could easily visit and work with their suppliers. The HSR system, in this way, became an extension of China's promotion of regional urban constellations and its ever-present drive toward greater urbanization.[72]

China also became active in export markets, actively soliciting connections in as many as twenty-eight nations for Chinese technology and expertise. Some of the ambitions were outsized, including one remarkable concept for an HSR line originating in Beijing that would pass through Siberia, Alaska, and Canada before reaching the continental United States. In order to make the project happen, a 200-kilometer underwater tunnel would need to be constructed beneath the Bering Strait. The designers envisioned a trip that would take less than two days, although the need to reach an agreement with Russia, Canada, and the United States made the project more fantastical, akin to proposals in the United States to transport an iceberg from Alaska to Los Angeles.[73]

But while the high-speed rail system helped increase various access needs, whether for middle-class professionals traveling between cities or those engaged in manufacturing and logistics or other economic activities requiring more of a just-in-time or reduced time approach, it also increased the inequality of access between regions. Similar to some of the issues associated with the new metro systems, location (where people lived and where stations were located) and constituency (who would actually use such a system) were crucial parts of the built-in inequality linked to the HSR. The displacement of the eighty thousand migrants in Shenzhen for its brand-new metro exemplified that problem.[74]

Today, China continues its path of constructing as rapidly as possible its modernist transportation system of automobiles, highways, and rail systems and its promotion of bike riding as recreation rather than as a way of meeting daily-life needs. Some of the new transportation systems have come with a green identity—electric cars, sleek new metro systems, bus rapid transit, transit-oriented development, and its global leadership in high-speed rail. But the greening of transportation is also contravened by the continuing expansion of a car culture and a loss of streets and alleyways as places to walk, especially for those without cars who may be without an urban hukou. While everybody still walks in China and many still ride their bikes, wealthier residents would now like to own their cars or travel by rail at 300 kilometers per hour. In doing so, they may shorten distances and experience a unified but less diverse China in which megacities continue to eclipse rural places and where former rural residents living in the cities may experience both the upsides and downsides of urbanization.

Strategies for Change

In Hong Kong, in the fall of 2014, the several weeklong protests known as Occupy Central had an unintended but revealing transportation—and air pollution—consequence. The Occupy Central protesters managed to cut off the most important east-west traffic corridor on Hong Kong Island, and a major north-south road in Kowloon. Car traffic slowed to a crawl and many bus routes were canceled or diverted. Car commuters instead switched to riding the MTR, or walking or cycling. Mong Kok, Causeway Bay, and Central (a short distance from Admiralty) where the demonstrations took place also happened to be the locations of three government roadside air quality monitoring stations. They showed that concentrations of NO_2, which is produced by internal combustion engines, substantially declined during the street occupations. In Central's case, emissions clearly rose again on October 14, the day the police reopened Queensway, one of the roads blocked to traffic.

In Los Angeles, in the early morning of June 15, 2003, a Sunday, several thousand bike riders and walkers assembled at the entrance of the I-110 (Pasadena) freeway to embark on a most unusual event in the automobile era in Los Angeles. After two years of negotiation between community groups and the California Department of Transportation (CALTRANS) as well as city transportation agencies and the California Highway Patrol, the agencies reluctantly signed off on a plan that included a closure of the I-110 freeway between its entrance point in Pasadena to just before the multiple freeway interchanges in downtown Los Angeles. For four hours, no cars would be allowed on this stretch of the freeway that passed along much of the same route of the cycleway proposed a hundred years earlier. Instead, the freeway would be open for the exclusive use of bike riders, walkers, skateboarders, and rollerbladers, among others seeking to experience what turned out to be a magical moment for Los Angeles. ArroyoFest, as the

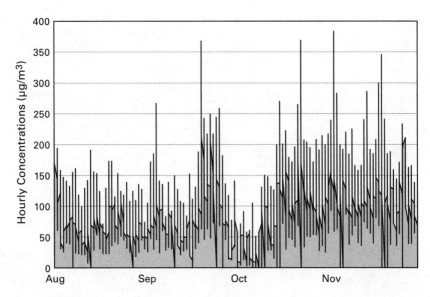

Figure 6.4
Hourly concentrations of nitrogen dioxide at Central roadside air quality monitoring station, August to November 2014. Source: Hong Kong EPD.

event was called, was designed to imagine a very different type of experience along those eight miles. After the event, residents living adjacent to the freeway noted how quiet it had been, how one could breathe more deeply, and how one could hear birds singing. Event organizers said it gave power to the imagination to envision a post-freeway, post-automobile era in a city historically identified with the car more than any other global city.[75]

In China, officials in the Guangming New District in the northwest section of an expanding Shenzhen met with officials from the Local Governments for Sustainability group (ICLEI) in early December 2015 about the possibility of the Guangming District becoming part of the ICLEI. Songming Xu, the district's Deputy Director, said at the meeting that Guangming District was determined to go green by "driving its growth with a new type of urbanization and industrialization that values both economic development and environmental conservation."[76]

The initial plan to develop Guangming District in 2007 as a green or "low carbon" New Town section of Shenzhen was to reorganize this heavily migrant district, which included an older industrial area, into a high-tech

Figure 6.5
ArroyoFest, June 2003. Source: Virginia Renner, UEPI.

development zone, an ecological development zone, and, for the remainder, a traditional industrial development zone. Plans included a green transport system, a transit-oriented development framework for new developments, an innovative stormwater capture system, and an overall "green city" orientation that could further serve as a tourist destination. The traditional industrial zone was still home to 400,000 low-income migrants, who lived adjacent to but essentially separated from those in the new high-tech and green development sectors; their presence was seen as marginal to Guangming District's coveted urban facelift and green remake.[77]

But two and a half weeks after the ICLEI meeting, at 11:40 AM on December 20, 2015, Guangming District experienced an event in its industrial district area that undermined the New Town's environmental identity. A construction waste site collapsed, unleashing a torrent of mud that spread over more than 380,000 square meters and, in some places, was more than 10 meters deep. Thirty-three buildings were covered with the mud, including fourteen factories and three dorm buildings where many of the migrant workers in the industrial zone worked. Most of the casualties were migrant workers from rural Henan Province. The event indicated the crucial problem of how policies got implemented—in this case, establishing a construction waste dump in order to tackle the problem of illegal dumping throughout the city—even if its "green city" reputation were to be established. It also pointed to the problem of environmental disparities and inequities that persisted and even deepened as part of China's development outcomes (as well as those of Los Angeles and Hong Kong).[78]

In Hong Kong, the search for a green identity has also remained uneven. Hong Kong has had the advantage of a world-class public transit system where, as noted earlier, 90 percent of all trips are transit-based. As early as the 1970s and 1980s, as the rail system began to expand, Hong Kong policymakers sought to establish a preference for rail and a disincentive for automobile use. Implementation of that preference has relied on a set of coordinated policies that helped establish an integrated public transit system, such as ease of transfer between bus and rail.

The issue of road congestion and concern about the increase in the number of cars and motor vehicles on the road has led to efforts to control the growing use of private cars while simultaneously seeking to expand the percentage of rail passenger journeys. Policies included a 1982 increase in the first-registration tax for cars, from 35–45 percent to 70–90 percent of the cost of the vehicle. That change coincided with a decline in car sales by as much as 70 percent, to a low rate of 25 vehicles purchased per thousand people during the mid-1980s. The registration tax continued to be maintained at a high level during the 1980s (35–100 percent of the vehicle cost) but was then reduced in 1994 (to 40–60 percent) after pushback from car and antitax advocates. Car sales increased again to 50 cars per 1,000 population, forcing the government to raise the initial registration tax back to the 35–100 percent rate in 2004. A higher fuel tax was also set in 2002 at HK $6.06 per liter for unleaded petrol, at the time almost eight times the fuel tax rate in the United States. Efforts to develop an electronic road pricing policy to reduce congestion in core commercial corridors were also discussed as early as the 1980s (and then again in the late 1990s and most recently in early 2016). These back-and-forth policies indicated that higher taxes and fees had the potential to reduce car purchases, even as their uneven implementation failed to slow increased car use over time.[79]

The problem of road congestion has continued to plague Hong Kong's transportation policy. While the low level of 10 percent car passenger journeys among all transport modes has been maintained, the number of cars on the road has increased. This is partly due to the overall increase in the number of passenger journeys in all modes: the increase in bus and rail trips has not offset the increase in the number of cars on the road. Policies have also vacillated in how aggressive they are in discouraging car ownership and use. The first-registration tax has only increased slightly since 2004, although in 2007 incentives through lower rates were introduced to encourage the purchase of "environment-friendly vehicles" such as hybrids. That policy was discontinued in 2015 due to the rapid increase in such vehicles, with the benefits of greater fuel efficiency more than offset by the increase in the number of car purchases. The fuel tax has also remained the same and is now at a lower rate than the various fuel taxes at the federal and state level in the United States.[80] A pilot scheme to implement electronic road pricing in Central and its adjacent area had its public consultation completed in March 2016.[81]

While policies on car use remain in flux, Hong Kong's public transit system continues to achieve strong results in terms of percentage and volume of use, efficiency of service (although bus service in central corridors continues to experience delays), and affordability (particularly with rail). What has not emerged, however, is a more active approach toward facilitating and expanding the pedestrian experience and the use of bikes for more than recreational purposes in limited areas. At the ground level, the roads still belong to vehicles, while above or below Hong Kong's streets, residents look to the footbridges and subways to get to where they need to go.

In Los Angeles, the streets continue to belong to vehicles, even as events like ArroyoFest have sought to introduce a different dynamic that could lead to policy change. ArroyoFest, originally planned as a one-time event, eventually resulted in CicLAvia, a bike ride through Los Angeles on streets closed off to cars. The first CicLAvia took place on October 10, 2010, stretching several miles from Hollywood to downtown Los Angeles, with an additional leg of the route to Boyle Heights and East LA. CicLAvia became an enormously popular happening, taking place as often as four times a year and spreading to multiple neighborhoods. As many as 150,000 bike riders and a far smaller number of walkers have participated at each event, with the event itself becoming more institutionalized through support from the City of Los Angeles. And similar to the reduced air pollution that air-monitoring equipment identified during Hong Kong's Occupy Central events, a UCLA research team found that the streets closed off to cars for the bike ride during the October 2014 event had reduced ultrafine particles by 21 percent and $PM_{2.5}$ by 49 percent.[82]

ArroyoFest and CicLAvia helped empower what had been a fledgling bike advocacy movement to become a more effective lobbying force pushing for an expanded bike infrastructure with more bike lanes, more bike paths, more bike sharrows, and a modest bike share program. The events also shifted much of the discourse about biking in Los Angeles, which has witnessed new bike-focused media coverage and policy initiatives and greater numbers of bike riders touting the health, environmental, and recreational value of biking in Los Angeles.

With the election of Eric Garcetti as mayor in 2013, efforts to establish a more bike- and pedestrian-friendly streetscape, which had begun to be discussed by Garcetti's predecessor, Antonio Villaraigosa, shifted to the policy arena. Garcetti turned for advice to former New York City Transportation Commissioner Janette Sadik-Khan, who emphasized the importance of linking traffic safety to efforts to expand bike and pedestrian infrastructure. Sadik-Khan, perhaps best known for her work to turn New York City's Times Square into a car-free pedestrian-friendly space, was among a new group of

Figure 6.6
CicLAvia ride, October 10, 2010. Source: CicLAvia.

transportation officials in the United States and Europe and in other parts of the world who had begun to reorient transportation policy away from the near exclusive focus on the automobile and its roadways.[83]

Garcetti's efforts culminated in the release of his "Mobility 2035" plan, approved by the LA City Council in August 2015. Mobility 2035 identified a range of objectives and a targeted timeline to expand bike infrastructure and support the regional transportation agency's ambitious rollout of new rail lines. It also identified new approaches to traffic calming and transportation demand management, and promoted two related high-profile plans ("The Great Streets Strategic Plan" and "Vision Zero"). These plans focused on the reduction and eventual elimination of traffic fatalities by 2025,[84] and the reorientation of existing major streets or boulevards into new types of transportation corridors with plazas, parklets, and bicycle corrals, along with other cultural and design features.[85] "The old model of a car-centric, different-neighborhood-for-every-task city is in many ways slipping through our fingers whether we like it or not," LA Mayor Garcetti told the *New York Times* about his new policy. "We have to have neighborhoods that are more self-contained. People want to be able to walk or bike or take transit to a movie."[86]

While the emphasis on pedestrian and bicycle mobility makes sense for Los Angeles since nearly 50 percent of all trips are under three miles, the resources available for walk and bike infrastructure have continued to be dwarfed by the expansion of car- and truck-related infrastructure and the highly visible effort to develop the new—and expensive—rail system. Similar to China, Los Angeles's push for its rail system has been partially justified as a way to get people out of their cars and into public transit (as well as to bike and walk). Reduction of car use has become a key policy goal for both Hong Kong and China as well as Los Angeles through its Mobility 2035 plan. Car use, measured by vehicle miles traveled, has also declined in Los Angeles since 2002, but not due to specific policy changes. Similar to Hong Kong and China, road congestion in Los Angeles has gotten worse, not better, particularly in areas where the jobs-housing mismatch is most pronounced, such as West Los Angeles. At the same time, the number of car registrations has continued to climb, albeit modestly, since 2002. Thus, while average vehicle miles traveled has declined, the number of cars on the road (and resulting road congestion) has continued to increase.[87]

The issue of choosing a mode of transportation—whether by car, rail, bus, bike, or walking—is fundamentally reflected in the question of how streets and urban space are used and what constitutes livable streets and a sustainable urban space. To walk in Los Angeles (the place where nobody is supposed to walk) or on the streets of Hong Kong (where walking is a necessity) or to walk or bike in China's megacities remains a challenge. In these global cities, the streets still belong to the vehicles, even as policymakers tentatively explore a shift in policy and in the culture of car-based mobility.

7 Spaces of the City

Occupying Spaces

On September 28, 2014, riot police in Hong Kong fired tear gas into a crowd of demonstrators in an attempt to disperse a weeklong pro-democracy student strike outside of government headquarters. The tear gas had the opposite effect. Televised images spurred thousands more onto the streets, and by the end of the day, eighty-seven canisters of tear gas had been fired, the police were in retreat, and three major sites across the city were occupied by protesters. Thus began seventy-nine days of protest known to the world as Occupy Central, or the Umbrella Movement.[1]

In Los Angeles, three years earlier, a group of demonstrators, following the lead of the occupation of Zuccotti Park in New York City, had met downtown in Pershing Square to identify a public space to establish the Los Angeles version of Occupy Wall Street. On October 1, 2011, the demonstrators regrouped on the lawn in front of Los Angeles City Hall and declared themselves the Occupy LA encampment. They pitched tents, held General Assemblies, and established a twenty-four-hour living environment, with food, sanitation, lending libraries, music, and poetry readings, among other activities.

In mainland China, occupying as a tactic and other overt forms of public demonstration are largely forbidden, particularly since the 1989 events at Tiananmen Square. However, in many cases where local residents seek to bring about change, such as protests against a polluting facility, people turn out to walk or "stroll together," often using social media to help identify the time and place. These strolls become virtual public demonstrations, or what Jeffrey Hou calls "self-made public spaces."[2]

In Hong Kong, there has been a more liberal tradition of peaceful and usually uneventful protest marches that receive prior police permission. However, occupying space as a disruptive act of civil disobedience has become a novel protest tactic regarded by the authorities with a great deal of alarm. The Occupy Central protesters' main demand was the withdrawal of the government's stipulations for electoral reform in 2017, which would have allowed the public to vote for Hong Kong's Chief Executive, but restricted the slate of candidates to two or three people vetted by a nominating committee loyal to Beijing.[3] While the protesters failed in this objective, one of the key byproducts of the protests was to challenge Hong Kong's experience and perception of public space.

In Los Angeles, the Occupy LA group resisted making any formal demands, but as for their counterparts in New York and Occupiers elsewhere, a core focus was the huge gap in wealth and power between what the Occupiers called the "1%" as opposed to the "99%," consisting of everyone else. Although the notion of public space was not central to the Occupiers, they warmly greeted the thousands of bike riders riding in the streets through the city as part of the CicLAvia event, held eight days after the Occupiers had taken over the southern front lawn at City Hall. The bike riders had identified a pit stop at City Hall and held their own event, with speakers and music, across the street from the Occupiers. As bike riders and protesters mingled, they discovered their affinity for public action, for creating an open space for protest, and for reimagining the spaces within the city.[4]

In Hong Kong, the main protest site was located outside of the new Central Government Complex on a prime waterfront site in Admiralty. The base of British naval forces in colonial times, Admiralty is an eastward extension of Hong Kong's central business district built in the 1970s. While centrally located and highly accessible by public transportation, the ground level is hostile to pedestrians, with its landscape of inactive frontage, safety railings, and uncrossable roads. Pedestrians navigate its gleaming office towers and high-end shopping malls through a network of elevated skyways and underground tunnels. Both the government headquarters and Tamar Park, the grassy lawn outside it where the student strike began, are accessible only from the MTR station via a pedestrian footbridge over Harcourt Road, an east-west traffic corridor carrying twelve lanes of traffic. Although the architecture of the government headquarters is supposed to

evoke an "open door" to symbolize openness and transparency, its inaccessible siting has, ironically, meant that there are very few ways in or out of
the complex.[5]

Somewhat similarly, in Los Angeles one cannot enter City Hall from the
lawns at the northern and southern ends of the building where the occupiers
resided. The immediate area around City Hall has limited foot traffic, and the
only visible public activity is a Thursday Farmers' Market on the south lawn
that is primarily used by government workers on their lunch break. The
Occupy encampment created an entirely new experience of the space around
City Hall, while the CicLAvia event added an even greater sense of public
gathering. The two events combined had created a Los Angeles agora.[6]

In Hong Kong, aside from Admiralty, the other two protest sites occupied major roads in Causeway Bay and Mong Kok, both busy and densely
populated shopping districts. As part of Hong Kong's older urban core,
Causeway Bay and Mong Kok had a tight, gridded street network which
enabled a vibrant and hectic street life. Part-time pedestrianization schemes
introduced in 2000 had enabled a culture of street performance to emerge
in minor streets. However, the changes posed different sorts of challenges
to pedestrians. The public space was of poor quality due to heavy congestion and vehicle-dominated traffic planning. Prior to the protests, the
blockaded roads Yee Wo Street and Nathan Road had an annual average
daily traffic load in 2013 of 22,430 vehicles and 32,420 vehicles, respectively.[7] Pedestrian crowding is exacerbated by narrow obstructed sidewalks,
a shortage of public open space, and inconvenient pedestrian crossings.
Long stretches of Nathan Road, for example, can only be crossed via underground passages. Tall buildings on either side of narrow, heavily trafficked
roads have produced a "canyon effect," trapping vehicle exhaust and
increasing pedestrians' exposure to pollutants. One side effect from the
Occupy Central events was reduced air pollution due to reduced car and
bus traffic, as described in the previous chapter.

In addition to temporarily improving air quality, Occupy Central also
challenged perceptions of the uses of public space. As days turned into
weeks, the occupied roads were transformed into tent settlements hosting
all manner of political and creative activities. The protesters set up art
installations, speaker's corners, an outdoor study room, "democracy class-
rooms" staffed by volunteer lecturers, an organic herb garden, a book lending library, and religious shrines, among other things.[8]

In both Hong Kong and Los Angeles, the protesters acted out their communitarian ideals. In Hong Kong, volunteers set up stations for charging mobile phones, collected and sorted trash for recycling, distributed food and supplies, and cleaned public bathrooms. Makeshift steps, monitored by safety volunteers, were built over concrete road medians. Office workers, shoppers, and tourists were frequently seen lunching at or strolling through the protest sites. In Los Angeles, the front lawn tents identified a temporal community and a place for intense political discourse as well as cultural happenings. As in several other Occupy sites, self-identified committees became de facto support networks, facilitating the needs of daily life.

These types of self-governing, collaborative behaviors in Hong Kong and Los Angeles—and in the walking or strolling events in China—are highly unusual in cities, where most formal public spaces are rigidly managed, nonexistent, or forbidden as gathering places.[9] The protesters managed to provoke conversations about how people can reclaim their city from vehicles, as well as overcome protest restrictions and demonstrate how occupying public space can enhance quality of life, build community, and establish informal democratic rights. Occupying public space can also identify where "new forms of the social and political can be *made*," as Saskia Sassen put it—and as the two Occupy events, the strolling together in China, and similar outpourings, from the *indignados* in Spain to the Arab Spring, all briefly demonstrated.[10]

Evolving Spaces

Hong Kong

The transformation of open space during Occupy Central in Hong Kong was seen as remarkable precisely because public open space is in short supply in such a densely vertical city, and is often inflexible and of poor quality.

Hong Kong's overall population density in 2014 was 6,690 people per square kilometer.[11] However, since the built-up areas of Hong Kong occupy only 24 percent of its land mass, the actual density is much higher. In some neighborhoods, densities can exceed 110,000 people per square kilometer.[12] These densities reflect a high-rise urban morphology where residential and commercial towers routinely reach thirty to forty stories, even in suburban areas. It is also indicative of small living spaces. In 2014, the average household size was 2.9 people, while homes with a floor area of less than 70 square meters (about 750 square feet) made up 80 percent of the housing stock.[13]

In contrast, about 40 percent of Hong Kong's land mass is protected country parkland, established in 1977 to provide affordable recreational opportunities for a population living in crowded conditions.[14] The country parks are generally located within 30 minutes of urban centers. However, within urban areas, open space is very limited. According to various government sources, there are somewhere between 2.6 square meters and 3.5 square meters of urban open space per person in Hong Kong.[15] By contrast, the city of Los Angeles, which is considered park-poor within a United States context, has 37.9 square meters of public parkland per person, or 35.1 square meters if the Angeles National Forest areas are excluded since they are more comparable to Hong Kong's country parks.[16]

This uniquely dense urban morphology is the result of natural and artificial constraints on urban land. Hong Kong has a hilly topography, and has historically relied upon sea reclamation to provide a source of flat land. About a third of its built-up area is reclaimed land. Artificial constraints play an equally important role. Dense development was established during the earliest days of colonial rule. When Hong Kong was ceded to Britain in 1842 to serve as a free port, the primary concern of the government was to facilitate commerce. It was politically unwilling to raise merchant taxes, yet running a colony was expensive and required regular subsidies from London. The government therefore turned to the sale of land for revenue. In 1843, all land in Hong Kong was declared Crown land to be auctioned on seventy-five-year leases.[17] Today, land sales and lease conversion premiums alone still account for 15 to 20 percent of government revenues. According to some estimates, after including profits tax, stamp duties, government rents, rates, and property taxes, this figure may be as high as one-third of revenues.[18]

In the nineteenth century, streets were laid out in a grid pattern (distorted by topography in hilly areas) in order to use land as profitably as possible. Little land was set aside for open space. The major exceptions were the Hong Kong Botanical Gardens, which was established in the 1870s for the purpose of researching the local botany, and the Happy Valley Race Course, which was built on a filled-in swamp deemed too unhealthy for residential use. These spaces mainly served the European colonial population. In 1890, the government permitted a piece of land formerly occupied by a military barracks in the Chinese working-class district of Sheung Wan to be used by market stalls and fortune tellers, with the site formalized as the Chinese Recreation Ground in 1923. Overall, the shortage of recreational public and open space created a business opportunity for merchants, who built private gardens (sometimes on the rooftops of commercial buildings) and charged admission.[19]

The government's revenue-raising methods drove up the price of land. Housing was built shoulder to shoulder and sometimes back to back, poorly ventilated, and severely overcrowded, resulting in outbreaks of bubonic plague beginning in 1894. The need to attract business investment repeatedly constrained government attempts to impose stricter building and sanitation requirements on private properties throughout the late nineteenth century. The earliest open space regulations, passed as part of the 1903 Buildings Ordinance, were limited to requiring backyards, alleyways, and spaces around buildings.[20]

Modern zoning was formally introduced in 1939 with the passage of the Town Planning Ordinance (TPO), although its implementation was delayed until several years after World War II. The TPO empowered the government to draw up statutory zoning plans identifying the intended land uses for particular plots of land, including open recreational space. This gave government planners the tools to reserve land for open space in new development areas, but in already built-up areas planners were usually reluctant to diverge from existing land uses. Statutory plans were a statement of intent; they could not be enforced against existing nonconforming land uses, which could stay in place until redevelopment.[21]

In the postwar decades, Hong Kong experienced phenomenal growth, fueled by waves of refugees from the north and the rise of an export-driven light manufacturing economy. Housing, which was already overcrowded and war-damaged, came under enormous pressure, and squatter settlements sprang up around the urban fringe. This demand for housing prompted the government to relax building density controls in 1956; they were further revised in 1962 to allow the replacement of three-story shop houses with high-rises, setting the stage for the intensely vertical urban form seen today.[22]

Nor was the city allowed to sprawl outward. The leasehold land system ensured that unauthorized building was not legitimized as this would entail giving up land sale revenues. Squatter areas were eventually cleared and redeveloped, with the inhabitants resettled in public housing estates that the government reluctantly began to build in the 1950s, expanding into a massive subsidized housing program by the 1970s. High-rise New Towns with factories and both public and private housing were planned in the New Territories from the early 1970s onward in order to disperse the population. The government therefore bore the full responsibility for allocating land, and its priority was to fulfill the demand for new housing, factories, shops, roads, schools, and hospitals.[23]

Open space remained a relatively low priority, with wooded slopes (green belt land) and beaches also identified as forms of open space. An Urban Council report published in 1969 further identified that the amount of open space included such dubious categories as car parks, public toilets, and changing facilities. In the old urban core, the government took over some dilapidated buildings upon the expiration of nonrenewable leases to create a number of scattered pocket parks. To find space for larger urban parks, the government looked to reclaimed land, the urban fringe, and defunct military sites.[24]

In 1981, the government adopted the Hong Kong Planning Standards and Guidelines (referred to here as the Planning Standards), a nonbinding set of administrative guidelines for the Planning Department. For the first time, it set a target amount of per capita open space: 15 ha (one hectare [ha] equals 2.47 acres) per 100,000 residents in urban areas and 20 ha per 100,000 in New Towns, split between "local open space" and "district open space."[25] Local open space was defined as small open spaces for passive use (such as gardens for sitting or walking, and children's playgrounds), located within a few minutes' walk of people's homes or workplaces. District open spaces were defined as spaces containing facilities for active recreation, such as swimming pools, basketball courts, and soccer pitches. The standards were later revised upward to 20 ha per 100,000 for both urban areas and New Towns, which is where they stand today.[26] Large parks, classified as regional open space, were to be defined as bonus space and could only count toward half of an area's district open space requirements. As of 2008, all but two of Hong Kong's eighteen administrative districts met the open space standards overall, but eleven districts still had a shortfall of either local or district open space, indicating that space and recreational facilities were unevenly distributed.[27]

Today, tensions between development and open space allocation are ongoing. Under intense public pressure to provide more affordable housing as quickly as possible, the government is currently in the process of rezoning for housing development land that was previously zoned for open space, green belt, and community usage.[28] This practice has brought about disputes between government planners attempting to quickly find sites to satisfy the demand for housing and local residents worried about negative impacts on local roads, schools, and wind circulation.[29]

As the following sections will detail, public and open space planning in Hong Kong, similar to Los Angeles's own land use history, faces challenges due to privatization, vehicle-dominated transport planning, sterile and homogenizing urban morphologies, and management problems.

Los Angeles

In the 1970s, a popular television newscaster opened his Los Angeles broadcast with the words, "From the mountains to the sea." His slogan sought to capture the vast expanse of the Los Angeles region, endowed with its open spaces and connections to Nature in its multiple forms. In contrast, Los Angeles has often been depicted in Hollywood films and popular culture as an urban (or suburban) dystopia, both spread out and polluted. Rather than open spaces, there are endless real estate developments, vast highways systems, and a river that runs through the city that nobody can find. Yet both images are accurate. The current site of the City of Los Angeles was in fact located near where the Los Angeles River enters the downtown area. It was settled by Spanish colonists whose approach to development had been influenced by the Laws of the Indies adopted by Spain in 1573, which called for settlements to be inland to avoid pirate raids, to be near fresh water, and to be close to native settlements to provide a pool of labor.[30] Land use regulations begun under Spanish rule were expanded during the approximately twenty-five years when Los Angeles was part of Mexico (1821–1847). These included requirements that residents help pay the labor costs of constructing infrastructure that benefitted their properties, such as streets, alleys, and *zanjas* (pipes and ditches that drew water from the river).[31]

After Los Angeles became a municipality (and California became part of the United States), the arrival of the railroads and subsequent land booms of the 1880s dramatically changed the contours of development and the built environment. The mixing of existing farms, new residential areas, and nascent commerce and industry caused controversies over uses of land. The city government began to restrict certain types of activities from specified districts or to restrict them to certain districts. Saloons, cemeteries and mortuaries, slaughterhouses, backyard cows, and commercial laundries all became subjects of debate and regulation.[32]

At the turn of the twentieth century, discussions of appropriate land uses and land use policies reflected a number of factors. These included the promotion of Los Angeles as a "residential city" different from the more crowded and industrial East Coast cities; the discovery of oil, which led to concerns over drilling and accompanying industrial activities; the racism against Chinese-owned laundries; and the rise of city beautiful and garden city approaches to urban planning. Los Angeles was promoted for its sunshine and connection to healthy, outdoors living, with its weather contributing to the design of homes. Yet tension emerged over how development, such as the early oil drilling and manufacturing, conflicted with LA's pastoral image. "Some of the most charming, picturesque and healthful portions of the municipality are being menaced by the oil district," complained the *Los Angeles Times*, the city' preeminent booster whose owners helped facilitate the region's countless real estate speculations and imported water schemes.[33]

In 1904, Los Angeles adopted the United States' first zoning ordinance, which was further elaborated in 1908–1909. It divided the city into residential districts where any types of dwellings were allowed (houses, apartments, hotels), with commercial activities permitted in the rest of the city. Industrial activities were permitted in specified districts near the river, although business elites worried that manufacturing activities could be too restricted.[34] Subsequently, in 1917, Los Angeles passed a new zoning ordinance with five zones: for single-family houses, for all dwellings, for commercial uses, for industrial uses, and for any uses. Early zoning and subdivision rules borrowed from standards and restrictions used in large private and planned residential developments in other communities in Southern California such as Beverly Hills, Culver City, and Palos Verdes Estates. Generalizing the development standards of higher-end subdivisions into public policy was promoted by the larger real estate interests. One prominent realtor argued that the purpose of the city zoning ordinances inspired by the real estate industry was "to protect good residence neighborhoods from trade uses that would destroy values."[35]

Another real estate and population boom in the 1920s led to the building of large numbers of detached houses as well as duplexes, courtyard apartments, and larger apartments.[36] During the 1920s and 1930s, the federal government recommended subdivision, zoning, and community design standards to help shape and idealize the low-density suburban form. In 1934, Los Angeles adopted a yard ordinance that essentially banned row houses and streets lined with uninterrupted multifamily apartments. The city also adopted its first minimum parking requirements (for multifamily dwellings). These rules led to a decline in the diversity in housing forms and helped cement a connection between driving and the built environment that shaped future development.[37]

The region sometimes took on the appearance of a shapeless sprawl of homes and commercial boulevards in subdivisions that opened up in concert with streetcar lines and then freeways. Nevertheless, major clusters of housing did form around industrial and employment centers, from oil drilling and refining, to film production, to airplane production and aerospace. A comprehensive revision of the zoning code in 1946 further set the pattern for postwar growth. Responding to the need for housing after World War II's material shortages and a major migration to the Los Angeles region, the code sought to balance residential, commercial, and suburban development with agricultural preservation in the San Fernando Valley. The pressures of the former ended up overwhelming the latter, and nearly 100,000 single-family homes were built in the valley between the late 1940s and late 1950s.[38]

With the increasing suburbanization of housing and transportation, the loss of agriculture, and the dispersal of industry and jobs, local elites raised concerns about a vanishing downtown Los Angeles and the need for a renewed focus on downtown redevelopment strategies. Instead of promoting new housing, redevelopment took the form of the construction of new cultural centers and office buildings. When Mayor Fletcher Bowron and his Housing Authority head Frank Wilkinson laid plans in the late 1940s for an expansion of affordable housing to accommodate returning veterans and the population growth caused by new migrations to Los Angeles, these plans were condemned as "socialistic." Informal efforts by behind-the-scenes power brokers led to the defeat of Bowron in 1953 by an obscure congressman, Norris Poulson, who was literally pushed into the race by those power brokers with promises of ample campaign funds and even a "Cadillac to strut around in" once elected.[39]

The conflict between new housing and commercial and cultural development was magnified when two areas in or near the downtown area—Bunker Hill and Chavez Ravine—witnessed the displacement of low- and moderate-income Latinos and loss of open space to make way for the downtown redevelopment plans and subsequently for the construction of a new baseball stadium for the recently relocated Los Angeles Dodgers team. These episodes as well as ongoing discriminatory housing policies became a flashpoint not only for the city's Latino population but for the African-American population that had migrated to the region after World War II. Already by the late 1940s, private deed restrictions and discriminatory Federal Housing Authority mortgage rating had prevented nonwhites from purchasing houses in most neighborhoods in the Los Angeles region, leading to segregation and the predominance of white homeowners. Along with federal housing legislation, a Supreme Court decision barring racially discriminatory housing covenants had in theory removed barriers for people of color to buy real estate and move into a wider range of neighborhoods. But a mix of legal resistance, violence against nonwhite home buyers, white flight, racial steering, and discriminatory lending maintained significant segregation even as racial patterns shifted within the region. California state legislation to address such housing discrimination was even overturned by a statewide referendum vote in 1964. This result further intensified the anger about discrimination and racial antagonisms, and played a role in the Watts Riots explosion the following year.[40]

Concerns about growth also played a role in Los Angeles's development patterns. Passage of the California Environmental Quality Act (CEQA) in 1972, rising housing costs and rents due to inflation, increasing density, and frustration over traffic from commercial development on boulevards all diversified and strengthened a "slow-growth" movement in the 1970s and '80s. The City of Los Angeles's first general plan, released in 1974, envisioned dozens of high-rise "centers" linked by mass transit, but slow-growth politics led to community plans and downzonings that scaled back density and increased the ability of neighborhood groups to challenge and slow down proposed projects.[41]

Starting in the 1990s, the construction of transit and concerns about air pollution and climate change led to a shift in some areas toward more mixed-use zoning and transit-oriented development under a smart growth framework supported by state and regional policy. In addition, large-scale migration from Latin America and Asia starting in the 1970s further led to a transformation from below of the built environment through the twin phenomena of "Latino urbanism," in which residents reshaped the traditional single-family house and subdivided dwellings, as well as the rise of majority nonwhite "ethnoburbs" to the south and east of the City of Los Angeles.[42]

Nevertheless, Los Angeles still suffered from the lack of public and open space within the confines of its built environment. As pressures mounted to change and ultimately reinvent this more diverse and complex city, planning officials launched in 2013 what they called a "re:code LA" process to comprehensively update the zoning code for the first time since 1946. This process raised questions about the future of land use in Los Angeles, as well as about how such decisions would be made. In the midst of those debates, the problems of affordability, lack of open space, gentrification trends, and deepening income and class divides continued to loom large.

Privatizing Spaces

Hong Kong

Public spaces can be seen as spaces for free expression, social mixing, and the fostering of civic values such as social tolerance and the exchange of ideas. In contrast, when open and public spaces are privatized, such spaces can seem exclusionary and inaccessible, whether as a result of physical location and design or of management policies or image making that make particular groups feel unwelcome.

With its history as a former colony, inclusiveness and civic engagement was never a top priority for land use and open space planning in Hong Kong. Most nineteenth-century open spaces catered toward European users. Even when the Chinese population was permitted to use them, so-called Western standards of behavior were enforced. Open spaces were designed and managed with quiet walking, sitting, nature appreciation, and team sports in mind, rather than the festival-like entertainments favored by working-class Chinese residents. Such events were permitted only at the Chinese Recreation Ground in the Sheung Wan neighborhood.[43]

In the postwar era, the drive to expand the definition of open space in order to rationalize its limited availability in urban areas led the government to accept activities in the countryside requiring private resources or memberships, such as golf courses, tea gardens, and sailing, as part of the city's recreational offerings. Inclusiveness and access was not a consideration.[44]

Part of the motivation for increasing the open space provision in the 1970s was to avert further political instability, even as the authorities deemphasized the role of open space in political participation. Hong Kong's current Planning Standards use the terms "open space" and "recreational open space" interchangeably, and narrowly define the function of open space as promoting "mental and physical well-being of the individual and the community" and allowing "the penetration of sunlight and air movement, as well as for planting areas for visual relief."[45] As such, the Planning Standards make no clear distinction between public and private open space. Since the early 1980s there has been a policy of encouraging the private sector to provide open space under planning gain arrangements. Developers can be required by lease conditions or planning permission to incorporate open space or other public facilities into their developments.

This policy was built into the Planning Standards' definition of "countable" open space. While open space in schools and private clubs were excluded, private open spaces within large housing or commercial developments could be counted, as long as certain criteria were fulfilled: the space should serve an identifiable population; it should be accessible to the intended users; it should be of an appropriate size and physical nature; and some public or private entity should be responsible for its management. Notably, the Planning Standards also states that open space "could be provided at both the ground level and on [a] podium."[46]

This policy dovetailed with an emerging mode of urban development similar to the growth of gated communities in Los Angeles, except instead of spreading horizontally over tracts of fenced-off land, "gating" in Hong Kong developed vertically. Hong Kong's intense crowdedness made privacy a valuable commodity for those who could afford it. Historically, the city's most privileged residents secluded themselves in relatively geographically inaccessible locations, such as hilltops and the southern coast of Hong Kong Island.[47] In keeping with their attitudes toward public open space, European colonists also used discriminatory land leases, zoning requirements, and legislation to exclude ethnic Chinese residents from the city's poshest residential areas, notably Victoria Peak and its environs. These discriminatory ordinances remained in place until 1946.[48]

From the 1960s onward, the construction of large-scale residential developments became increasingly common. For the upper middle classes, walled multitower complexes with private security guards offered some of the privacy previously available only to the very rich.[49] These complexes, which increasingly offered residents recreational facilities such as gardens, tennis courts, and swimming pools, are essentially gated communities with high-rise condominiums instead of single-family detached houses. Yet, as stated earlier, the government permitted the recreational open spaces within them to be counted toward the Planning Standards' targets.

In the 1980s and 1990s, vertical gating emerged as part of a multilevel, grade-separated urban design strategy for New Town centers, as planners and developers exploited building regulations in order to utilize land to its maximum. The Buildings Ordinance amendment of 1962, which is still in place today, sought to improve natural lighting and ventilation in buildings by encouraging the construction of tall, narrow towers rather than the hulking tenements permitted in the 1950s. The new regulations stipulated an inverse relationship between site coverage and building height: the narrower the tower, the taller it could be built. However, in order to maximize valuable shop frontage at ground level, the nonresidential lower floors of a building could cover as much as 100 percent of the site, up to 15 meters.[50] This set the stage for Hong Kong's present-day dominant building typology, the podium-and-tower block.

As developers became increasingly wealthy and ambitious, podiums evolved into full-sized air-conditioned shopping malls with residential or office towers on top. Several malls could be connected to each other and to public transport hubs through footbridges, which freed up the ground level for the flow of traffic. In retail/residential developments, the shopping mall rooftops serve as amenity gardens for residents. Residential tower entrances typically open onto the gardens, which provide a security buffer between their homes and the shopping mall and the street below. In retail/office developments, podium gardens are usually made accessible to the public during the shopping mall's open hours. Again, these private and privately managed open spaces are counted toward the Planning Standards. In addition to open space, podiums also commonly house exclusive recreational facilities such as clubhouses, gyms, pools, and children's play areas. Developers sold this new type of vertically segregated gated community as mass-market middle-class housing and also for development in New Town centers, large-scale urban renewal projects, and new development areas on reclaimed land.[51]

Figures released by the Planning Department in 2008 indicated that private open space (including fully private spaces and privately owned public open spaces) accounted for about 10 percent of the total open space in urban areas.[52] That year, a public backlash against the privatization of public space emerged, due to two key cases. The first involved Times Square, a shopping mall in Causeway Bay. It had a covered ground-floor piazza which had been dedicated to "public passage" in return for bonus floor area under the Buildings Ordinance (a separate incentive scheme not to be confused with planning gain). These areas are usually classified as pedestrian circulation space and not public open space. Times Square's deed of dedication, however, also stipulated that the piazza could be used for passive public recreation, even as the shopping center continued to be responsible for its maintenance and was allowed to use some of the space for exhibitions. Under such blurred arrangements, the owner continued to treat the area as private land, charging rent for commercial exhibitions and at one point leasing a corner to a chain coffee shop. The seating was deliberately uncomfortable, and people were frequently chased off by security guards for lingering too long. This situation remained unchanged until 2008, when the media brought the situation to public attention, and the government was pushed to sue the owner to recover millions of dollars' worth of rental fees erroneously charged since 1993.[53]

The second case involved a podium-and-tower-style residential/retail development called Metro Harbour View. Planning conditions required its podium rooftop garden to be open to the public between 7 a.m. and 10 p.m. Residents who bought condominiums there under the impression that the open space was private objected to having to pay for its upkeep if the general public were allowed to use it. They complained of noise and nuisance, and considered public use to be a security risk since the garden allowed access to the residential towers' entrance lobbies. In response, the government announced that it would consider granting a waiver to convert the garden to private use. However, this solution only provoked further controversy and questions about why this particular development deserved a waiver while others did not.[54]

As a response to these two cases, the government began to publish an online list of all public amenities available in private developments built since 1980. It also announced that it would halt the practice of requiring public open space in private developments, especially residential ones, unless there was a shortage of open space (according to the Planning Standards) in the district.[55]

In 2011, the government followed up by issuing a set of voluntary Design and Management Guidelines for public open space in private developments which attempted to address the design, accessibility, and management of privately managed public open spaces.[56] In 2014, government auditors surveyed thirty-six privately managed public open spaces and found that nearly all had problems or limitations: ten had low levels of patronage; three were located on podiums difficult to access from the ground level (in one case, the owner had disabled the elevators and could not be compelled to reinstate them); two were locked; two were poorly maintained; six had opening hours of less than thirteen hours a day; and one was used to store construction materials, thereby preventing people from using it.[57]

Ultimately, the reforms seemed to be geared toward avoiding further controversy. The Design Guidelines devoted special attention to the range of permissible activities in privately managed public open spaces, especially profit-making activities. This focus was a direct reaction to the Times Square controversy, as the government came under heavy criticism for allowing a developer to capture rents from land for which it had already been compensated through bonus floor area. The guidelines therefore recommended that any commercial uses had to be approved by the relevant government department, and that the government would charge market rate fees to the owner, who would in turn charge the user.[58] This caution was not necessarily conducive to improving the vitality of such spaces, since it could deter owners from providing desirable place-making services such as al fresco dining.

Moreover, as the controversies revolved around the subset of private open space that was accessible to the general public, there has been virtually no discussion of the much larger amount of privately owned open space that is purely for private use. By ending the policy of situating public open space within private developments without reexamining the policy more broadly, the government has ensured that future privately owned open spaces will be reserved purely for private use and yet will continue to count toward the Planning Standards' targets.

Los Angeles

In Los Angeles, shifting definitions of public and private space have caused conflict and affected the changing use of space, including public spaces that are privatized or those that are privately owned and operated public spaces. Public space itself is a highly contested issue. The 1930 Olmsted-Bartholomew report discussed in chapter 6 sharply criticized Los Angeles's approach to public spaces, particularly its open spaces such as parkland, beaches, and recreational spaces. Despite Los Angeles's enormous natural resources—national forest and park land, coastal lands and beaches, among other places—the lack of open spaces remains critical. At the same time, the privatizing of public spaces has long included efforts to isolate and marginalize displaced persons, such as the homeless and mentally ill or rebellious young people, and to also provide convenient and safe private homes for the middle-class and the wealthy. Privately owned or managed public spaces have also played an important role in the development of large commercial centers such as downtown or the Wilshire corridor.

The most elaborated private spaces in Los Angeles have been the gated or walled-off communities. Researchers have identified various types of gated communities in Los Angeles and other parts of the United States, including areas characterized as elite communities, leisure communities, or security zones. California has been the leading location for such communities, and Southern California has often provided the model for these various types of gated settlements.[59]

Private elite housing in Los Angeles has long been associated with the estates, mansions, and private lands of its wealthiest residents. Large private estates date back to the massive Spanish land grants in the eighteenth and nineteenth centuries. The first land grant was established in 1784 in the area extending from Palos Verdes to the Harbor area. As Los Angeles transitioned in the 1880s and 1890s to its Anglo-dominated, real-estate-driven form of politics and development, the one-time ranchos transitioned into the largest land estates owned by its emerging Anglo elite. Henry Huntington, for example, took over the San Marino Ranch in 1903, around the time he began to build his Southern California electric rail empire. But instead of creating his own version of a walled estate, Huntington subdivided parts of the estate to create the City of San Marino, which established itself as the wealthiest and most exclusive city in the region, if not the United States. Houses were built as mini-mansions, and residential guidelines reinforced the notion of exclusivity. The policy for open space was more private than public—for example, park use on the weekend was free to residents, but nonresidents had to pay a fee. San Marino was almost entirely white (99.7 percent in the 1970 census), and its politics extremely conservative. During the 1980s, the demographics of San Marino began to change dramatically, however, as wealthy Chinese buyers entered the housing market, keeping the city wealthy but now part white and part Chinese.[60]

The notion of exclusivity for gated communities in the Los Angeles region expanded to include not just the rich but the near-rich and upper-middle-class. This expansion originated with the development of the Rolling Hills enclave in the 1930s and 1940s, and subsequently the Hidden Hills gated community in the San Fernando Valley in the 1950s. The Rolling Hills development was proposed by real estate developer Frank Vanderlip Sr., who owned 16,000 acres in the Palos Verdes Peninsula, part of the original 70,000-acre Rancho Palos Verdes estate and cattle grazing land that he had purchased in 1913. A couple of decades later, Vanderlip solicited Beverly Hills architect A. E. Hanson to develop an exclusive community on part of this land. Hanson began work in 1932, in the middle of the Depression. Hanson's concept was to develop the city's large lots into single-family vacation dude ranches with a minimum lot size of one acre of land and with "one-story luxury homes with room enough to keep and ride horses," as he described it. A gatehouse was built in 1935 and all building exteriors and the gate to the lot were painted white. Homes gradually began to sell, and by 1957 the enclave of high-priced homes (though not mansions) had become sufficiently established that residents voted to incorporate into the city of Rolling Hills, which was then proclaimed the first incorporated fully gated city.[61]

Hanson then shifted his attention in 1950 to an area in the San Fernando Valley situated next to nature preserves and park land, where he established the upscale community of Hidden Hills, perhaps the best-known of the elite gated communities. Marketed for its wealth and exclusivity, the enclave came with bridle trails and three-rail fences painted white, while there were no sidewalks or street lights within the confines of the gated area. Rated as the richest place in the United States in 2011 and 2012 and "the cream of gated living," according to *Forbes* magazine, the Hidden Hills focus on extreme privacy became especially attractive to high-profile celebrities. A long list of celebrities who owned homes in

Hidden Hills was in fact a feature of the Wikipedia site for the gated settlement.[62]

Leisure-oriented gated communities, or what geographer Renaud le Goix has characterized as "commoditized suburban communities for the upper and middle classes," have also been strongly associated with Southern California, with its mild weather and emphasis on the outdoors. These communities often have themes: for example, they might be for seniors or retirees, with highlighted attractions such as a private golf course, bowling alley, or other kinds of amenities for that population. Developers Ross Cortese (who built two Leisure World communities in Orange County) and Del Webb (who built the massive Sun City complex in Phoenix, Arizona) were the pioneers in developing an exclusive country club environment for those over the age of fifty-five. The focus was especially retirees who could experience "comfortable surroundings with amenities," as one Del Webb official put it. When Cortese established the first Leisure World, his 1962 ad in *Life* magazine touted it as "the most revolutionary new concept in housing since World War II." Although Cortese sought to attract a homogenous white middle-class population, the retirement or leisure community culture began to attract a more diverse (albeit still middle-class or upper-middle-class) constituency as Southern California's own demographics underwent change in later years.[63]

By the 1980s and 1990s, gated communities had become increasingly linked with real estate speculation and development. By 2002, more than one in eight families lived in gated communities in the United States, with Southern California continuing to lead the way. The development of the locked "security zone" extended the concept to low-income residents fearful of crime and wanting greater security. In his book *City of Quartz*, Mike Davis labeled this emphasis on security and safety "Fortress L.A.," or the militarization of space, both public and private. The use of private security guards and physical security services became common, as did shopping malls ringed with staked metal fences that led to the "destruction of accessible public space." With these kinds of changes, privatizing spaces, as embodied most directly in gated communities, became, in Le Goix's words, "predators of public resources."[64]

Complementing these changes has been a redefinition of what actually constitutes public space, as is reflected in the growing phenomena of "privately owned public spaces" or "POPs." The POPs had their origin in the development of plaza areas associated with commercial developments which sought to differentiate between desirable and undesirable users of the space. The design of these spaces, as earlier noted, only reinforced the distinctions about who was wanted and who was discouraged to use the space. Anastasia Loukaitou-Sideris described spaces that were part of the redevelopment of downtown Los Angeles in the 1960s and 1970s as having such characteristics as "rectangular, concrete form, fixed benches, modern sculpture, water elements, limited color palette and other design conventions that are intended to keep unwanted guests out." Privately owned and managed public spaces have further allowed the area in front of or part of a commercial development complex to meet corporate goals of safeguarding these areas from the undesirables. Such strategies reflect a form of "spatial injustice."[65]

Streets and Sidewalk Spaces

Hong Kong

In such a dense city with limited official public open space, streets in Hong Kong provide multiple economic, social, and recreational functions and have long been hives of human activity. Photographs of street life from the 1950s through the 1970s show hawkers offering goods and services, from vegetables to shoe-shining. Laundry was hung to dry. Children played there. Inexpensive eateries (*dai pai dong*) laid out tables and chairs in the street. Goods were delivered on trolleys, bicycles, and people's heads. Performances and parades took place.[66]

This level of vibrancy (or urban chaos, for its critics) was not appreciated by policymakers. Street activities were and are to a large extent still seen as unsanitary, nuisances, and road obstructions. While street hawking was once tolerated as a form of social welfare, since the 1970s the government has adopted a policy of regulating them out of existence, as described in chapter 5. New licenses have not been issued since then, and strict conditions have been imposed on the transfer of existing ones. Street markets and dai pai dongs have in turn been relocated into indoor municipal centers.[67]

As older hawkers retire and pass away, their number continues to shrink. As of 2013, there were 6,400 licenses, including 28 dai pai dongs.[68] As social attitudes have changed over time, their public image has shifted, and they are now considered iconic local symbols worthy of promotion by the Tourism Board. After a government review in 2009, certain classes of hawkers can more easily transfer licenses to their children, or new licenses can be granted if local district councilors are supportive. However, new licenses are still not issued on a regular basis, and hawkers are still offered financial compensation to surrender their licenses. The attempt to remove unlicensed street food stands that led to the Fish-Ball riots of 2016 was just the latest example of the squeeze on street hawker activity.[69]

Two other examples illustrate policymakers' failure to consider the value of vibrant streets. In 1993, the famous Mid-levels Escalators were built to carry commuters between the Central Business District and the residential neighborhoods on the hillside above. Three years later, the government deemed the project a failure because it went over budget and did not measurably reduce traffic congestion.[70] However, it had the unintended consequence of catalyzing economic regeneration by transforming the streets it passed through into a trendy dining and bar district thanks to increased pedestrian accessibility. These benefits were not taken into account, and no further escalator projects were considered for the next twenty years.

The second example of policymakers' disregard for lively streets is the Transport Department's experiment with pedestrianization in the early 2000s. Several pilot part-time pedestrianization schemes were implemented in crowded shopping districts in order to reduce congestion on sidewalks. However, the government seemed unprepared for the eruption of street activity that ensued. After decades of absence from Hong Kong's streets, street performance came roaring back in popularity. Besides the buskers, there were broadband Internet subscription booths, product promoters, religious and political protesters, crafters and artists. Existing hawker control regulations were outdated (for example, there are no regulations pertaining specifically to buskers). The enforcement capacity of the Food and Environmental Hygiene Department (FEHD) and the Hong Kong Police Force was overwhelmed.

In Mong Kok, where residents live above the pedestrianized street, noise and nuisance complaints mounted. After twice cutting back the pedestrian zone's evening hours, the Yau Tsim Mong District Council voted in 2013 to ask the Transport Department to cancel the Mong Kok scheme on weeknights.[71] Modernizing street management was beyond the purview of the local District Council, whose powers are only advisory,[72] but high-level officials did not see it as a priority. The street was given back to vehicles as a form of crowd control. Siloed administration again meant that other benefits of pedestrianization were ignored, including better air quality, place making, and cultural enrichment.

In a larger context, Hong Kong's mode of urban design since the 1980s has not been successful at creating vibrant streets. The podium-and-tower style of building in New Towns, on reclaimed land, and in urban renewal projects has resulted in deadened streets in the newer parts of the city. As podiums became superblocks and the urban fabric less permeable, the viability of street-facing shops declined. Many large-scale podium-and-tower developments turn blank walls to the street, while human activity is absorbed indoors. In that sense, the privatization of public space is a much broader problem than just the creeping private ownership of formally defined open spaces. This is especially significant in a city where people rely so heavily on public and quasi-public spaces for their recreational needs, given that their homes are too cramped to accommodate guests.[73]

In a way, despite a vastly different urban form, Hong Kong has parallels with the Los Angeles urban identity in which residents live in cul-de-sacs or gated communities, drive everywhere, and primarily encounter fellow citizens in controlled, privately owned spaces such as malls and gyms. A Hong Kong resident may live in a tower above a shopping mall, which connects directly to the metro system, which brings her to another shopping mall in the central business district, which is connected to her office building through a series of skyways. She ultimately may go days at a time without venturing out onto the street. Streets then become not a part of the rhythm of daily life but a part of the civil and transportation engineering emphasis on moving people between locations without contributing to a sense of place.

Los Angeles

Streets and sidewalks in cities like Los Angeles and Hong Kong are important links between the private and public realms. Sidewalks in particular have veered between being places of disrepute (for example, through associations with the gutter) and spaces for pedestrians and for social and economic life. Sidewalks continuously raise questions of what types of people and activities belong in visible areas of cities.[74]

When Los Angeles became a pioneer in mass motorization in the 1920s, the city was an innovator in applying traffic rules to its streets. These included laws limiting when and where pedestrians could cross roadways. As discussed in the previous chapter, by declaring pedestrians in streets to be "jaywalkers," Los Angeles and its auto advocates also needed to ensure there was space on sidewalks for people to move. Uses like shop merchandise displays and sidewalk vendors that had previously been tolerated as economic attractions were now seen as obstacles to walking.[75] The city restricted the use of sidewalks for displays and vending activities, and restricted the space and times of day allowed for deliveries. While these regulations may have uncluttered the sidewalks, they also removed the social activity, visual interest, and commercial attractions along the way. Ironically, by clearing space for pedestrians, limits on sidewalk uses transformed sidewalks into sterile places where people didn't want to walk.

Today, Los Angeles has begun to reenvision streets, moving from car-dominated corridors to a "complete streets" model, while trying to enhance sidewalks as places for walking and lingering. The city's Mobility Plan 2035 anticipates creating "pedestrian enhanced districts" where walking is emphasized. The complete street design guide meant to help implement the mobility plan analyzes sidewalks as comprising a frontage zone where business owners can place decorations, signs, and displays near their entryways; a pedestrian zone that must be kept clear for pedestrians to walk (or roll in wheelchairs); and a frontage zone for bike racks, sign poles, and other infrastructure. Outdoor seating and dining would be encouraged in the amenity and frontage zones as long as adequate space is left to ensure pedestrian passage. By acknowledging that sidewalks serve multiple uses, new rules can help allow some of the vitality that used to exist on sidewalks before the dominance of cars.[76]

People-friendly designs, however, are undermined by the poor physical condition of Los Angeles's 11,000 miles of sidewalks. Through much of the nineteenth and twentieth centuries, sidewalk maintenance was considered to be the responsibility of adjacent property owners. In 1970, the City of Los Angeles took over sidewalk repair and received federal grants to help pay to fix areas that had been torn up by tree roots as well as from other wear and damage. When the funding ran out, a backlog grew of sidewalks in need of repair. Today, 40 percent of the city's sidewalks are considered to be in poor condition. Disabled residents sued the city over its failure to maintain sidewalks, reaching a settlement in early 2015 that committed Los Angeles to spend $1.4 billion over thirty years to fix sidewalks. The city will focus on repairs near government buildings and in heavily used areas. Yet this large pool of money will still not be enough to fully fix or maintain sidewalks, so Los Angeles plans to require commercial property owners to take responsibility for sidewalk maintenance. Under a "fix and release" plan, LA would fix sidewalks in residential areas one more time, and then turn over future responsibility to owners. Pedestrian and safety advocates have criticized these plans, questioning why the city should pay for repairs of the roadways used by vehicles but not for the sidewalks where people need to walk safely.[77]

Much of the pedestrian-oriented activity that is returning to Los Angeles sidewalks is a result of immigrants who are used to vibrant public space. Activities include the sale and purchase of food and nonfood items on sidewalks (the Los Angeles equivalent of Hong Kong's street hawkers), which is prevalent in parts of the city, although it is not legal. Despite its reputation as a hotbed of street food, Los Angeles is, in fact, the only one of the ten most populous cities in the United States to not allow sidewalk vending of food.[78] Los Angeles had banned sidewalk vending in business districts starting in the 1930s, and it was forbidden citywide in 1980. Sidewalk vending did not reemerge as a significant activity in Los Angeles until the 1970s and '80s, when many new immigrants from Latin America and Asia turned to the informal economy, including vending, for work.[79]

In 1994, the ban on sidewalk vending was amended to allow the establishment of up to eight Sidewalk Vending Districts.[80] Complicated regulations, including a requirement that 20 percent of surrounding landowners and residents sign the application in favor of a new district and the assignment of vendors to fixed locations, made it difficult to establish vending districts, however. Only one vending zone, in McArthur Park, was ever created.

In 2012, advocates for vendors, food access, and vibrant streets began meeting under the auspices of the Los Angeles Food Policy Council to consider how sidewalk vending could be legalized in the city. This mobile food task force evolved into a campaign endorsed by more than forty organizations, which led to a City Council motion to legalize sidewalk vending in late 2013. More than three years of debate ensued in the City Council without a resolution over whether a permit system for legal vending should be citywide, whether vending should be allowed only in special districts, or whether to pass a hybrid model with a citywide permit system but stricter limits in certain areas. Nevertheless, the street vending campaign picked up significant support among community-based organizations and food and pedestrian advocates, creating momentum for the renewal of street and sidewalk life.[81]

The crisis in lack of affordable housing combined with the persistence of poverty has also turned sidewalks in Los Angeles into places to live. In Los Angeles County, homelessness rose 12 percent between 2013 and 2015. With the number of homeless encampments increasing and expanding beyond skid row in downtown, the City of Los Angeles has struggled over the "appropriate" uses of sidewalks, even as the need of the homeless to use these public spaces as shelter has only grown. The increasing number of people living on sidewalks and in other open areas has led to further policy debates about a "homelessness crisis" and has transformed the debate over sidewalk policy into an issue of crisis management. For the homeless, the role and purpose of streets and sidewalks as public spaces have also become a matter of survival.[82]

Water Spaces

Hong Kong

Surrounded on three sides by water and a limited land base, the Hong Kong coastline has the potential to provide valuable recreational space and relief from urban stress. Appreciation for "blue space" has become increasingly prevalent among governments, urban planners, public health experts, and open space advocates around the world, and research has identified that living near the coast may bring psychological benefits.[83]

Hong Kong's most recognized water space is Victoria Harbor, the channel between Hong Kong Island and the Kowloon Peninsula that had made Hong Kong a prized trading base. Until the late 1970s to early 1980s, it was a working harbor with docks, warehouses, and shipyards, and therefore unavailable for recreational use. When containerization was introduced in the 1960s and 1970s, the docks were removed to Kwai Chung in northwest Kowloon.[84] Following the demolition of a nineteenth-century railway terminus, the Tsim Sha Tsui waterfront (in the center of the harbor's north shore) was converted into a cultural complex and waterfront promenade.

However, most of the rest of the harbor's waterfront was turned over to large private housing complexes, roads, and infrastructural uses. On the north side of Hong Kong Island, new roads were built along the waterfront to alleviate inland traffic congestion. Coastal sites were then used to accommodate infrastructure such as sewage screening plants, pumping stations, public cargo works areas, refuse transfer stations, wholesale produce markets, and bus and tram depots.

One notable example of spontaneous waterfront recreation was the "Poor Man's Nightclub," also known as *Dai Dat Dei* (big piece of land), which used to exist on a vacant coastal lot in Sheung Wan. In its heyday, it was a bustling nighttime flea market known for hawker stalls selling household objects, dai pai dongs, and street performances.[85] It became the second incarnation of the nearby Chinese Recreation Ground, which had been made into a park after being redeveloped for reclamation in the 1970s.[86] Patronage began to decline in the 1980s, and the site was vacated in 1992 to make way for reclamation works for the Western Harbor Tunnel Crossing. There was a short-lived government-sponsored attempt to revive it as a tourist attraction in the early 2000s, but since the site was now relatively isolated, with only footbridge access, it was not commercially viable.[87]

The 1990s saw the emergence of social movements against harbor reclamation, which had already reduced the harbor's size by half. Advocacy by the Society for the Protection of the Harbor, first organized in 1995, changed public attitudes and led to the passage of the Protection of the Harbor Ordinance in 1997, which identified the harbor's status as a "special public asset."[88] The society then sued the government seven times to enforce the ordinance. In 2004, the Court of Final Appeal (the highest court) ruled that the government was required to convincingly demonstrate an "overriding and present need" for public purposes. As a result, the society was able to reduce the scale of the government's initial 584 ha reclamation plan to the 30 ha necessary for its Central and Wanchai Reclamation project.[89]

Having established the principle of the harbor as a "public asset," the debate shifted to qualitative issues. It became clear to urban design experts and advocates that the usual infrastructure-and-superblock-centric mode of development would hamper public access to the planned waterfront park and produce a sterile environment. Since this would be the last chance to create a high-quality waterfront, civil society groups began to campaign for the reduction of development intensity on the new waterfront, arguing that plans for a new government headquarters (outside of which Occupy Central took place) and large commercial buildings would bring too much traffic to the area, resulting in a need for high-capacity, pedestrian-unfriendly, and polluting roads.[90]

In 2007, as part of its broader campaign for greater open space and public access, Designing Hong Kong filed a rezoning application to the Town Planning Board. It proposed reducing building density, establishing ground-level pedestrian access to the waterfront, splitting up oversized commercial developments, reducing road widths, constructing a tramway, converting part of the People's Liberation Army's military base to open space, and preserving the Queen's Pier *in situ*.[91] The Town Planning Board, whose members are appointed by the government, predictably rejected the application.

To shift the development paradigm, advocates like Designing Hong Kong continue to face many challenges. Access remains an intractable problem, with wide, multilane roads and impermeable blocks between the city and the shore. Fragmentation across government departments remains another difficulty. With various zonings, transport networks, and land ownerships scattered across the waterfront, innovative ideas, such as a proposal by cyclists to build a bike path along the entire north coast of Hong Kong Island, have become caught in bureaucratic tangles between the Transport and Housing Bureau, the Development Bureau, the Highways Department, the Leisure and Cultural Services Department (LCSD), the Transport Department, and the Hong Kong Police Force.[92]

Yet some important policy changes have occurred over the past two decades due to public pressure. The government has committed to the principle that waterfronts should be for recreational use. Aside from the Central Waterfront, waterfront promenades have been built on several shores on both sides of the harbor and even in the New Towns. In Kwun Tong, an old industrial district in the process of being transformed into a second central business district, the Energizing Kowloon East Office (EKEO) has been working to foster waterfront vibrancy through programming and flexible management. Space underneath the elevated Kwun Tong Bypass has been opened up for public use, with activities from art displays to BMX biking to loud musical performances encouraged rather than prohibited.[93]

These various struggles represent a microcosm of the issues affecting public open space in Hong Kong. They include transport planning and traffic management, pedestrian-friendly approaches, diversity and vitality, the privatization of open space, the prioritization of land revenues, the political influence of the property sector, and problems of quasi-representative government. How these issues are addressed will also shape the urban environment in the decades to come.

Los Angeles

When Jenny Price arrived in Los Angeles in the mid-1990s, fresh from completion of her PhD dissertation, she decided she wanted to become an urban nature advocate and urban nature writer rather than an academic. She immediately focused on the Los Angeles River and joined other advocates, including poet Lewis MacAdams, to promote the idea that this was indeed a river despite its sixty-year history of channelization. While the water engineer managers at the LA County Flood Control District and the US Army Corps of Engineers sought to limit access to the river, Price saw that the river could afford a modest reconnection to urban nature. She began to provide tours along the edge of its fifty-one-mile path, with an occasional foray past the "Do Not Enter" signs to have her tour participants begin to experience the river as part of a revived urban nature.

Over time, Price shifted her focus to broader issues of land use, open space, and public access. Her concern about the issue of access to urban nature loomed especially large and led her to consider access to the beaches along the ocean front, such as in the wealthy enclaves along the coast like Malibu, as another preeminent issue of the Los Angeles urban environment. To Price, the LA River and the beaches and ocean front at Malibu constituted "two of the great public spaces in Los Angeles that for decades have not functioned as great public spaces. [And] even though they seem really different, they're linked in that way."[94]

Price used her skills as a journalist to document some of the more egregious ways that wealthy homeowners along the coast, including several famous celebrities, had established numerous barriers, many of them illegal or deceptive, to prevent people from finding a path to the beach. The California Coastal Commission (established in 1972 after an earlier campaign by open space advocates had led to the passage of a proposition about coastal access) required public access as a condition for development along the coast. However, Price was able to document that such access requirements were not being met through such subterfuges as fake "No Parking" signs, fake "No Trespassing" signs, bagged parking meters (to suggest they were not working), illegal curb markings, and many more tricks. This was particularly the case along the part of the coast known as "Billionaire's Beach" and on roadways such as Malibu Road that had countless illegal signs suggesting there was no public access.[95]

Price, along with colleague Ben Adair, created a smartphone app called Our Malibu Beaches to identify where fake barriers were located and how people could assert their right to access and utilize the beach. Part whimsical but also serious about the importance of public access to the coast, the Our Malibu Beaches app quickly caught on and represented the most recent effort in a long, contested history in Los Angeles about coastal access and discriminatory practices of coastal and waterfront areas.[96]

The issue of coastal access has also reflected some of the racial and economic divides that have long plagued the region. In the 1920s, African-Americans' efforts to increase access to coastal beaches evoked a hostile response. Up to then public beaches had largely been seen as off-limits to people of color. As historian Douglas Flamming writes, when subdividing or selling their beach-front properties, white property owners sought to "place Caucasian restrictions on their properties barring negroes from ownership and occupation." One area called Bruce's Beach in the coastal enclave of Manhattan Beach was an exception, in that access had initially been made possible by a nondiscriminatory approach taken by the black and white owners of the land. But Manhattan Beach officials, concerned about African-American access, paid the owners $75,000, ostensibly to turn the area into a public park; they then leased the land for $1 a year to a private individual. That new owner, working in tandem with city officials, excluded blacks from the beach. The change in beach access led a young African-American college student to protest by going to the beach and getting arrested in her swimsuit. The subsequent controversy, made visible through the black media, forced Manhattan Beach officials to relent. The property was then taken back, and full public access was renewed.[97]

Today, the waterfront, whether along the coastline or along the Los Angeles River, still remains contested space. Despite plans to make the LA River more connected rather than separate from the communities that border it, many of the areas along the river are kept off limits, either by legal requirements or by barriers, due to the potential for dangerous flood waters cascading through the channelized river during rainy periods. And while access to the beaches has evolved so that restrictions are no longer allowed, efforts are still made to keep access a private matter at the choicest locations. The ability to turn these water spaces in Los Angeles into places for connection and renewal, whether through policies or community action, remains a challenge.

The Connection to China

For China, the evolution of urban public space has been bound up with the changing relationships between rural and urban areas. The pattern of urban transformation since the 1980s has generally involved redevelopment of the urban core along with continuing expansion at the urban edge, with expropriation of rural farmland at the edges of the metropolitan region. Once that land had been converted for urban settlement, it also extended the boundaries of the city.[98] The transformation in Guangdong Province of Shenzhen and Guangzhou is emblematic of those changes.

Prior to 1979, the Shenzhen area consisted of several small villages and communes similar to the hundreds of such villages in Guangdong. These villages and communes constituted Bao'an County, which was 2.9 square kilometers in size and had a population of 300,000, most of whom were engaged in agricultural or fishing activities. In 1979, Shenzhen became a designated municipality, and the next year the area adjacent to Hong Kong became one of the four Special Economic Zones (SEZ) set up to stimulate export production and foreign trade. Shenzhen's population, economic activity, and geographic boundaries expanded enormously in the next three decades. Its annual GDP growth reached 35 percent, its population grew to 10.5 million by 2014 in the city and 18 million in the metropolitan region, and its boundaries increased its urban footprint to more than 1,700 square kilometers.

Shenzhen's growth further identified the enormous class (and often ethnic) divides that have characterized China's cities. Its industrial expansion was made possible by the huge influx of migrant workers from multiple locations throughout China. Nearly all the migrant transplants living within the city were without an urban hukou and its attendant benefits, significantly reinforcing the class divide between longtime city dwellers and those formerly rural residents now living within the metropolitan region.[99] The new migrant enclaves thus established what anthropology professor Li Zhang identifies as "a new dimension of spatial segregation, one of the most visible aspects of the changing urban space in the reform era."[100]

While Guangzhou's annual growth rate of more than 6 percent was not as dramatic as Shenzhen's, it still witnessed increasing pressures on resources and available land, due in part to the influx of its migrant workers. Official population estimates of ten million "registered residents" in 2011, for example, underestimated the migrant population, which added an additional five million people within the city. "Anyone who is a mayor will know the intensity of the problem we are facing," Guangzhou mayor Wang Qingliang told a national planning seminar that year, arguing that population pressures could best be addressed by limiting the influx of low-skilled workers and "attracting better qualified residents." This position was subsequently adopted as the city's formal policy, with goals to limit population growth by 2020.[101]

As the migrant labor pool rapidly expanded, a new kind of urban area, known as the urban village or "village in the city," had emerged. These places, almost entirely populated with rural migrants and providing an essential source of affordable housing not otherwise available to them, were seen by authorities as "dirty, chaotic, and inferior" environments (*zang, luan, cha*). These areas conflicted with the efforts by officials to modernize their cities and showcase to the world their tall skyscrapers, new roadways, contemporary dwellings, and corporate identities, as well as their outward, global-oriented face.[102]

The migrant residents of the villages in the city, despite their lack of key urban amenities such as health care and education, were nevertheless able to establish their own translocal networks and minieconomic and entrepreneurial activities. They in turn injected "a new diversity into [Guangzhou's] overwhelmingly Cantonese culture," argue Margaret Crawford and Jiong Wu. While village activities had a certain degree of autonomy, their power was "potential rather than actual … a genuine 'bottom-up' urbanism in a city where top-down mandates play an ever-increasing role in city building."[103]

Those top-down mandates came into play as the villages-in-the-city expanded in size and population, creating a growing tension between the continuing need for a low-cost migratory labor supply (and places to house migrants) and the goals of urban modernity and the desire to be seen as a

global city. Various plans began to be implemented in Shenzhen, Guangzhou, and several other cities in China to "clean and deport" particular residents in order to "maintain order and crack down on crimes" and address
such problems as noise and sanitation in order to make those places "clean,
smooth, and quiet." But the "clean and deport" strategy was more ad hoc
than part of a broader urban planning initiative. In its place emerged new
policies focused on "demolition-redevelopment," which included dismantling existing structures and promoting real estate development strategies
to essentially remake the village in the city into a direct part of the modern
city. In Shenzhen that took the form of the City Planning Bureau's "Master
Plan of Villages-in-the-City Redevelopment"; while in Guangzhou, redevelopment was intensified with the efforts to further modernize the city in
anticipation of the 2010 Asian Games. In both cities, efforts were sometimes resisted, but the logic of modernization, based on the priority of real
estate development, continued to prevail. By 2016, much of this process of
redevelopment had succeeded, with migrant enclaves more scattered, new
efforts to change the hukou system in discussion, and redevelopment seen
as the linchpin of a full transition to the modern, global city.[104]

One form that the quest for modernity took was the gated community,
which became prevalent in cities like Shenzhen, Guangzhou, Shanghai,
and Beijing. Chinese gated communities, or "sealed residential quarters,"
contrasted with those in Los Angeles and Hong Kong due to their size (often
12 to 20 hectares of land) and population (between 200,000 and 300,000
residents). While Los Angeles's gated communities often consist of low-rise,
single-family homes, with as few as twelve to fifteen homes per hectare,
and Hong Kong's gated communities are vertical, consistent with the city's
density and limited land base, the Chinese gated communities have high
buildings (six to ten stories) and are often sprawling. They may look more
like "a city within a city" than the "greenfields" nature-themed gated communities of the United States. For example, China's largest property developer, the China Vanke company (or Vanke), which is based in Shenzhen
and operates in more than thirty cities, established its reputation by constructing a range of gated communities (as well as high-rise apartment towers and single-family residences), including Vanke City (or "Vanke Garden
City") in the Shenzhen suburbs.[105]

While China's gated communities provide a mechanism for an emerging
middle class to become homeowners (*yezhu*), they also represent strategies

to establish new types of middle-class identities. These are reflected in the theme-based gated communities where developers promote the idea that they are building not just houses but an "entirely serviced small town outside the city." Borrowing from the language of "new urbanism," Chinese developers like Vanke emphasize such themes as "low density," "greener," "ecological," "luxury," and "livable." But such themes also incorporate the ideas of opulence and ostentation as well as exclusivity. Even parking, at such a high premium in China's cities, is part of the identity of exclusivity. "In present urban China," James Wang and Qian Liu point out, "most neighborhoods with purposely built parking spaces are gated-communities that have a controlled access and are independent from outside by a closed perimeter of walls or fences." By distinguishing themselves from rural and traditional old neighborhoods as well as from the migrant enclaves, China's gated communities, and more broadly new urban development, represent what has been called a "civilized modernity."[106]

Like China's push for marketization, the dominant forms of urban development have had major implications for the urban environment. In Shenzhen, for example, during the first decades of redevelopment, "urban planning and design prioritized speed and price over any other value, including environmental impact," noted architecture critic and long-time Shenzhen resident Mary Ann O'Donnell. "The Chinese expression for land reclamation, 'yi shan tian hai' or 'move mountains and fill the sea,' literally describes the step-by-step transformation of the Shenzhen Bay coastline," O'Donnell writes of one noted development. "First, raze a mountain—and many Shenzhen hills no longer exist except as place names—and, second, reclaim coastal land, creating flat, relatively inexpensive building sites. The point, of course, is that as the city has prospered and natural features such as mountains and coastlines have been restructured, their market value has increased exponentially."[107]

Beyond severing the city's connection to the natural environment, the urban development strategies have generated the types of urban environmental problems—air pollution, poor water quality, contaminated food—associated with the marketization and urban modernity experienced in the global city. As the cities have grown, urbanization has come to replace industrialization as the driving force in China's development. "The political economy of urban expansion as well as the ideology of urbanism,"

You-tien Hsing has argued, "[have combined] to dominate the logic of China's transformation."[108]

As urban environmental problems loom larger, China's response has been to highlight its own quest for "urban sustainability," including by designing self-contained "eco-cities" which have been modeled and planned in hundreds of places across China. While some eco-cities reside within or adjacent to major metropolitan areas, their primary characteristic has been the idea of building a city from scratch, complete with green building design and walkable places. In cities like Shenzhen, itself essentially built from scratch in its transformation from rural to urban and now grappling with many of China's urban environmental problems, plans have been developed to create more sustainable models of development. Along those lines, real estate developer Feng Lun (head of Vantone Holdings and president of the China Real Estate Chamber of Commerce) has promoted what he calls the "Great City" concept—a "green, relational, economical, all-encompassing, and technological" approach to building new cities.[109]

Yet the problem of building from scratch or remaking a highly polluted city into a "green city" is that plans go awry or get postponed, while the urban environmental problems remain deeply embedded in the prevailing logic of development and marketization. In 2014, China released a new Urbanization Plan that emphasized the goals of continuing to accelerate the shift from rural to urban (a rise of the urban population by 1 percent each year to 2020, to 60 percent of China's overall population) while putting in place sustainability guidelines for land use and construction associated with that growth. The goal of a new policy framework of "sustainability" at the national level is noteworthy, yet implementation success remains a local or province-based challenge. For the national leaders, development (and economic growth) remains the priority; at the local level, expansion and the private/market interests that make development happen are what dictates whether policies get implemented. And whether that translates into an urban environmental success story remains more problematic, as the recent history of China's urban expansion indicates.[110]

Strategies for Change

For Hong Kong, non-revenue-generating land use in a city with a constrained land supply and high-density development has meant that open space has usually been sacrificed to other priorities. Barring a complete overhaul of the tax structure, Hong Kong is highly unlikely to reduce its dependence on land revenues. Hence, its infrastructure-driven, land-value-maximizing, economic growth model will likely remain dominant for the foreseeable future.

For Los Angeles, efforts have been made to densify certain corridors, such as downtown, Hollywood, and Wilshire Boulevard. Yet the primary drivers continue to be real estate and financial interests that influence land use patterns. Strategies for change, such as the rezoning process currently under way, become critical to identifying what types of development can reorient or at least modify what one observer a century ago characterized as Los Angeles's "chief stock in trade"—its real estate development.[111]

For China, pressures to improve the condition of migrants remain one of the key determinants of how spaces are going to be designed or reconstructed. The tension between the modern and the premodern or traditional, combined with the environmental problems that accompany urbanization, continue to unsettle the rapid-fire process of urbanization so essential to the Chinese leadership's strategies of growth and development. How to unravel that association and undo that tension is at the heart of any future strategies for change.

In Hong Kong, throughout the city's history, strategies for change have come about as a result of public pressure, whether from the Urban Council of the mid-twentieth century pushing for more parks and playgrounds, or from recent advocacy over waterfront development. These efforts cannot be separated from Hong Kong's overall political situation. Top officials, preoccupied by political conflicts over the government's own legitimacy, have little political capital to spend on issues deemed to be of secondary importance, such as public and open space. The low level of trust between the government and opposition parties in Hong Kong has spawned a risk-averse attitude in government as decision makers become leery of attracting any controversy. Bureaucratic inertia results, because officials are unwilling to try anything new. Interdepartmental and interbureau collaboration becomes problematic. Since public space implicates many different spheres of responsibility, from air pollution to social interaction and traffic management, it becomes even more challenging to effect changes.

Urban planners and NGOs advocating for change related to spatial issues often face the prospect of not only trying to build broad public support for their ideas but of convincing government departments and bureaus one at a time. For example, in 2014, five months before the Occupy Central protests, an alliance of planners, traffic engineers, policy researchers, and scientists proposed a pedestrian scheme for Des Voeux Road Central that would also provide air quality benefits. They seized upon the air quality data recorded during the Occupy Central protest period, arguing that commuters had been able to adapt and that people still got to work on time. The number of traffic complaints had actually dropped during the first two weeks of the protest, and traffic congestion only worsened after the reopening of Queensway. A pedestrianization approach that included planned bus rerouting and functional tram service could bring about the same benefits without going through an impromptu street action. Yet the Des Voeux Road Central plan also needed to navigate various departmental jurisdictions. Through sustained organizing and effective presentations, the alliance of planners was successful in obtaining support from the Environmental Protection Department and the Transport and Housing Bureau as well as from local District Councillors and business owners. Nevertheless, the Transport Department's concern over the scheme's impact on traffic created yet another bureaucratic hurdle.[112]

All too often, policy change occurs only under extreme public pressure or crisis. Minimally responsive, crisis-driven government has encouraged a new generation of social activists resorting to more radical tactics. These mostly young, loosely organized activists have periodically assembled large protests against urban-spatial issues such as heritage conservation, railway construction, New Town development, and agriculture. Instead of working within the system (e.g., through the Town Planning Board, the court system, or traditional media campaigns), they have adopted civil disobedience as a tactic to force a government response. In hindsight, the Queen's Pier protests of 2007 were part of a trend that would eventually find expression in the Occupy Central events of 2014.

In Los Angeles, the momentum for changes in land use and the very concept and legitimacy of public space in a car-dominated and real-estate-driven city have come from below, both in the form of community action and in the major demographic changes that have created a different-looking city that is more diverse and more oriented toward public space. The 2003 ArroyoFest event that closed a freeway for bike riders and pedestrians on a Sunday morning led in turn to the street closures for the CicLAvia bike rides

Figure 7.1
Des Voeux Road Central: before and after, a photomontage. Source: Hong Kong Institute of Planners.

and pedestrian walk events that now take place quarterly. While initially organized with limited resources by a long-time transportation and environmental community activist (as well as an environmentally oriented event planner), CicLAvia had unanticipated massive participation (more than 100,000 participants in the very first event) that got the attention of city officials, including the mayor. As a result, the city offered financial and logistical support for a quarterly CicLAvia while also identifying a constituent base for a new approach to transportation and street planning, which culminated in the Mobility 2035 plan.

The impact of the wave of immigrants—the newest Angelenos—on the uses and design of streets and public space has been noteworthy in influencing strategies for change in Los Angeles. As documented by the Latino Urbanism advocates, immigrants' use of the built environment has led to important changes in land use and public space that have been central to the *LA Times*'s Christopher Hawthorne's characterization of a potential "Third Los Angeles" that is not dominated by cars and real estate. Such a transition remains more potential than realized, a testament to the deeply embedded car culture and the power of the real estate developers that have long determined the shape and nature of the built environment—and its urban environmental implications.[113]

In China, the role of social movements remains constrained. National policymakers, however, can be sensitive to the ad hoc, spontaneous protests that erupt over environmental impacts on people's lives, as well as to the increasing urban middle-class discontent about such issues as air pollution, food safety, and transportation gridlock. The rush toward modernization, moreover, has created some nostalgia related to the loss of traditional aspects of the urban (and rural) built environments and the related social and cultural associations with the alleyways, shops, public dancing (primarily by seniors, or the "dancing grannies"), and other aspects of neighborhood life.

China's 2014 New Urbanization policy document identified environmental impacts that needed to be addressed, notably air pollution and sprawl, and reflected an effort to address that discontent. However, the document—and urban policy more generally—still prioritized continuing urban growth and expansion as part of its overall economic model. Spaces have become increasingly privatized, whether due to new commercial developments or gated communities, the identity-driven focus on consumption, and the rapid increase in automobile ownership and use. These have all become essential components of urbanization, modernization, and the remaking of the global city.

Environmental protest has become one of the few outlets for registering opposition to particular problems or development impacts. But the ability of environmental discontent to translate into a new type of urban movement—and an urban politics—still appears limited, despite the enormity of China's urban environmental challenges. It is a challenge that environmental movements in Los Angeles and Hong Kong face as well.

8 Social Movements and Policy Change

Making Change Happen

How can environmental and social change happen in Los Angeles with its fragmented political system and geographically dispersed urban and suburban clusters? How can it happen in Hong Kong with its colonial history, long-standing market orientation, and difficult and complex relationship with China that limits opportunities for political decision making? And how can it happen in China, with its rapid-fire urbanization, its massive environmental disruptions embedded in its development push, and its global role associated with its export-oriented production, huge energy and resource appetites, and climate impacts?

We open this concluding chapter with a profile of a Los Angeles community organizer, a Hong Kong environmental leader who has become an influential policymaker, and a Chinese academic and environmental activist. The focus of these three environmental players—organizing and mobilizing, working an inside-outside game, and representing a deep attachment to place—indicates the potential as well as constraints for environmental change in Los Angeles, Hong Kong, and China.

In the Los Angeles region, the organizer, Angelo Logan, grew up in the city of Commerce, southeast of downtown Los Angeles. Commerce is one of the densest inner-city suburbs that had transitioned during the 1960s from a white working-class community to a low-income Latino immigrant community. By the time Logan entered the local public school, the community was 95 percent Latino and subject to enormous environmental stresses from the crisscrossing freeways, high-traffic roads, and four massive railyards with their diesel emissions and twenty-four-hour operations. Logan was very much a product of the working-class community of

Commerce. A one-time artist, he had dropped out of school in tenth grade to take a job as a maintenance mechanic in an aerospace factory, and had flirted with the gang culture on the Eastside when he was a teenager. Restless and inquisitive, he had done labor organizing in the auto plant while beginning to immerse himself in social and environmental justice advocacy and community organizing. With his talents as a communicator and a quick learner, he eventually became a recognized figure within those southeast communities that were grappling with some of the worst environmental hazards in the region but mostly lacked effective political leadership and community organization.

In 2001, Logan decided to move back to his former hometown where he founded a community-based organization, East Yard Communities for Environmental Justice. East Yard quickly emerged as a powerful new force for environmental change, first in Commerce, then expanding to West Long Beach near the ports, and ultimately throughout Southern California and the United States. East Yard draws upon a strong youth base, and represents a kind of coming-of-age political consciousness and a new type of community identity associated with the desire to change the conditions and circumstances of everyday life. Logan's effectiveness has been based on his understanding of the intricacies of community organizing, which enables him to help develop campaigns at the regional and national levels. He has also effectively focused on where and how environmental change can happen. Logan is a visionary, but he is also realistic about the magnitude of his challenges. "We're up against huge forces: the railroads, the shippers, the Wal-Marts, companies with lobbyists and money and PR firms," Logan told the *Los Angeles Times* in 2009. But Logan also insists that the work of East Yard and other environmental and social justice groups is "changing the mentality," making it possible for people in the community, who once felt powerless, to now demand that their home become "the cleanest and greenest city" and to force those in power to respond to their anger. In recognition of his effectiveness as an organizer, *Grist* magazine, characterizing Logan as "a coalition builder with serious but unassuming chops," named him one of the "50 People You'll Be Talking About in 2016."[1]

In Hong Kong, Christine Loh, the consummate inside/outside player, honed her skills as an environmental change leader through a variety of roles to become the most widely recognized environmental figure in Hong Kong. With law-related degrees from the University of Hull and the City

Figure 8.1
Angelo Logan. Source: East Yard Communities for Environmental Justice.

University of Hong Kong, Loh spent her early career in the fields of multinational commodities trading and business strategic management. In her twenties, Loh became focused on global environmental issues and briefly served as the Hong Kong unremunerated chairperson of Friends of the Earth (HK). In 1992, she was appointed to Legislative Council (LegCo), and was elected to LegCo again in 1995 through direct election. She helped establish the Citizens Party in 1997, which combined a democratic and an environmental agenda, and was reelected to LegCo in 1998. As a member of LegCo and through the Citizens Party, she became an effective inside player on specific issues, notably the protection of Hong Kong Harbor.[2]

Concerned about the limits of her (or anyone's) role as a LegCo member, Loh decided not to run for reelection in 2000 and instead founded Civic

Exchange.[3] Her stature rose even further as an environmental leader, albeit now as an "outside player" focused on evidence-based research to highlight critical environmental issues. In recognition of her worldwide status and her credibility and growing authority on environmental issues, *Time* magazine named her one of its "heroes of the environment" in 2007. Civic Exchange research reports became major documents detailing the importance, challenges, and potential solutions of the urban environmental issues facing Hong Kong. Loh was also a prolific writer, with books, articles, and op-ed pieces that touched on everything from Hong Kong legislative politics, the SARS epidemic, a manual for democratic participation, a history of the Communist Party in Hong Kong, and writings on the global environment, among other topics. She was insistent that research in an area where one was passionate needed to hold up to rigorous scrutiny. Staff members recalled that they would often redouble their efforts to be as thorough as possible even if Loh did not directly say anything to them about the standards they anticipated she would expect them to meet.[4]

After twelve years at Civic Exchange, Loh decided to establish herself as an insider once again. In 2012, she was invited to join the government as Under Secretary for the Environment in the administration of Chief Executive Leung Chun-ying. She assumed the role as the most important inside player in addressing several core environmental issues, such as air pollution, environmental impacts from the ports, transportation, and the design of the built environment. She immersed herself in the details (and the limits) of policymaking. She could point to important accomplishments, such as the development of Hong Kong's Clean Air Plan, or efforts to extend Hong Kong's port-related environmental initiatives to the port of Shenzhen. But policymaking on the inside was fraught with conflicting and competing pressures, and Loh continued to emphasize the importance of outside players and solid research to expand the opportunities and spaces for change.[5]

In China, the capacity to make environmental change has been most effective but also often most disheartening at the local level, as the inspiring yet cautionary story of Xiao Liangzhong reveals. Xiao Liangzhong grew up in Chezhu Village on the banks of the Jinsha River (also known as the Upper Yangtze) in Yunnan Province. A member of the Bai ethnic group, he went to school with members of several other ethnic groups in the area with their shared but also specific identities—identities also shaped by their

Figure 8.2
Christine Loh. Source: Civic Exchange.

environments. These experiences, he would later say, contributed to his research about how the myriad multiethnic groups and their minority identities needed to be understood in the ways they coexisted with each other and with their places.

Liangzhong became an anthropologist, writer, and editor based in Beijing, although much of his focus concerned the region extending from Tibet, Sichuan, and Yunnan and the different minority groups and farming villages and communities interspersed throughout those areas. His family—grandmother, parents, and several siblings—still resided in Chezhu. In his many trips back to the region, he became immersed in the major environmental, cultural, and community issues that would impact his village and many others. These included plans to build a series of huge hydroelectric plants along the Jinsha River, plans that were an extension of the massive

South-North Water Transfer Project. One of those proposed projects—to dam the Leaping Tiger Gorge world heritage site—would soon emerge as one of the most important environmental battles in China.[6]

For Liangzhong, the Leaping Tiger Gorge battle was not simply about its massive quantifiable impacts—100,000 villagers displaced and 13,333 hectares of productive farmland lost—but the eroding of long-standing cultural and social ties. "Ecological issues [for Liangzhong] do not only concern protection of the natural environment, but also cultural ecologies and social rights," the well-known academic Wang Hui wrote in his moving portrait of Liangzhong. Liangzhong's advocacy embodied Hui's critique of developmentalism as an erosion of rural societies in the rush for urban expansion, huge energy projects, globalization forces, and market-dominated decision making and goals.[7]

Liangzhong, intense, passionate, knowledgeable, and dedicated to his cultural and place-based roots, became the linchpin for opposition to the Leaping Tiger Gorge project. He worked closely with sympathetic journalists, and was effective in mobilizing people in the region. He connected with but also sought to shift the role of the NGOs that had become involved in the issue by insisting that their focus had to include the cultural and ecological diversity of the communities involved. And he appealed to national leaders about the manipulations and illegal activities of local officials in conjunction with the project developers.

Liangzhong became the very best example of the action researcher—focused on the research that is needed to inform action, and deeply engaged in action that informs and shapes research. He was a whirlwind of activity, identifying with his people and the place in which he grew up and to which he remained attached. As his reputation grew, he maintained a commitment to a China that could value rather than undermine cultural and ecological diversity and multiplicity. "Even if you had enough gold to fill the Jinsha River Valley," he declared, "it wouldn't be enough to buy up this free-spirited river, nor would it be enough to buy up our homeland."[8]

Exhausted and driven, Liangzhong collapsed and died of cardiac arrest at the age of only thirty-two. He never saw the outcome of the campaign he had led and inspired. Tiger Leaping Gorge was spared, although plans continued to be pushed for other huge hydroelectric projects along other sections of the river. One thousand villagers attended his memorial, including many who had never met him. After his death, the villagers created a stone monument along the mountainside with large characters that read "Son of Jinsha River" in respect for what Liangzhong stood for and sought to achieve.

These three environmental figures—the organizer, the insider/outsider, and the place-based advocate and action researcher—represent important environmental players in their own right, each engaged in bringing about environmental and social change. Logan helped shut down a major polluting facility and helped bring about the adoption of cleaner technologies at the Ports of Los Angeles and Long Beach and for the trucks passing through his community and the region. Among her many accomplishments, Loh played a crucial role in the development of Hong Kong's Clean Air Plan, both as research director and policymaker. Liangzhong demonstrated that communities—including the ethnic minorities often marginalized in the push for urban, industrial, and energy development—could successfully engage in environmental protest and demonstrate why place matters. Yet all three recognized the enormity of the challenges to bring about such change. At the same time, they understood that change was not due to individual effort but arose from social, political, and cultural mobilization

Figure 8.3
Xiao Liangzhong. Source: Chen Meiqun.

and the movements it generated, the policies that could be adopted, and the shift in the discourse about the environment they helped create.

Change from Below

In Los Angeles, in October 1998, a conference was held that sought to link historical social movements and urban struggles in Los Angeles with an array of contemporary community-based organizations and urban initiatives. The gathering brought together activists, researchers, grassroots leaders, organizers, and legislative representatives. One of the conference papers, entitled "Progressive Los Angeles: From Liberty Hill to a Living Wage—The Struggle for a Livable City," described a hidden history of social movements in Los Angeles, including those that had sought to connect the issues of the urban environment, such as air, water, land use, food, and transportation, with the issues of daily life. Reflecting the dispersed Los Angeles region itself, conference participants recognized the challenges of establishing such linkages, including the issue silos and the class, racial, and ethnic barriers that prevented a more holistic approach toward a social change agenda. A conference organizer (and coauthor of this book) told the *Los Angeles Times* at the time that "if you took all the movements and activists, you could fill Dodger stadium. That doesn't mean you've captured that as a force."[9]

In Hong Kong, five years later, on July 1, 2003, more than 500,000 demonstrators marched to protest the introduction of Article 23 of the Basic Law. The Basic Law had provided the framework for Hong Kong's relationship with mainland China, and Article 23, first introduced in 2002, was essentially a political or antisubversion law, intended for Hong Kong to insure political stability and prevent any anti-China activities to take root. The huge demonstration went beyond the specifics of Article 23 to identify concerns about transparency and the very ability of social movements to operate in the one country, two systems framework. The timing of the demonstration, coming on the heels of the SARS outbreak and the failure to provide accurate and timely information as the epidemic spread, provided a subtext for the demonstrators as well. What was perhaps most striking about the 2003 demonstration was that years after the 1997 restructuring of the Hong Kong–China relationship and the formation of the Special Administrative Region, Hong Kong still remained a "city of protests." This reputation, which has ebbed and flowed over the years, would dramatically reappear with the Occupy Central events in 2014.[10]

In China, protests have assumed many forms, particularly after the 1989 Tiananmen Square events. They have included massive and spontaneous eruptions of civil discontent about a polluting facility or local government corruption; social-media-inspired informal gatherings to block an unwanted facility; the ubiquitous collective strolls that represent a discreet form of mobilization, organized by social media, to avoid the restrictions on formal demonstrations; or even the use of photography classes for organizing purposes.[11] Despite the proliferation of protest actions and strategies, the creation of nongovernmental organizations has been uneven at best, given the role of the State and its myriad set of rules and operational requirements that have potentially limited their capacities. More broadly, the idea of a grassroots, independent set of actors that functions outside the structure of the State has been challenged as a "Western" and anti-Chinese concept, associated with a type of "Western" discourse.[12] Yet the enormity of environmental problems, including those in an urban context, has nevertheless created opportunities for action and policy change.

In Los Angeles, some of the earliest environmental battles in the period before World War I involved the fights about water and urban development. Those struggles linked questions of environment, land use, labor, and political power. The successful completion of the Los Angeles aqueduct in 1913 coincided with the diminished role of the "managed growth" approach of the Socialist Party and its labor movement allies in shaping urban and environmental policy. From the 1920s through the 1940s, environmental problems multiplied, related to the horizontal mode of real estate development and population growth, oil drilling and refinery development, and the expansion of the region's manufacturing base. Environmental opposition tended to be localized and problem-specific, although that began to change as the problems of air pollution came to dominate the region. Air quality protests became more forceful during the 1950s and 1960s, anticipating the rise of a grassroots environmentalism that began to influence the policy process at the regional and state level.

By the late 1960s and the 1970s, environmental groups had become important actors in shaping public discourse as well as legislation and regulations. At the national level, a series of environmental laws and regulations were established in the early 1970s, influenced in part by the growth of

environmental organizations skilled in the use of litigation, policy development, and lobbying. The election of Tom Bradley as mayor of Los Angeles
in 1973 and Jerry Brown as governor of California the following year provided further opportunities for some of these groups (characterized as the
"mainstream environmental movement") to help initiate their legal, lobbying, and environmental policy agenda in California and Los Angeles. At the
same time, local protests about hazardous facilities in low-income neighborhoods and communities began to coalesce into a more action-oriented
form of protest identified as the "environmental justice" movement.

These two movements and the organizations associated with them have
at times worked together on specific issues. But they have also clashed over
certain specific policy agreements and goals. Several of the mainstream
environmental organizations such as the Environmental Defense Fund
have worked closely with a number of industries on air pollution and promoted the concept of market solutions, including a pollution trading program at the regional and state levels. The community-based environmental
justice groups sharply criticized this approach, arguing that purchase of
pollution "credits" could be used to continue (or even intensify) polluting
activities, for example by oil refineries in low-income communities that
suffered from the highest levels of such pollution.[13]

However, when the different wings of the environmental movements
have worked together around particular, community-identified goals, as in
the case of the ports and goods movement activities, important changes
have resulted. With the ports, environmental justice groups like East Yard
and the Coalition for a Safe Environment worked effectively with lawyers
from the Natural Resources Defense Council to bring about key policy
changes. This included the Clean Air Action Plan at the ports and the Clean
Trucks program (in conjunction with the labor-focused Clean and Safe
Ports Coalition). The environmental groups have also been effective in
advocating for and often identifying the opportunities for greener technologies as well as for the policies that promote or mandate such technologies. These actions from below have been effective in beginning to change
the environmental policy discourse and shifting the direction of policy and
regulatory bodies, such as the South Coast Air Quality Management District
(for Southern California), the California Air Resources Board (for state government), and the US Environmental Protection Agency (for the federal
government). Through pressures from the environmental justice groups

especially, these agencies have begun to incorporate the problem of differential community impacts and social equity into their policy deliberations.[14]

Despite some important community victories and policy changes, such as at the ports, Los Angeles's environmental issues and protracted problems have remained as challenging as ever. For every effort to make Los Angeles more bike-, transit-, and pedestrian-friendly, such as the Mobility 2035 plan or the CicLAvia events, cars still rule and Los Angeles remains sprawling even as a "more elegant density" is promoted. The region still has major congestion problems, and is nearly always ranked the most traffic-congested of US metropolitan areas. It has the worst air pollution in the United States, even though its air pollution levels have declined (and though other places, like the city of Houston or areas in California's Central Valley, now vie for the number-one spot). Los Angeles has felt the loss of a sense of place, even as new communities and ethnic groups have established their own hybrid-like sense of place. And industry opponents and their lobbyists and political allies continue their efforts to neutralize or roll back environmental policies and regulatory actions, even though they have lost some battles and been forced to take a more defensive posture.[15]

For every initiative to increase food access, provide healthier food, reduce the distance from farm to table, or reinvent an urban agriculture, the food system in California has continued to be dominated by its global (export- and import-oriented) focus, its industrial inputs and organization, and its fast food culture, which is still widespread despite reduced domestic profits and sales as well as menu changes by companies like McDonald's. Nevertheless, the food justice movement continues to grow, even as it seeks to extend its influence.

For all the efforts to reinvent a free-flowing Los Angeles River, reduce the Bay pollution, create new landscape ethics and decrease water consumption, and shift from building new infrastructure to focusing on how water is managed and used, changes have been slow in coming. The Los Angeles River still remains largely off limits and is managed as a storm channel, although local officials including the mayor, in response to more than thirty years of river advocacy, have committed to a river revival. A new water ethic, embracing major reductions in water use, has begun to take root and has contributed to the effective response to California's most recent drought, although a greater shift may be called upon in the expectation of more severe droughts to come. Moreover, impacts from climate change threaten to unravel any of the environmental gains described above, without a deeper focus on the land use patterns that still influence where water comes from and how it is used; without addressing how sprawl patterns are extended; and without confronting why car use, including single-driver commuters, remains the dominant transportation mode.

The actions from below—and the deliberations and actions at the policy level—have brought about important policy breakthroughs, as illustrated in this book. A Clean Air Action Plan at the ports has begun to address Los Angeles's most serious air pollution problem in the region—namely, particulate matter (or small particles) in the ambient environment. Efforts of groups like Friends of the Los Angeles River and Heal the Bay and even the "Our Malibu Beaches" app have brought an important focus on two of Los Angeles's most important water-related environmental assets. They have also helped establish policy initiatives involving stormwater runoff, public access, and water resource management. Other policy initiatives in the areas of food, transportation, the built environment, and environmental planning in its broadest sense have contributed to a changed focus on the urban environment. For Los Angeles, the importance of environmental action through change from below, the value of a continuing and deeper change of consciousness, and the need for effective policy change are as compelling as ever.

In Hong Kong, environmental action and policy initiatives have been bounded by its colonial history, its complex relationship with mainland China, and its resource constraints. After World War II and through the 1970s, protests targeted such issues as public housing conditions, a telephone tax increase, and a marriage ban for hospital nurses. Labor actions during the 1950s and 1960s, such as the 1952 tram workers' protests, and student protests during that time revealed an intense political divide regarding Hong Kong's relationship with China, associated with left-wing (pro-Maoist) and right-wing (pro-Chiang Kai-shek) factions. Riots broke out in 1956 and again in 1966 and 1967 led by pro-Maoist groups that tapped into the anger against colonial rule and identification with and support for China. Colonial authorities responded with a number of harsh measures restricting the right to protest and form organizations, and allowing the retention of protestors for up to a year without a trial. During the 1970s, some spaces for social and political action began to open up, such as mobilization by residents about housing conditions. By the early 1980s, the focus of several of the Hong Kong groups and political factions began to turn toward the transition from colonial rule to the relationship with China. This was particularly true after the crucial meeting in 1982 between Margaret Thatcher and Deng Xiaoping, at which Deng made it clear that the 1997 handover and the one country, two systems approach were non-negotiable. According to Thatcher, Deng also warned that China would take over the colony sooner if there were "very large and serious distur-bances in the next fifteen years" before the handover.[16]

During this period, there was little by way of environmental organiza-tion or action, even as some critical environmental problems, such as air pollution and lack of open space, became more prominent. In 1980, in one of the few actions, a handful of environmentalists registered their opposi-tion to a plan for the development of a nuclear power plant at Daya Bay, about fifty kilometers from central Hong Kong. Opposition to the plan, however, remained minimal until 1986 when the Chernobyl nuclear power plant accident raised fears about Hong Kong's vulnerability to an accident so close to the city. Before 1986, China had had only a limited development of nuclear power (six domestic plants tied to the military) and had resisted efforts to bring in foreign nuclear power developers and their expensive technologies. But with the emerging industrial development in Guangdong in the 1980s and the creation of the SEZ at Shenzhen, the Chinese govern-ment decided to move forward with plans for the $3.7 billion Daya Bay facility, using nuclear technology from French and British companies. Of the 1,800 megawatts of power to be generated, 70 percent was to be ear-marked for Hong Kong. In March 1986, the mainland government and the developers signed a letter of intent, again with only limited dissent in Hong Kong. Then, on April 26, the Chernobyl nuclear plant accident transformed the debate and elevated fears in Hong Kong about safety issues. Within weeks, a massive protest movement emerged that included as many as 107 organizations which formed the Joint Conference for the Shelving of the Daya Bay Nuclear Plant. Among other protest actions, one million signa-tures were gathered on a petition that asked for a change in the plans, most prominently a desire to relocate the plant away from Hong Kong.[17]

The primary focus of the protests against the Daya Bay nuclear plant was safety, including mistrust of China's technical capabilities for managing the facility. As awareness of Hong Kong's vulnerability increased, it also helped forge a stronger sense of an independent Hong Kong identity, both with respect to the British colonial government and Hong Kong's relationship to China. A broader environmental critique of nuclear technology (and Hong Kong's own limited resource capacities, whether for energy, water, or food) failed to take root. Instead, environmental considerations tended to be subsumed within the emerging democracy movements, as the 1997 deadline for the transfer to China came closer. The Chinese government ignored the protests, refused to consider a plant relocation, and moved ahead with construction of the Daya Bay facility. China has subsequently expanded its nuclear footprint to include three other nuclear plants in operation or on the drawing board in Guangdong, including a plant just one kilometer north of the Daya Bay facility.[18]

The Daya Bay 1986 protests pointed to the importance of the connection to China for social movements in Hong Kong through much of the 1980s and 1990s. That focus was intensified by the 1989 Tiananmen events, which inspired a huge demonstration in Hong Kong. Although tensions about the China transition dominated the political discourse, some of Hong Kong's environmental groups took a less confrontational approach to their issues, such as seeking cooperative relationships with Hong Kong's business elites that played a central role in shaping Hong Kong's policies during the transition to the one country, two systems framework. As distinct from the pro-democracy groups, environmental organizations tended to be apolitical, focused as much on a "green lifestyle" approach as on a push for policy and structural change.[19]

That dynamic began to shift again as Hong Kong's environmental problems such as air pollution became more prominent and as environmental and grassroots groups began to focus on the Hong Kong–China connections around such issues as food, water, and transportation. Hong Kong-based groups such as China Water Risk and the food-oriented Partnerships for Sustainable Development complemented the Hong Kong–China focus of the international NGOs such as Greenpeace East Asia (which was launched in 1997 and initially focused on the energy mix of Hong Kong's largest utility, China Light and Power). At the same time, local grassroots environmental protests and community actions in Hong Kong, focused on issues ranging from port-related pollution to the displacement of an agricultural village in the New Territories to make way for a high-speed rail link between Hong Kong and Guangzhou, provided a more activist orientation that complemented the policy focus of groups such as the research think tank Civic Exchange. With the appointment of Christine Loh as Undersecretary for the Environment, policy innovation and change regarding the environmental problems facing Hong Kong seemed more possible, with the recognition that Hong Kong's connection to China needed to be part of any environmental agenda.

In China, a range of environmental groups, including NGOs, international groups and funders, issue-based advocates, and local ad hoc groups, have formed around particular environmental impacts. Each of these has played an important role in elevating environmental issues at the policy level, in relation to impacts from particular facilities, and also as part of the public discourse. Even as that role has been constrained due to the organizational restrictions posed by regulatory policies and by the generally apolitical nature of much of the organized environmental activity in China, some local protests have involved thousands of participants and brought about specific, even if limited, changes.

The government's attitude and approach toward social movements has been a key factor in the evolution of Chinese environmentalism. In 1998, Premier Zhu Ronji issued an order through the State Council that stipulated a set of "Regulations on the Registration and Management of Social Organizations" for any nonprofit organization, including environmental organizations. Article 4 of the order sought to insure that groups would not "damage national interests [and] the interests of society," and would not "violate the prevailing social morals."[20] Discussions by state authorities on how to manage the proliferation of NGOs and other civil society organizations had intensified in the wake of the 1989 Tiananmen events. Prior to 1989, environmental organizations, including environmental NGOs (ENGOs), were small in number and tended to be more research- and education-oriented. That situation began to change after the mid-1990s, as environmental problems became increasingly apparent. By the end of 1999, more than 1,800 national and 165,000 regional organizations had already registered, although environmental organizations constituted a small percentage of that pool. By 2012, however, environmental issues—and environmental protests—had become a critical part of the political landscape. That year was described by *China Dialogue* as "China's year of environmental protests," citing research that identified the growth of environmental protests at an annual rate of 29 percent since the mid-1990s.[21]

Among local protests, the spontaneous and sometimes explosive demonstrations against the siting of paraxylene (PX) petrochemical production plants have been notable. Protests against the production of PX, which is used in the manufacture of plastic bottles and polyester, first occurred in Xiamen in 2007 in Fujian Province, and led to a decision by the provincial and municipal governments to relocate the proposed plant. Anti-PX protests then spread to a number of other places around the country, including Dalian, Chengdu, Ningbo, Kumming, Maoming (in Guangdong Province), and Shanghai. The large Maoming and Shanghai protests of 2014 and 2015 were met with police repression that only intensified opposition to the proposed plants. The Maoming project was also noteworthy in that the city, described as the "oil town of the south," was already host to a huge number of petrochemical plants, including the production center for Sinopec, the state-owned oil and petrochemical company. Thanks to social media, information about previous anti-PX protests, including the successful one in Xiamen, and long-standing health concerns about cumulative impacts fed the local protest movement. Similarly in Shanghai, the proposal to shift a PX plant from the Pudong development zone to a chemical industrial zone in the Jinshin district of the city generated huge protests over concerns about the impacts from paraxylene production in an area already subject to cumulative hazards.[22]

Despite the steady stream of these types of events, environmental NGOs had little or no role in the local protests. Moreover, the complex set of regulations that had been put in place in 1998 and their strong bias toward protection of government interests and the broader social order (as well as several subsequent regulations) had limiting effects on China's organized environmental groups. On the one hand, it led many organizations to seek to avoid the registration process itself by identifying themselves as businesses or some other private or commercial enterprise. More importantly, it created constraints, both formal and informal, on the type of environmental advocacy carried out by many of the environmental organizations, whether registered or not. Some NGOs were identified as extensions of the government (GONGOs, or government-organized and operated NGOs). Others intentionally sought to avoid political controversy by portraying themselves as partners rather than opponents to government, for example by lobbying the central authorities and avoiding more controversial environmental issues. Already by 2000, the apolitical and nonconfrontational nature of many of the environmental NGOs came to be characterized as "female mildness" by the press and stood in contrast to the more confrontational protests about local issues directed especially against local industries and their allies in local government.[23]

Despite the widespread concern about environmental issues, the environmental NGOs in China have played a relatively minor role in building an organized and nationally based social movement. Groups that have had some of the biggest impacts, such as Ma Jun's Institute of Public and Environmental Affairs (IPEA), have made transparency about environmental impacts a major issue (such as through IPEA's China Pollution Maps) and have helped increase the government's reporting and monitoring, including real-time information.[24] Some protracted protests, such as the struggle against the PX petrochemical plants, have led to important, even if limited and localized, environmental victories. Perhaps the most important changes have been the development of several national and globally oriented environmental policies. These include the decision to begin to shift away from coal and achieve reduced carbon emissions by 2030; the release of a strengthened Environmental Protection Law (the first revisions since 1989); the strengthening of environmental regulations in such areas as air and water pollution; and the change in government language, which has included writing "ecological civilization" into China's constitution and calling it part of the "Chinese dream." But policy change only becomes effective when implemented at the local level, and the failure to do so has continued to be a problem that remains widespread.[25]

Today, negative environmental impacts in China, including in the largest urban centers, still remain incredibly visible and life-threatening. The danger, for the government, is that any environmental advocacy runs the risk of turning into a deeper, more political focus that challenges the very heart of government policy in the reform era—its developmentalism and marketization strategies that this book has explored. By 2014–2015, new and more severe restrictions on both international and domestic NGOs were put in place that could be interpreted widely for any activities seen as "subverting state power," "spreading rumors," and/or damaging "the national interest or society's public interest." Because they are currently apolitical, unable to operate at a national level, and subject to operational constraints, environmental social movements in China have yet to realize their full potential.[26]

A "Right to the City": A Conclusion

This book originated in discussions between urban environmental researchers and activists based in Hong Kong and Los Angeles about the environmental impacts from global trade, from the ports of Hong Kong and Los
Angeles/Long Beach, and from the systems of production and distribution
that make the connection to China prominent in addressing such impacts.
We decided to turn those discussions into a book project in part so we could
explore whether and how the conditions in these two global cities and their
counterparts in China are similar and how they are different, noting the
importance of social and environmental change and how the agendas for
change in these different places might be shared.

Environmental impacts are among the most critical issues in Los Angeles, Hong Kong, and China. They reinforce such problems as income disparities and immigrant abuses, and they can deeply affect daily life. Even
the differences in each of these places are noteworthy in understanding the
history, advocacy, policy interventions, and changes (or barriers to change)
related to such environmental issues as the air we breathe, the water we use,
the food we eat, the spaces we occupy, and how we get from one place to
the next. These environmental issues are both global and local in their origins and in how they are experienced. Climate change, the most far-
reaching and potentially devastating global issue, cannot be separated from
each of those issues we have explored and how each is rooted in local,
national, and increasingly international decisions.

The generations that have come of age since 1997 in Hong Kong, or
since the 1990s in Los Angeles, or since the Reform period in China, have
all begun to assert in some capacity their "right to the city," to borrow
Henri Lefebvre's term.[27] This concept suggests the right to participate in
actively reshaping urban places and the global city. But right to the city
advocacy remains an incomplete process. The political demands of the
Occupiers in Hong Kong were about electoral reform, not about reshaping
the places where they live, work, and play, even though many of the protestors have identified the environment and related issues of social and economic justice as issues of concern. The Occupiers and the bike and
pedestrian activists in Los Angeles (who rallied together for a brief moment
near the steps of City Hall in 2011) have not extended their arguments to
advocate for such social, economic, and environmental justice issues as
homelessness, real estate development and gentrification, or unsafe and

highly polluted streets and freeways in low-income communities. And the environmental protests in China have not extended to a call for bottom-up public planning to address urban environmental impacts for more transformative changes in urban and rural environments or for an equity agenda and political set of goals.

By calling for changing the conditions of daily life and creating more livable and just urban places, these various advocates have nonetheless begun to trace the outlines of a broad critique of a political economy which prioritizes development and profits over human well-being. During the Occupy events in Hong Kong and Los Angeles, many protesters said that despite their perception of life in Hong Kong or Los Angeles as soulless and stressful, their experience at the protest site gave them a renewed sense of community. The spontaneous protests that have taken place in China in factories, villages, or around air pollution or chemical plants have created temporary communities, even if they lack a larger regional or national agenda.

Much of this activism, at this point, tends to be reactive and temporary. Protests often fail to address the broader context for change. They may raise the cost of government inaction and even help identify specific policy changes, but they have yet to directly shape broader targets for policy change such as the role of industries, financial interests, real estate and land speculation, or other market forces in shaping the urban environment of the global city.

This book has argued that beyond any immediate conflict or single environmental problem, approaches need to be developed with a capacity to bring about what André Gorz has characterized as "structural reform."[28] Such approaches would involve changes that could lead to opportunities for a deeper, more systemic transformation as well as a change in consciousness among the participants engaged in any particular struggle.

This book also reflects our own action research agenda. We have come to better understand more of the details and the enormity of the environmental problems that our respective global cities are confronting and the history and intricacies as well as limits and opportunities of any environmental change agenda. If the end point of that agenda is constructing an "ecological civilization," it would need to be not just a dream for China, or an aspiration for Los Angeles or Hong Kong, but a change at both the global and local level. Such a change may still seem distant, but its need is immediate.

Notes

Chapter 1

1. Peter Hall, *The World Cities* (New York: McGraw-Hill, 1966).

2. John Friedmann, "The World City Hypothesis," *Development and Change* 17 (1986), 70.

3. In his discussion of neoliberalism, David Harvey distinguishes between its theoretical and practical applications. According to Harvey, the classic nineteenth-century definition of liberalism, which dated back to Adam Smith and John Locke, argued that "individual liberty and freedom can best be protected and achieved by an institutional structure, made up of strong private property rights, free markets, and free trade: a world in which individual initiative can flourish." But the practice of neoliberalism, which emerged in the 1970s with the formation of groups like the Business Roundtable and attacks against the welfare state, environmental laws and regulations, and a more active public sector, promoted instead privatization, limited government intervention, and a dismantling of many welfare state and social safety net programs. It reached its political consolidation with the elections of Reagan and Thatcher. See Sasha Lilley, "An Interview with David Harvey," *MR Zine* (*Monthly Review*), June 19, 2006, http://mrzine.monthlyreview.org/2006/lilley190606.html; see also David Harvey, *A Brief History of Neoliberalism* (Oxford: Oxford University Press, 2007). For the Thatcher quote, see "Margaret Thatcher: A Life in Quotes," *Guardian*, April 8, 2013, http://www.theguardian.com/politics/2013/apr/08/margaret-thatcher-quotes.

4. Michael J. Enright, Edith E. Scott, and Ka-mun Chang, *Regional Powerhouse: Greater Pearl River Delta and the Rise of China* (Singapore: John Wiley & Sons [Asia], 2005). Hong Kong began to emphasize its designation as Asia's world city in part to trump the growing role of Shanghai, perceived to be its main competitor in the Asia-Pacific region. See John Rennie Short, *Global Metropolitan: Globalizing Cities in a Capitalist World* (London: Routledge, 2004), 23; Friedmann, "The World City Hypothesis," 72.

5. Saskia Sassen, *The Global City: New York, London, Tokyo* (Princeton: Princeton University Press, 2001), 3–4.

6. Reyner Banham, *Los Angeles: The Architecture of Four Ecologies* (Berkeley: University of California Press, 2009), 195.

7. Robert Gottlieb and Margaret FitzSimmons, *Thirst for Growth: Water Agencies as Hidden Government in California* (Tucson: University of Arizona Press, 1989).

8. John McPhee, "Los Angeles against the Mountains," *New Yorker*, September 26, 1988, http://www.newyorker.com/magazine/1988/09/26/los-angeles-against-the-mountains-i; Mike Davis, "The Case for Letting Malibu Burn," *Environmental History Review* 19, no. 2 (Summer 1995), 1–36.

9. David Harvey, "Possible Urban Worlds," in Steef Buijs, Wendy Tan, and Devisari Tunas, eds., *Megacities: Exploring a Sustainable Future* (Rotterdam: 010 Publishers, 2010), 168.

10. James Rojas, "Latino Urbanism in Los Angeles: A Model for Urban Improvisation and Reinvention," in Jeffrey Hou, ed., *Insurgent Public Space: Guerrilla Urbanism and the Remaking of Contemporary Cities* (London: Routledge, 2010), 36–44.

11. Robert Gottlieb, Regina Freer, Mark Vallianatos, and Peter Dreier, *The Next Los Angeles: The Struggle for a Livable City* (Berkeley: University of California Press, 2006), xviii.

12. Christine Loh, *Nimble and Nifty: Transforming Hong Kong* (Hong Kong: CLSA, 2002).

13. Y. C. Jao, "The Rise of Hong Kong as a Financial Center," *Asian Survey* 19, no. 7 (July 1979), 674.

14. Manuel Castells, Lee Goh, and R. Yin-Wang Kwok, *The Shek Kip Mei Syndrome: Economic Development and Public Housing in Hong Kong and Singapore* (London: Pion, 1990).

15. Crystal Tse, "Hong Kong Is Slowly Dimming Its Neon Glow," *New York Times*, October 13, 2015, http://www.nytimes.com/2015/10/14/world/asia/hong-kong-neon-sign-maker.html. Tse points out that the volume of neon signs began to decrease by the turn of the twenty-first century and that the lights also dimmed, with LED replacing neon. Nevertheless, Hong Kong's reputation for its twenty-four-hour bright lights across its tall towers has remained, similar to Los Angeles's own "Hollywood Nights" with its sea of neon-lit billboards along Sunset Boulevard.

16. Hong Kong Government, *White Paper: Pollution in Hong Kong—A Time to Act* (Hong Kong: Government Printer, 1989).

17. China has seven of the ten largest container ports in the world since 2012, including Shanghai, Shenzhen, Hong Kong, and Guangzhou, but Hong Kong's ranking slipped to no. 5 in 2015. Kenneth Rapoza, "The World's Ten Busiest Ports,"

Forbes, November 14, 2014, http://www.forbes.com/sites/kenrapoza/2014/11/11/the-worlds-10-busiest-ports/.

18. One interesting connection involves China's role in Los Angeles and California's interest in high-speed rail. The state-owned China Railway in 2015 established a joint venture with a US group to develop a Los Angeles to Las Vegas high-speed railway. China has also sought to become a supplier of some of the trains that would be used for an ambitious (and controversial) California high-speed rail system that is several years from construction. See "China, US Reach Agreement on High Speed Rail before Xi's Visit," *Bloomberg News*, September 16, 2015, http://www.bloomberg.com/news/articles/2015-09-17/china-u-s-reach-agreement-on-high-speed-rail-before-xi-visit.

19. Hugo Martin, "Los Angeles County Reports Record 45.5 Million Visitors in 2015," *Los Angeles Times*, January 11, 2016, http://www.latimes.com/business/la-fi-los-angeles-visitors-for-2015-20160111-story.html; Ferdinando Guerra, "Growing Together: China and Los Angeles County," Los Angeles Economic Development Corporation and Kyser Center for Economic Research, June 2014, http://laedc.org/wp-content/uploads/2014/05/2014-Growing-Together-China-and-LA-County.pdf.

20. The urban growth in the Pearl River Delta region has been breathtaking. Between 1990 and 2000, more than 40 million people, primarily immigrants from China's poorer provinces, settled in the PRD. Growth rates since then have slowed, but the population continues to climb, with an estimated 100 million-plus in the province as a whole. Wendell Cox, "China's Shifting Growth Patterns," *New Geographies*, April 23, 2015, http://www.newgeography.com/content/004904-chinas-shifting-population-growth-patterns.

Chapter 2

1. The idea of the doll's journey as representative of the goods movement system was first introduced by Andrea Hricko from the University of Southern California in 2005 and has subsequently been used by community activists and researchers as a way to characterize its components and its impacts. See Paul Welitzkin, "China's Toy Story," *China Daily Asia*, March 11–13, 2016; see also Rone Tempest, "Barbie and the World Economy," *Los Angeles Times*, September 22, 1996, http://articles.latimes.com/1996-09-22/news/mn-46610_1_hong-kong.

2. Joan Magretta, "Fast, Global, and Entrepreneurial: Supply Chain Management, Hong Kong Style: An Interview with Victor Fung," *Harvard Business Review* (September-October 1998), 103–114, http://hbr.org/1998/09/fast-global-and-entrepreneurial-supply-chain-management-hong-kong-style/ar/1.

3. Choi's toy manufacturing business was one of the first Hong Kong companies to relocate to Guangdong and establish a system of small subcontractors feeding its

large manufacturing facilities. The company took a hit when one of its subcontractors was found responsible in 1978 for using lead paint on one of the toy products earmarked for Mattel. But Choi escaped any serious impact when he laid plans to establish his own factory to handle some of his supplier needs. See Shu-Ching Jean Chen, "Choi's Toys," *Forbes*, January 18, 2008, http://www.forbes.com/global/2008/0128/026.html. See also "Hong Kong's 50 Richest: #13 Francis Choi," *Forbes* (as of January 2014), http://www.forbes.com/profile/francis-choi/. Early Light Industrial Co., Ltd., is a subsidiary of Early Light International (Holdings) Ltd.; see the company website at http://www.earlylight.com.hk/en/manufacturing.html; and Daniel Poon, "Toy Industry in Hong Kong," *HKTDC* [*Hong Kong Trade Development Council*] *Research*, February 10, 2014, http://hong-kong-economy-research.hktdc.com/business-news/article/Hong-Kong-Industry-Profiles/Toy-Industry-in-Hong-Kong/hkip/en/1/1X000000/1X001DGH.htm.

4. Andrea Hricko, "Ships, Trucks and Trains: Effects of Goods Movement on Environmental Health," *Environmental Health Perspectives* 114, no. 4 (April 2006), A204–A205; Ferdinando Guerra, "Growing Together: China and Los Angeles County," Los Angeles Economic Development Corporation, June 2014, http://laedc.org/wp-content/uploads/2014/05/2014-Growing-Together-China-and-LA-County.pdf; Joseph Bonney, "Port Woes Threaten Toy Importers with Unhappy Holidays," *Journal of Commerce*, October 21, 2014, http://www.joc.com/port-news/us-ports/port-los-angeles/port-woes-threaten-toy-importers-unhappy-holidays_20141021.html.

5. Testimony of Dr. Geraldine Knatz, executive director of the Port of Los Angeles, regarding "The Marine Vessel Emissions Reduction Act of 2007," before the US Senate Committee on Environment and Public Works, August 9, 2007.

6. On the "maritime world economy," see Carolyn Cartier, "Cosmopolitics and the Maritime World City," *Geographical Review* 89, no. 2 (April 1999), 278–289.

7. Hong Kong Special Administrative Region, Marine Department, "Port of Hong Kong Statistical Tables," 2015, http://www.mardep.gov.hk/en/publication/pdf/portstat_ast_2014.pdf.

8. Hong Kong Special Administrative Region, Census and Statistics Department, *Hong Kong Annual Digest of Statistics* (2015).

9. Hong Kong Special Administrative Region Government, "The Port of Hong Kong—Today and Tomorrow," press release, August 21, 2012, http://www.info.gov.hk/gia/general/201208/21/P201208210154.htm.

10. According to a dispatch from Lord Stanley, Secretary of State for War and the Colonies, to Sir Henry Pottinger, governor of Hong Kong (June 1843–May 1844), dated June 3, 1843, Hong Kong was occupied "for diplomatic, commercial and military purposes." Quoted in George B. Endacott, *A History of Hong Kong*, 2nd ed. (London: Oxford University Press, 1964), 255.

11. Ibid., 194.

12. David R. Meyer, *Hong Kong as a Global Metropolis* (Cambridge: Cambridge University Press, 2000), 115–116.

13. Endacott, *A History of Hong Kong*, 253.

14. Hong Kong Government Information Services, *Hong Kong 1957* (1958), 7.

15. Baruch Boxer, "Ocean Shipping in the Evolution of Hong Kong," research paper no. 72, University of Chicago, Department of Geography, 1961, 61.

16. Ross Robinson, "Size of Vessels and Turn-around Time: Further Evidence from the Port of Hong Kong," *Journal of Transport, Economics, and Policy* 12, no. 2 (May 1978), 161–178, http://www.bath.ac.uk/e-journals/jtep/pdf/Volume_X11_No_2_161 -178.pdf.

17. Hugh Farmer, "The Development of Containerization at the Port of Hong Kong," *Industrial History of Hong Kong Group*, June 30, 2014, http://industrialhistoryhk .org/development-containerization-port-hong-kong/; Tzu-nang Chiu, *The Port of Hong Kong: A Survey of Its Development* (Hong Kong: Hong Kong University Press, 1973).

18. Dong-Wook Song, "Regional Port Competition and Cooperation: The Case of Hong Kong and South China," *Journal of Transport Geography* 10 (2002), 99–110; Alexander McKinnon, "Hong Kong and Shenzhen Ports: Challenges, Opportunities, and Global Competitiveness," Hong Kong Centre for Maritime and Transportation Law, City University of Hong Kong, Working Paper Series, 2011, http://www.cityu .edu.hk/slw/HKCMT/Doc/Working_Paper_3_-_Shenzhen_-_Final_(v6).pdf; Hutchison Port Holdings Limited website, accessed June 27, 2016, https://www.hph.com/ en/globalbusiness/gbpage-1.html#.

19. Richard C. M. Yam and Esther P. Y. Tang, "Transportation Systems in Hong Kong and Southern China: A Manufacturing Industries Perspective," *International Journal of Physical Distribution and Logistics Management* 26, no. 10 (1996), 46–59, http://202.116.197.15/cadalcanton/Fulltext/20977_2014318_11419_6.pdf.

20. Hong Kong Special Administrative Region, Information Services Department, "The Port," Hong Kong fact sheet, 2015, http://www.gov.hk/en/about/abouthk/ factsheets/docs/port.pdf.

21. World Shipping Council, "Top 50 World Container Ports" (2011–2014 numbers), accessed June 27, 2016, http://www.worldshipping.org/about-the-industry/ global-trade/top-50-world-container-ports; Joanne Chiu, "Hong Kong's Port Is Caught in a Storm," *Wall Street Journal*, February 16, 2016, http://www.wsj.com/ articles/hong-kong-ports-steady-drop-persists-1455654601.

22. Hong Kong Special Administrative Region, Census and Statistics Department, "Hong Kong Merchandise Trade Statistics—Domestic Exports and Re-Exports:

Table 1: Annual Value of Merchandise Trade," December 2015, 17, http://www
.statistics.gov.hk/pub/B10200032015MM12B0100.pdf.

23. Port of Los Angeles, "About the Port," accessed June 27, 2016, http://www
.portoflosangeles.org/idx_about.asp; Port of Long Beach, "Facts at a Glance,"
accessed June 27, 2016, http://www.polb.com/about/facts.asp; Grace M. Lavigne,
"Port Congestion Worse at LB than LA, CargoSmart Says," *Journal of Commerce*,
November 4, 2014, http://www.joc.com/port-news/us-ports/port-los-angeles/port-
congestion-worse-lb-la-cargosmart-says_20141104.html; Bill Mongelluzzo, "Megaships Dealing Worst Congestion Hand to LA-LB, NY-NJ," *Journal of Commerce*, July 1,
2015, http://www.joc.com/port-news/us-ports/port-new-york-and-new-jersey/mega
-ships-dealing-worst-congestion-hand-la-lb-ny-nj_20150701.html; Chris Kirkham
and Andrew Khouri, "Bigger Ships, Bigger Problems," *Los Angeles Times*, June 2,
2015, http://www.latimes.com/business/la-fi-big-ships-ports-20150602-story.html.

24. Clarence Matson, "The Port of Los Angeles: Historical," in *The Story of Los Angeles Harbor: Its History, Development, and Growth of Its Commerce* (Los Angeles: Department of Foreign Commerce and Shipping, Los Angeles Chamber of Commerce,
1935), 1.

25. Ibid., 10.

26. Marc Levinson, in his history of containerization, argues that containers essentially turned ports into "mere 'load centers,' places through which large amounts of
cargo flowed with hardly a break." Marc Levinson, *The Box: How the Shipping Container Made the World Smaller and the World Economy Bigger* (Princeton: Princeton
University Press, 2006), 236.

27. Tom Bradley's comment is in Steven Erie, *Globalizing L.A.: Trade, Infrastructure,
and Regional Development* (Stanford: Stanford University Press, 2004), 92.

28. Dan Weikel, "Freighters Enter the Age of the Mega-ship," *Los Angeles Times*, June
15, 1999.

29. The "trading center of the world" comment was made by Mark Pisano of the
Southern California Association of Governments in 1998, based on anticipated
infrastructure development. Dan Weikel, "L.A., Long Beach Ports Will See More
Cargo Volume," *Los Angeles Times*, December 18, 1998, http://articles.latimes
.com/1998/dec/18/business/fi-55161.

30. Bill Mongelluzzo, "The Port Moves Inland," *Journal of Commerce*, September 13,
2010; Chris Kirkham, "Inland Empire Sees Surge in Warehouse Jobs, but Many Are
Low-Pay, Temporary," *Los Angeles Times*, April 17, 2015, http://www.latimes.com/
business/la-fi-inland-empire-warehouses-20150419-story.html#page=1.

31. Remarks by Zhou Wong of the Anhui Leadership Delegation at Occidental College, September 17, 2007.

32. United States Department of Agriculture Agricultural Marketing Service, "Impact of Panama Canal Expansion on the U.S. Intermodal System," January 2010, https://www.ams.usda.gov/sites/default/files/media/Impact%20of%20Panama%20Canal%20Expansion%20on%20the%20U.S.%20Intermodal%20System.pdf; Natalie Kittroeff, "L.A. Ports Face a Sea of Rivals," *Los Angeles Times*, April 27, 2016; Martha Matsuoka et al., "Global Trade Impacts: Addressing the Health, Social, and Environmental Consequences of Moving Freight through Our Communities," Occidental College and University of Southern California, March 2011, http://scholar.oxy.edu/cgi/viewcontent.cgi?article=1410&context=uep_faculty.

33. HKND is wholly owned by a young Chinese telecom industry billionaire named Wang Jing, who had also invested in Nicaragua's telecom infrastructure. Wei Tian, "Logistics Project of the Century," *China Daily*, July 5, 2013, http://africa.chinadaily.com.cn/weekly/2013-07/05/content_16735118.htm; Tim Maverick, "Nicaragua Canal: China's Boondoggle or Brilliant Strategy," *Wall Street Daily*, January 24, 2015, http://www.wallstreetdaily.com/2015/01/24/nicaragua-canal/; Michael D. McDonald, "China's Building a Huge Canal in Nicaragua, but We Couldn't Find It," *Bloomberg Business*, August 19, 2015, http://www.bloomberg.com/news/articles/2015-08-19/china-s-building-a-huge-canal-in-nicaragua-but-we-couldn-t-it; "China Eyes Cheniere's LNG," *Oil and Gas 360*, May 22, 2015, http://www.oilandgas360.com/china-eyes-chenieres-lng/.

34. A number of the ships that had used the Panama Canal (known as Panamax ships), were also being scrapped in favor of a new generation of post- or "neo-Panamax" ships, which increased capacity from 4,500 to 13,200 containers and measured 1,400 feet in length. Despite delays in expanding the Panama Canal, the southern and East Coast US ports still moved aggressively to expand their own port capacity, with ten of the ports prepared to spend more than $11 billion to deepen dredging, build much higher cranes, and upgrade their terminals. Dan Molinski, "Ports, Shipping Companies Retool before Panama Canal Expansion," *Wall Street Journal*, February 17, 2014; see also Walt Bogdanich et al., "The New Panama Canal: A Risky Bet," *New York Times*, June 22, 2016, http://www.nytimes.com/interactive/2016/06/22/world/americas/panama-canal.html?_r=0. Andrea Hricko, "Progress and Pollution: Port Cities Prepare for the Panama Canal Expansion," *Environmental Health Perspectives* (December 2012), 470–473, http://ehp.niehs.nih.gov/120-a470/.

35. *Hong Kong Annual Digest of Statistics* (2015), 315.

36. The number of container trucks crossing the border increased almost fourfold between 1991 and 2001. To give a sense of the traffic volume, there were about 12,100 container trucks and some 13,000 other goods vehicles passing the border checkpoints every day in 2001. Data gathered from Hong Kong Special Administrative Region, Transport Department, *Monthly Traffic and Transport Digest*, December 2001, http://www.td.gov.hk/en/transport_in_hong_kong/transport_figures/monthly_traffic_and_transport_digest/200112/index.html.

37. The idea of a Port Rail Line was first suggested in a Freight Transport Study published by the Transport Department in 1994. It was mentioned again in the Second Rail Development Strategy, published in 2000 by the Transport Bureau.

38. This is known as the McClier Report, published in 2001 by a team led by the Chicago-based McClier Corporation. See Port and Maritime Board, "Competitive Strategy and Master Plan to Strengthen Hong Kong's Role as the Preferred International and Regional Transport and Logistics Hub," 2001, http://www.logisticshk .gov.hk/board/bes.pdf.

39. Bill Barron, Simon Ng, Christine Loh, and Richard Gilbert, "Sustainable Transport in Hong Kong: Directions and Opportunities," Civic Exchange, Hong Kong, 2002.

40. Caitlin Gall and Marcos Van Rafelghem, "Marine Emission Reduction Options for Hong Kong and the Pearl River Delta Region," Civic Exchange, Hong Kong, 2006.

41. A special feature of the San Pedro Bay Ports Clean Air Action Plan was covered in another Civic Exchange publication: Marcos Van Rafelghem and Rob Modini, "Lessons for Hong Kong: Air Quality Management in London and Los Angeles," Civic Exchange, Hong Kong, 2007. See also Simon K. W. Ng et al., "Marine Vessel Smoke Emissions in Hong Kong and the Pearl River Delta," final report, Hong Kong University of Science and Technology, Atmospheric Research Center, March 2012.

42. Simon Ng et al., "Study on Marine Vessels Emission Inventory," Institute for the Environment, Hong Kong University of Science and Technology, final report submitted to the Hong Kong Special Administrative Region Environmental Protection Department, February 2012; Grigory Kravtsov, "Ship Emissions Blamed for Worsening Pollution in Hong Kong," CNN, December 20, 2013, http://www.cnn .com/2013/12/19/world/asia/hong-kongs-worsening-ship-pollution/.

43. Mike Kilburn, Simon Ng, et al., "A Price Worth Paying: The Case for Controlling Marine Emissions in the Pearl River Delta," Civic Exchange, September 2012. See also Hak-Kan Lai et al., "Health Impact Assessment of Measures to Reduce Marine Shipping Emissions," final report, University of Hong Kong, Department of Community Medicine, School of Public Health, July 2012.

44. Zheng Wan et al., "Pollution: Three Steps to a Green Shipping Industry," Nature 530, no. 7590 (February 17, 2016), http://www.nature.com/news/pollution-three -steps-to-a-green-shipping-industry-1.19369; Veronica Booth and Christine Loh, "Reducing Vessel Emissions: Science, Policy and Engagement in the Hong Kong– Pearl River Delta Region," Civic Exchange, Hong Kong, April 2012.

45. International Agency for Research on Cancer, "Diesel Engine Exhaust Carcinogenic," June 12, 2012, http://www.iarc.fr/en/media-centre/pr/2012/pdfs/pr213_E .pdf; T. Tran Hein et al., "Methodology for Estimating Premature Deaths Associated

with Long-Term Exposure to Fine Airborne Particulate Matter in California: Staff Report," California Environmental Protection Agency Air Resources Board, December 7, 2009, http://www.arb.ca.gov/research/health/pm-mort/pm-mortdraft.pdf.

46. Estimates about increases in truck traffic have varied considerably, and partly reflect economic conditions in China and the US, as well as the political maneuvering by goods movement advocates about the need to expand I-710's capacity. California Department of Transportation (Caltrans), "Technical Memorandum: I-710 Corridor Project EIR/EIS Travel Demand Modeling Methodology," February 26, 2010, http://www.dot.ca.gov/dist07/resources/envdocs/docs/710corridor/docs/I-710_Travel%20Demand%20Modeling_Report_Revise_Feb262010_FINAL.pdf; Dan Weikel, "Two Options Considered for Reconstructing Part of Congested 710 Freeway," *Los Angeles Times*, March 17, 2015, http://www.latimes.com/local/california/la-me-california-commute-20150317-story.html.

47. East Yard Communities for Environmental Justice, "I-710 Corridor Project: Community Alternative 7 (CA 7)," accessed June 27, 2016, http://eycej.org/campaigns/i-710/. For a discussion of highway expansion and increased rather than reduced congestion, see for example the US Department of Transportation Federal Highway Administration, "Induced Traffic: Frequently Asked Questions," accessed June 27, 2016, https://www.fhwa.dot.gov/planning/itfaq.cfm#q3.

48. See the Final Environmental Impact Report, Southern California International Gateway Project, accessed June 27, 2016, http://www.portoflosangeles.org/EIR/SCIG/FEIR/feir_scig.asp; and the Intermodal Container Transfer Facility (ICTF) Joint Powers Authority website at http://www.ictf-jpa.org/; Ryan Zummallen, "A Call for Zero Emissions as Railroads Seek to Expand," *Long Beach Post*, January 26, 2011, http://lbpost.com/news/11004-a-call-for-zero-emissions-as-railyards-seek-to-expand.

49. California Air Resources Board, "ARB Health Risk Assessment Guidance for Rail Yard and Intermodal Facilities," 2006, http://www.arb.ca.gov/railyard/hra/1107hra_guideline.pdf; Andrea Hricko et al., "Global Trade, Local Impacts: Lessons from California on Health Impacts and Environmental Justice Concerns for Residents Living near Freight Yards," *International Journal of Environmental Research and Public Health* 11 (2014), 1914–1941.

50. Wolfgang Babisch, "Transportation Noise and Cardiovascular Risk: Updated Review and Synthesis of Epidemiological Studies Indicate that the Evidence Has Increased," *Noise and Health* 8, no. 30 (2006), 1–29; P. Lercher et al., "Ambient Neighborhood Noise and Children's Mental Health," *Occupational and Environmental Medicine* 59, no. 6 (June 2002), 380–386, http://www.ncbi.nlm.nih.gov/pmc/articles/PMC1740306/; M. Sorensen et al., "Road Traffic Noise and Stroke: A Prospective Cohort Study," *European Heart Journal* 32, no. 6 (March 2011), 737–744.

51. Jim Steinberg. "2015 a Big Year for Warehouse Development in the Inland Empire," *San Bernardino Sun*, June 6, 2015, http://www.sbsun.com/business/

20150606/2015-a-big-year-for-warehouse-development-in-the-inland-empire.   See also Jones Lang Lasalle, "Perspectives on the Global Supply Chain: The Emergence of the Inland Port," Spring 2011, http://www.us.jll.com/united-states/en-us/ Research/The%20emergence%20of%20the%20Inland%20Port.pdf; Penny Newman, "Inland Ports of Southern California: Warehouses, Distribution Centers, Intermodal Facilities, Impacts, Costs and Trends," Jurupa Valley, CA; Center for Community Action and Environmental Justice, 2012, http://caseygrants.org/wp-content/ uploads/2012/08/Inland+Ports+of+Southern+California+-+Wareshouses+Distributio n+Centers+and+Intermodal+Facilities+-+Impacts+Costs+and+Trends.pdf.

52. See the Walmart website, http://careers.walmart.com/career-areas/transportation -logistics-group/distribution-center/, accessed July 29, 2016.

53. Jim Steinberg, "2015 a Big Year for Warehouse Development in the Inland Empire," *San Bernardino Sun*, June 6, 2015, http://www.sbsun.com/business/ 20150606/2015-a-big-year-for-warehouse-development-in-the-inland-empire; Erica E. Phillips, "Developers Dig into Distribution: Southern California's Inland Empire Illustrates Speculative Building Surge of Warehouse Hubs," *Wall Street Journal*, August 13, 2013, http://online.wsj.com/news/articles/SB10001424127887323446404 579010853015245422; Randall A. Bluffstone and Brad Ouderkirk, "Warehouses, Trucks, and $PM_{2.5}$: Human Health and Logistics Industry Growth in the Eastern Inland Empire," *Contemporary Economic Policy*, January 1, 2007.

54. South Coast Air Quality Management District, "Multiple Air Toxics Exposure Study in the South Coast Air Basin" (MATES II Study), March 2000; James W. Gau-derman et al., "Association between Air Pollution and Lung Function Growth in Southern California Children," *American Journal of Respiratory and Critical Care Medicine* 166, no. 1 (2002), 76–84.

55. National Bureau of Statistics of China, "China Statistical Yearbook 2014, Section 11-2 Total Value of Imports and Exports of Goods," 2014, http://www.stats.gov.cn/ tjsj/ndsj/2014/indexeh.htm.

56. Wang Yong, "WTO Accession, Globalization, and a Changing China," *China Business Review*, October 1, 2011, http://www.chinabusinessreview.com/wto -accession-globalization-and-a-changing-china/; Kevin Cullinane et al., "Container Terminal Development in Mainland China and Its Impact on the Competitiveness of the Port of Hong Kong," *Transport Reviews* 24, no. 1 (2004), 33–56.

57. There are a number of sources that annually identify the largest container ports in the world and that have listed the six China ports and the Hong Kong ports among the top ten since 2012. See, for example, "Top Fifty World Container Ports," *World Shipping Council*, 2016, http://www.worldshipping.org/about-the-industry/ global-trade/top-50-world-container-ports; and Wen Jiabao's tenth anniversary talk in People's Republic of China, Permanent Mission of China to the WTO, "China in the WTO: Past, Present and Future: The Tenth Anniversary of China's Accession to

the WTO, December 2001 to December 2011," accessed June 27, 2016, https://www
.wto.org/english/thewto_e/acc_e/s7lu_e.pdf.

58. Yuohong Wang et al., "The Role of Feeder Shipping in Chinese Container Port
Development," *Transportation Journal* 53, no. 2 (Spring 2014), 253–267.

59. Simon Ng, "Lower Sulphur Fuel Puts HK on Top in Asia Ship Emission Control,"
Voice of Hong Kong, July 3, 2016, http://www.vohk.hk/2016/07/03/lower-sulphur
-fuel-puts-hk-on-top-in-asia-ship-emission-control/.

60. Turloch Mooney, "China Regulates Sulfur Emissions from Ships," *Journal
of Commerce*, December 11, 2015, http://www.joc.com/regulation-policy/
transportation-regulations/international-transportation-regulations/china-regulates
-sulphur-emissions-ships_20151211.html; "China Comes to Grips with Emissions
from Ships," *World Maritime News*, June 9, 2015, http://worldmaritimenews.com/
archives/163212/china-comes-to-grips-with-emissions-from-ships/.

61. Information about the Fair Winds Charter discussions is from Simon Ng's notes.
See also Maersk Line, "Skies Clearing in Hong Kong," January 31, 2013, http://www
.maerskline.com/en-us/countries/int/news/news-articles/2013/01/skies-clearing.

62. Robert Gottlieb's personal communication with Jesse Marquez and Andrea
Hricko, July 2010.

63. Analilia Garcia et al., "THE (Trade, Health, Environment) Impact Project:
A Community Based Participatory Research Environmental Justice Case Study,"
Environmental Justice 6, no. 1 (2013), 17–26, http://lbaca.org/wp-content/
uploads/2013/07/LBACA-Trade-Health-Environment-Impact-Project.pdf;   Lydia
DePillis, "Ports Are the New Power Plants—at Least in Terms of Pollution," *Washington Post*, November 24, 2015, https://www.washingtonpost.com/news/wonk/
wp/2015/11/24/ports-are-the-new-power-plants-at-least-in-terms-of-pollution/.

64. Keia Lu Huang and Nectar Gan, "China's Leadership 'Infuriated by Tianjin
Government's Attempts to Underplay Death Toll of Blasts,'" *South China Morning
Post*, August 27, 2015, http://www.scmp.com/news/china/policies-politics/article/
1852945/tianjin-government-officials-detained-prosecutors-over.

65. "Deadly Tianjin Warehouse Explosion: Review Leads to China Moving Ten
Chemical Plants," *South China Morning Post*, February 15, 2016, http://www.scmp
.com/news/china/policies-politics/article/1913250/deadly-tianjin-warehouse
-explosion-review-leads-china.

66. "Deadly Tianjin Blasts Weigh on China's Shipments as Imports and Exports Fall
amid Weakening Demand," *South China Morning Post*, September 8, 2015, http://
www.scmp.com/news/china/policies-politics/article/1856249/deadly-tianjin-blasts
-weigh-chinas-shipments-exports.

67. Scott Cummings, "Preemptive Strike: Law in the Campaign for Clean Trucks," *UCI Law Review* 4 (2014), 939–1065, http://www.law.uci.edu/lawreview/vol4/no3/Cummings.pdf; Louis Sahagun, "Lawsuit Seeks to Block Shipping Terminal Plan," *Los Angeles Times*, June 15, 2001, http://articles.latimes.com/2001/jun/15/local/me-10836.

68. Deborah Schoch, "Port Project Suit Settled," *Los Angeles Times*, March 6, 2003, http://articles.latimes.com/2003/mar/06/local/me-port6.

69. Port of Los Angeles, "Alternative Marine Power," https://www.portoflosangeles.org/environment/amp.asp. Accessed July 29, 2016.

70. Deborah Schoch, "Accord Clears Way for '03 Plan to Clean Up Air at Port," *Los Angeles Times*, March 12, 2004, http://articles.latimes.com/2004/mar/12/local/me-china12.

71. Harbor Community Benefit Foundation, "Our Roots: History of the Port Community Mitigation Trust Fund," accessed June 27, 2016, http://hcbf.org/about/.

72. California Environmental Protection Agency Air Resources Board, "Commercial Marine Vessels," last modified October 28, 2015, http://www.arb.ca.gov/ports/marinevess/marinevess.htm.

73. San Pedro Bay Ports Clean Air Action Plan (CAAP), accessed July 29, 2016, https://www.portoflosangeles.org/environment/caap.asp. See also "Port of Los Angeles Clean Truck Program," accessed July 29, 2016, https://www.portoflosangeles.org/ctp/idx_ctp.asp; David Bensman, "Port Trucking Down the Low Road: A Sad Story of Deregulation," *Demos 3*, 2009, http://www.demos.org/sites/default/files/publications/Port%20Trucking%20Down%20the%20Low%20Road.pdf; Kristen Monaco, "Incentivizing Truck Retrofitting in Port Drayage: A Study of Drivers at the Ports of Los Angeles and Long Beach," Final Report, Metrans Project 06-02, February 2008, https://www.metrans.org/sites/default/files/research-project/06-02%20Final%20Report_0_0.pdf; Scott L. Cummings, "Preemptive Strike: Law in the Campaign for Clean Trucks," UCLA School of Law, 2014.

74. Ronald D. White, "Immigration Rallies Fuel Resolve of Port Truckers," *Los Angeles Times*, May 4, 2006, http://articles.latimes.com/2006/may/04/business/fi-truckers4; see also Chris Kutalik, "As Immigrants Strike, Truckers Shut Down Nation's Largest Port," *Labor Notes*, September 28, 2006, http://labornotes.org/2006/09/immigrants-strike-truckers-shut-down-nation%E2%80%99s-largest-port.

75. Jack Dolan and Tony Barboza, "Port of L.A.'s Bad Bet," *Los Angeles Times*, March 24, 2016.

76. Tony Barboza and Jack Dolan, "Port of L.A. Terminal Fails to Comply with Pollution-Reduction Measures," *Los Angeles Times*, February 2, 2016, http://www.latimes.com/local/california/la-me-0202-port-pollution-20160202-story.html; Tony

Barboza, "The Port of L.A. Rolled Back Measures to Cut Pollution—During Its 'Green' Expansion," *Los Angeles Times*, December 15, 2015, http://www.latimes.com/local/california/la-me-port-pollution-20151215-story.html; Bill Mongelluzzo, "LA Reviewing Environmental Impact of China Shipping Terminal," *Journal of Commerce*, October 15, 2015, http://www.joc.com/port-news/us-ports/port-los-angeles/la-reviewing-environmental-impact-china-shipping-terminal_20151015.html.

77. Dan Weikel, "Judge Tosses Environmental Report on L.A. Port Rail Yard," *Los Angeles Times*, March 31, 2016.

78. Wendy Laursen, "The Demise of the Fair Winds Charter," *The Maritime Executive*, January 5, 2015, http://www.maritime-executive.com/features/the-demise-of-the-fair-winds-charter; "Hong Kong Wants Ship Emission Standards Now," *The Maritime Executive*, March 19, 2015, http://www.maritime-executive.com/article/hong-kong-wants-ship-emission-standards-now.

79. White House, Office of the Press Secretary, "Fact Sheet: U.S.–China Climate Leaders Summit," September 15, 2015, https://www.whitehouse.gov/the-press-office/2015/09/15/fact-sheet-us-%E2%80%93-china-climate-leaders-summit.

Chapter 3

1. Scott Hamilton Dewey, *Don't Breathe the Air: Air Pollution and U.S. Environmental Politics, 1945–1970* (College Station: Texas A&M University Press, 2000), 93–94; Joshua Dunsby, "Localizing Smog: Transgressions in the Therapeutic Landscape," in E. Melanie DuPuis, ed., *Smoke and Mirrors: The Politics and Culture of Air Pollution* (New York: New York University Press, 2004), 170–200; Betty Koster, "A History of Air Pollution Control Efforts in Los Angeles County," Public Education and Information Division, Los Angeles County Air Pollution District, August 31, 1956, 36.

2. The Clean Air Network video is available at http://www.hongkongcan.org/eng/; Clean Air Asia, "Do Not Adapt to Air Pollution—Clean Air Asia Launches Hairy Nose Campaign," December 5, 2012, http://baqconference.org/2012/assets/Uploads/5-Dec.-Do-Not-Adapt-to-Air-Pollution-Clean-Air-Asia-launches-Hairy-Nose-Campaign.pdf, Accessed July 29, 2016.

3. Alan Taylor, "Beijing's Toxic Sky," *Atlantic*, March 4, 2015, http://www.theatlantic.com/photo/2015/03/beijings-toxic-sky/386824/; Julie Makinen, "Artists Finding Inspiration in China's Bad Air," *Los Angeles Times*, May 7, 2014, http://www.latimes.com/world/la-fg-c1-china-art-pollution-20140507-story.html; Nidhi Subbaraman, "Beijing's Smogpocalypse: China's Air Crisis by the Numbers," *NBC News Blog*, February 25, 2014, http://www.nbcnews.com/science/environment/beijings-smogpocalypse-chinas-air-crisis-numbers-n38251.

4. Michelle FlorCruz, "Chinese Province of Guizhou Selling Canned 'Fresh Air' in New Tourism Gimmick," *International Business Times*, March 21, 2014,

http://www.ibtimes.com/chinese-province-guizhou-selling-canned-fresh-air-new
-tourism-gimmick-1562954; Carol Driver, "China Sells BOTTLED AIR to Tourists as
Smog Described as 'Environmental Crisis,'" *Daily Mail*, March 25, 2014, http://www
.dailymail.co.uk/travel/article-2588788/China-sells-BOTTLED-AIR-tourists.html.

5. The comments of Fan Xiaqiu from Green Beagle are cited in Joy Y. Zhang and
Michael Barr, *Green Politics in China: Environmental Governance and State-Society Relations* (London: Pluto Press, 2013), 64–65.

6. "'Gas Attack,'" *Los Angeles Times*, July 27, 1943, 1; Ed Ainsworth, "Fight to Banish
Smog, Bring Sun Back to City Pressed," *Los Angeles Times*, October 13, 1946, 1.

7. John Anson Ford, *Thirty Explosive Years in Los Angeles County* (San Marino, CA:
Huntington Library, 1961), 121; Sarah S. Elkind, *How Local Politics Shape Federal
Policy: Business, Power, and the Environment in Twentieth-Century Los Angeles* (Chapel
Hill: University of North Carolina Press, 2011), 560; Marvin Breines, "The Fight
against Smog in Los Angeles: 1943–1957," PhD diss., University of California at
Davis, 1975.

8. Chip Jacobs and William J. Kelly, *Smogtown: The Lung-Burning History of Pollution
in Los Angeles* (Woodstock, NY: Overlook Press, 2008), 90–93.

9. A. J. Haagen-Smit, "Chemistry and Physiology of Los Angeles Smog," *Industrial
and Chemical Engineering* 44, no. 6 (1952), 1342–1346, http://cires1.colorado.edu/
jimenez/AtmChem/2013/Haagen-Smit_1952_IndEngChem_LA_smog_paper.pdf;
Kenneth Hahn, *Health Warning: Smog: Record of Correspondence between Kenneth
Hahn, Los Angeles County Supervisor, and the Presidents of General Motors, Ford, and
Chrysler on Controlling Air Pollution* (Los Angeles: Los Angeles County Board of Supervisors, May 1972).

10. South Coast Air Quality Management District, "History of Air Pollution Control," 2014, http://www.aqmd.gov/home/library/public-information/publications/
history-of-air-pollution-control#Last%20Stage%20III.

11. Alfred Sloan Jr., *My Years with General Motors*, ed. John McDonald with Catha-
rine Stevens (Garden City, NY: Doubleday, 1964).

12. Alice Hamilton, the leading critic at the time of TEL's health impacts, argued
that both occupational health and public health were at stake and thus needed to be
addressed "at the very outset," not after its health effects had been demonstrated.
Alice Hamilton, "What Price Safety: Tetraethyl Lead Reveals a Flaw in Our Defenses,"
Survey Midmonthly 54, no. 6 (June 15, 1925), 333.

13. David Rosner and Gerald Markowitz, "A 'Gift of God'? The Public Health Controversy over Leaded Gasoline during the 1920s," *American Journal of Public Health*
75, no. 4 (April 1985), 344–352; J. H. Shrader, "Tetraethyl Lead and the Public
Health," *American Journal of Public Health* 15, no. 3 (March 1925), 213–216.

14. Joseph C. Roberts, *Ethyl: A History of the Corporation and the People Who Made It* (Charlottesville: University Press of Virginia, 1983), 295–310. Cole's statement is also discussed in Gerald E. Markowitz and David Rosner, *Lead Wars: The Politics of Science and the Fate of America's Children* (Berkeley: University of California Press, 2014), 77. See also United States Environmental Protection Agency, "Milestones in Mobile Source Air Pollution Control and Regulations," 2012, http://www3.epa.gov/otaq/consumer/milestones.htm; and Robert Gottlieb, Maureen Smith, and Julie Roque, "Greening or Greenwashing: The Evolution of Industry Decision-Making," ch. 6 in *Reducing Toxics: A New Approach to Policy and Industrial Decision Making* (Washington, DC: Island Press, 1995), 171–177.

15. The 1971–1972 industry mobilization to prevent reduced automobile use anticipated a similar industry mobilization more than forty years later against climate-change-related legislation that would have mandated a targeted climate change and air quality goal of reducing petroleum use by 50 percent by 2030. That regulatory initiative had been developed by California Governor Jerry Brown and Mary Nichols, who was then his head of the California Air Resources Board. Adam Nagourney, "California Democrats Drop Plan for 50 Percent Oil Cut," *New York Times*, September 9, 2015, http://www.nytimes.com/2015/09/10/us/california-democrats-drop-plan-to-force-50-percent-cut-in-oil-use.html; Connie Koenenn, "Bent on Clearing the Air: Attorney Mary Nichols Hopes 'Amazing L.A.' Campaign Will Spur Residents to Join the Fight to Keep Los Angeles from Choking to Death," *Los Angeles Times*, November 20, 1991, http://articles.latimes.com/1991-11-20/news/vw-109_1_los-angeles-attorney/2.

16. Fred Lurmann, Ed Avol, and Frank Gilliland, "Emissions Reduction Policies and Recent Trends in Southern California's Ambient Air Quality," *Journal of the Air and Waste Management Association* 65, no. 3 (2015), 324–335, http://www.tandfonline.com/doi/abs/10.1080/10962247.2014.991856?journalCode=uawm20&

17. American Lung Association, "State of the Air 2015: Report Card: California," accessed June 28, 2016, http://www.stateoftheair.org/2015/states/california/.

18. United States Environmental Protection Agency, "Ozone Pollution," accessed July 29, 2016, https://www.epa.gov/ozone-pollution.

19. Tony Barboza, "Southern California Air Board Moves to Weaken Pollution Regulation," *Los Angeles Times*, March 4, 2016, http://www.latimes.com/local/lanow/la-me-southern-california-air-board-20160304-story.html.

20. Jan Morris, *Hong Kong* (New York: Random House, 1988), 53–54.

21. Bryan Walsh, "Choking on Growth," *Time*, Asia Edition, December 13, 2004, http://content.time.com/time/world/article/0,8599,2048019,00.html.

22. Hong Kong Special Administrative Region, Environmental Protection Department, "Hong Kong's Environment: Milestones in Hong Kong's Environmental Protection," 2003, http://www.epd.gov.hk/epd/misc/ehk03/textonly/eng/hk/1_2.html.

23. "Air Pollution Control Ordinance, 1987," *Historical Laws of Hong Kong Online*, accessed June 28, 2016, http://oelawhk.lib.hku.hk/items/show/3174.

24. Mike Kilburn, "Air Quality Report Card of the Donald Tsang Administration (2005–2012)," Civic Exchange, Hong Kong, January 2012.

25. Hong Kong Special Administrative Region, Environmental Protection Department, "A Study to Review HK AQO," 2010, http://www.epd.gov.hk/epd/english/ environmentinhk/air/air_quality_objectives/review_aqo.html.

26. "World Standard Air Quality Objectives Needed for Asia's World City," press release, Civic Exchange, October 5, 2006, http://www.civic-exchange.org/Publish/ LogicaldocContent/20061005AIR_PR_WorldStandardForAsiaCity_en.pdf; see also Hong Kong Legislative Council, "Official Record of Proceedings: May 19, 2011," 10763–10767, http://www.legco.gov.hk/yr10-11/english/counmtg/hansard/cm0519 -translate-e.pdf, Accessed July 29, 2016

27. Donald Tsang. "CE Speaks at 'Business for Clean Air,'" Joint Conference of Project CLEAN AIR and Project Blue Sky, November 27, 2006, accessed July 29, 2016, http://www.info.gov.hk/gia/general/200611/27/P200611270129.htm.

28. Hong Kong Special Administrative Region, Environmental Protection Department, "Hong Kong's Air Quality Objectives," http://www.epd.gov.hk/epd/english/ environmentinhk/air/air_quality_objectives/air_quality_objectives.html.

29. "Hong Kong Air Pollution: Real Time Air Quality Index (AQI)," http://aqicn.org/ city/hongkong/; Ma Jun et al., "Hong Kong's Role in Mending the Disclosure Gap," Institute of Public and Environmental Affairs, Beijing, and Civic Exchange, Hong Kong, March 2010.

30. University of Hong Kong School of Public Health, "Hedley Environmental Index," http://hedleyindex.sph.hku.hk/html/en/; Anthony J. Hedley et al., "Air Pollution: Costs and Paths to a Solution in Hong Kong: Understanding the Connections among Visibility, Air Pollution, and Health Costs in Pursuit of Accountability, Environmental Justice, and Health Protection," *Journal of Toxicology and Environmental Health, Part A* 71 (2008), 544–554; Eric Cheng and Rui Lou, "Hedley Index Sheds New Light on Hong Kong's Air Pollution," Wilson Center China Environment Forum, July 7, 2011, https://www.wilsoncenter.org/publication/hedley -environmental-index-sheds-new-light-hong-kongs-air-pollution; Clean Air Network, "'Hedley Environmental Index': Articles at Clean Air Network," accessed June 29, 2016, http://www.hongkongcan.org/v2/tag/hedley-environmental-index/.

31. Karl V. Canter and Ivan S. Deckert (Air Pollution Control District of Los Angeles County), "Control of Air Pollution from Oil Refineries in Los Angeles County," *Air Repair* 4, no. 4 (1955), 197–212; reprint available from the *Journal of the Air and Waste Management Association*, accessed July 29, 2016, http://www.tandfonline.com/ doi/pdf/10.1080/00966665.1955.10467669.

32. Rochelle Green et al., "Residential Exposure to Traffic and Spontaneous Abortion," *Environmental Health Perspectives* 117 (2009), 1939–1944; Rochelle Green et al., "Association between Local Traffic-Generated Air Pollution and Preeclampsia and Preterm Delivery in the South Coast Air Basin of California," *Environmental Health Perspectives* 117 (2009), 1773–1779; Rob J. Laumbach et al., "Sickness Response Symptoms among Healthy Volunteers after Controlled Exposures to Diesel Exhaust and Psychological Stress," *Environmental Health Perspectives* 119 (July 2011), 945–950, http://www.ncbi.nlm.nih.gov/pubmed/21330231.

33. Doug Houston, Wei Li, and Jun Wu, "Disparities in Exposure to Automobile and Truck Traffic and Vehicle Emissions near the Los Angeles–Long Beach Port Complex," *American Journal of Public Health* 104, no. 1 (January 2014), 156–164, http://ajph.aphapublications.org/doi/abs/10.2105/AJPH.2012.301120?journalCode=ajph; Gregory Rowangould, "A Census of the US Near-Roadway Population: Public Health and Environmental Justice Considerations," *Transportation Research Part D: Transport and Environment* 25 (December 2013), 59–67.

34. W. James Gauderman et al., "Effect of Exposure to Traffic on Lung Development from Ten to Eighteen Years of Age: A Cohort Study," *Lancet* 369, no. 9561 (2007), 571–577; Shishan Hu et al., "Observation of Elevated Air Pollutant Concentrations in a Residential Neighborhood of Los Angeles, California, Using a Mobile Platform," *Atmospheric Environment* 51 (May 1, 2012), 311–319; Rob McConnell et al., "Traffic, Susceptibility, and Childhood Asthma," *Environmental Health Perspectives* 114 (2006), 766–772; W. James Gauderman et al., "Childhood Asthma and Exposure to Traffic and Nitrogen Dioxide," *Epidemiology* 16 (2005), 737–743; Rob McConnell et al., "Childhood Incident Asthma and Traffic-Related Air Pollution at Home and School," *Environmental Health Perspectives* 118 (2010), 1021–1026; Cathryn Tonne et al., "Traffic-Related Air Pollution in Relation to Cognitive Function in Older Adults," *Epidemiology* 25, no. 5 (September 2014), 674–681; Annette Peters et al., "Triggering of Acute Myocardial Infarction by Different Means of Transportation," *European Journal of Preventive Cardiology* 20 (2013), 750–758.

35. "MATES-IV: Multiple Air Toxics Exposure Study in the South Coast Air Basin," Final Report (Diamond Bar, California: South Coast Air Quality Management District, May 2015), http://www.aqmd.gov/docs/default-source/air-quality/air-toxic-studies/mates-iv/mates-iv-final-draft-report-4-1-15.pdf?sfvrsn=7.

36. "Children's Health Study," Southern California Environmental Health Sciences Center, accessed June 29, 2016, http://hydra.usc.edu/scehsc/about-studies-childrens.html.

37. Carla Truax et al., "Neighborhood Assessment Teams: Case Studies from Southern California and Instructions on Community Investigations of Traffic-Related Air Pollution," THE Impact Project, September 2013, https://www.oxy.edu/sites/default/files/assets/UEP/Speakers/Neighborhood%20Assessment%20Teams%20-%20Case%20Studies%20from%20Southern%20California%20on%20traffic%20

related%20air%20pollution.pdf; Hu et al., "Observation of Elevated Air Pollutant Concentrations in a Residential Neighborhood of Los Angeles," 311–319. The UCLA/ USC mobile platform paralleled similar mobile-platform pollution-tracking instruments developed in Hong Kong at the Hong Kong University of Science and Technology.

38. Moving Forward Network, "Getting to Zero: What the EPA Must Do Now to Reduce Deadly Diesel Emissions in Port Communities and Freight Corridors—An Environmental Justice Policy Brief," January 4, 2016, http://www.edocr.com/ doc/362820/what-epa-must-do-now-reduce-deadly-diesel-emissions-port -communities-and-freight-corridor.

39. Lurmann, Avol, and Gilliland, "Emissions Reduction Policies and Recent Trends in Southern California's Ambient Air Quality," 334.

40. Anthony Hedley et al., "Cardiorespiratory and All-Cause Mortality after Restrictions on Sulphur Content of Fuel in Hong Kong: An Intervention Study," *Lancet* 360 (2002), 1646–1652; Hong Kong Special Administrative Region, Environmental Protection Department, "An Overview on Air Quality and Air Pollution Control in Hong Kong," last revised April 28, 2016, http://www.epd.gov.hk/epd/english/ environmentinhk/air/air_maincontent.html.

41. Christine Loh and James Patterson, "From Boomtown to Gloomtown: The Implications of Inaction," CLSA and Civic Exchange, Hong Kong, 2006.

42. Bill Barron, Simon Ng Ka Wing, and Ben Lin Chubin, "Owning Up to Responsibility for Manufacturing Contribution to the Pearl River Delta's Poor Air Quality," Institute for the Environment, Hong Kong University of Science and Technology, and Civic Exchange, March 2006, http://www.civic-exchange.org/en/publications/ 164986959.

43. Lisa Hopkinson and Rachel Stern, "One Country, Two Systems, One Smog: Cross-Boundary Air Pollution Policy Challenges for Hong Kong and Guangdong," *China Environment Series*, no. 6 (2003), Wilson Center, Washington DC, China Environment Forum, http://www.thesalmons.org/lynn/prd/3-feature_2.pdf.

44. Alexis Lau et al., "Relative Significance of Local vs. Regional Sources: Hong Kong's Air Pollution," Institute for the Environment, Hong Kong University of Science and Technology, and Civic Exchange, 2007, http://www.civic-exchange.org/ en/publications/164986956 (English and Chinese).

45. Simon K. W. Ng et al., "Policy Change Driven by an AIS-Assisted Marine Emission Inventory in Hong Kong and the Pearl River Delta," *Atmospheric Environment* 76 (September 2013), 102–112.

46. World Health Organization, "Ambient (Outdoor) Air Quality and Health," Fact sheet no. 313, updated March 2014, http://www.who.int/mediacentre/factsheets/ fs313/en/.

47. Tung Chee-hwa, "Quality People, Quality Home: Positioning Hong Kong for the 21st Century," 1999 Policy Address, 86–105, accessed June 29, 2016, http://www.policyaddress.gov.hk/pa99/english/espeech.pdf.

48. Ibid.

49. Anthony Hedley, "Air Pollution and Public Health," University of Hong Kong, School of Public Health, February 2009, http://www.legco.gov.hk/yr08-09/english/panels/ea/ea_iaq/papers/ea_iaq0212cb1-733-2-e.pdf.

50. "Audit Commission Controlling Officer's Environmental Report 2012," http://www.aud.gov.hk/pdf_e/coer12_e.pdf.

51. "China's Xi Says He Checks Pollution First Thing Every Day," *Daily Mail*, November 10, 2014, http://www.dailymail.co.uk/wires/afp/article-2828616/Chinas-Xi-says-checks-pollution-thing-day.html.

52. "China Blocks US Air Pollution Data as Apec Leaders Meet," *BBC News*, November 11, 2014, http://www.bbc.com/news/world-asia-china-29999784.

53. Vaclav Smil, *The Bad Earth: Environmental Degradation in China* (New York: M. E. Sharpe, 1984), 172; People's Republic of China, Ministry of Environmental Protection, "Law of the People's Republic of China on the Prevention and Control of Atmospheric Pollution," 2000, http://english.mep.gov.cn/Policies_Regulations/laws/environmental_laws/200710/t20071009_109943.htm.

54. Shanghai Project Research Team, "The Impact of Air Pollution on Mortality in Shanghai," ch. 14 in Ligang Song and Wing Thye Woo, eds., *China's Dilemma: Economic Growth, the Environment and Climate Change* (Canberra, Australia: ANU E Press and Asia Pacific Press, 2008), 298.

55. Yan-Lin Zhang and Fang Cao, "Fine Particulate Matter (PM2.5) in China at a City Level," *Nature: Scientific Reports*, published online October 15, 2015, accessed July 29, 2016, http://www.nature.com/articles/srep14884.

56. Shiwani Neupane, "Beijing's Blinding Pollution," *Columbia Journalism Review* (February 22, 2013), http://www.cjr.org/the_observatory/beijing_air_pollution_media_co.php?page=all. On the use of the term "airpocalypse," see for example, "China's 'Airpocalypse' Engulfs Millions," *CBC News*, October 21, 2013, http://www.cbc.ca/news/world/china-airpocalypse-engulfs-millions-1.2159056.

57. Yao Chen's comment is in Sarah Keenlyside, "Yao Chen Interview: Meet China's Answer to Angelina Jolie," *Telegraph*, August 24, 2014, http://www.telegraph.co.uk/culture/culturevideo/filmvideo/film-interviews/11050458/Yao-Chen-interview-meet-Chinas-answer-to-Angelina-Jolie.html. See also David Roberts, "How the US Embassy Tweeted to Clear Beijing's Air," *Wired*, March 6, 2015, http://www.wired.com/2015/03/opinion-us-embassy-beijing-tweeted-clear-air/; Tom Phillips, "Airpocalypse Now: China Pollution Reaching Record Levels, *Guardian*, November 8, 2015,

http://www.theguardian.com/world/2015/nov/09/airpocalypse-now-china
-pollution-reaching-record-levels.

58. Ministry of Environmental Protection of the People's Republic of China, "The State Council Issues Action Plan on Prevention and Control of Air Pollution Introducing Ten Measures to Improve Air Quality," September 12, 2013, http://english .mep.gov.cn/News_service/infocus/201309/t20130924_260707.htm.

59. Gonghuan Yang et al., "Rapid Health Transition in China, 1990–2010: Findings from the Global Burden of Disease Study 2010," *Lancet* 381 (2013), 1987–2015, http://iapbwesternpacific.org/download/countries/china/Rapid%20health%20 transition%20in%20China%201990%C2%AD2010.pdf; Zhu Chen et al., "China Tackles the Health Effects of Air Pollution," *Lancet* 382 (December 14, 2013), 1959–1960; Jiaojioa Lu, "Air Pollution Exposure and Physical Activity in China: Current Knowledge, Public Health Implications, and Future Research Needs," *International Journal of Environmental Research and Public Health* 12 (2015), 14887–97, http://www .mdpi.com/1660-4601/12/11/14887; Mun S. Ho and Chris P. Nielsen, *Clearing the Air: The Health and Economic Damages of Air Pollution in China* (Cambridge, MA: MIT Press, 2007); Heiko J. Jahn et al., "Ambient and Personal PM$_{2.5}$ Exposure Assessment in the Chinese Megacity of Guangzhou," *Atmospheric Environment* 74 (2013), 402–411.

60. Zhao Wen, "Parents Push Schools to Install Air Purifiers," *Shanghai Daily*, December 10, 2013, http://www.shanghaidaily.com/Metro/education/Parents-push-schools-to-install-air-purifiers/shdaily.shtml; Zhang Yue, "School Beats Heavy Smog with Online Classes," *China Daily*, January 2, 2014, http://www.chinadaily.com.cn/ china/2014-01/02/content_17209857.htm; Edward Wong, "Urbanites Flee China's Smog for Blue Skies," *New York Times*, November 22, 2013, http://www.nytimes .com/2013/11/23/world/asia/urbanites-flee-chinas-smog-for-blue-skies.html ?pagewanted=all.

61. Elizabeth Economy, "China's Environmental Enforcement Glitch," *Diplomat*, January 21, 2015, http://thediplomat.com/2015/01/chinas-environmental -enforcement-glitch/.

62. Li Keqiang, "Report on the Work of the Government," March 5, 2014, delivered at the second session of the Twelfth National People's Congress, Beijing, China, http://online.wsj.com/public/resources/documents/2014GovtWorkReport_Eng.pdf.

63. Liu Jinqiang, "China's New Environmental Law Looks Good on Paper," *China Dialogue*, April 24, 2014, https://www.chinadialogue.net/blog/6937-China-s-new -environmental-law-looks-good-on-paper/en; Sam Geall, "Interpreting Ecological Civilization (Part One)," *China Dialogue*, July 6, 2015, https://www.chinadialogue .net/article/show/single/en/8018-Interpreting-ecological-civilisation-part-one; Central Document Number 12, "Opinions of the Central Committee of the Communist Party of China and the State Council on Further Promoting the Development of

Ecological   Civilization,"   http://paper.people.com.cn/rmrb/html/2015-05/06/nw
.D110000renmrb_20150506_3-01.htm.

64. Emily Rauhala, "Airpocalypse, Again: With Winter Coming, China Hit by
'Doomsday' Smog," *Washington Post*, November 9, 2015, https://www
.washingtonpost.com/news/worldviews/wp/2015/11/09/airpocalypse-again-with
-winter-coming-china-hit-by-doomsday-smog/.

65. Friedrich Geiger, "Merkel Complained in 2010 about California Emission
Rules," *Wall Street Journal*, November 12, 2015. http://www.wsj.com/articles/merkel
-complained-in-2010-about-california-emissions-rules-1447349303.

66. In June 2016, VW finally entered into a settlement agreement with US EPA to
pay $10 billion to compensate its car buyers and another $4.7 billion to mitigate
pollution and support zero-emission technology. United States Environmental Protection Agency, "Volkswagen to Spend up to 14.7 Billion to Settle Allegations of
Cheating Emissions Tests and Deceiving Customers on 2.0 Liter Diesel Vehicles,"
press release, June 28, 2016, https://www.epa.gov/newsreleases/volkswagen-spend
-147-billion-settle-allegations-cheating-emissions-tests-and-deceiving.   See   also
Charles Fleming, "California Regulators Reject VW Repair Plan for Diesel Vehicles
Linked to Scandal," *Los Angeles Times*, January 12, 2016, http://www.latimes.com/
business/autos/la-fi-hy-carb-rejects-vw-diesel-plan-20160112-story.html; the VW
clean diesel commercial, "2015 Volkswagen Jetta TDI commercial 'No Compromise,'" can be seen on YouTube at https://www.youtube.com/watch?v=ZOjGc93olIY;
David McLaughlin and Alan Katz, "Volkswagen's Advertising of Rigged Vehicles
Probed by FTC," *BloombergBusiness*, October 14, 2015, http://www.bloomberg.com/
news/articles/2015-10-14/volkswagen-s-advertising-of-rigged-diesels-probed-by-ftc.

67. Hong Kong Special Administrative Region, Environmental Bureau, "A Clean Air
Plan for Hong Kong," March 2013, 3.

68. Hong Kong Special Administrative Region, Environmental Protection Department, "Air Pollution Control Strategies," 2016, http://www.epd.gov.hk/epd/english/
environmentinhk/air/prob_solutions/strategies_apc.html.

69. Edward Wong, "Beijing Issues Red Alert over Air Pollution for the First
Time," *New York Times*, December 7, 2015, http://www.nytimes.com/2015/12/08/
world/asia/beijing-pollution-red-alert.html.

70. Edward Wong, "Beijing Issues a Second 'Red Alert' on Pollution," *New York
Times*, December 17, 2015, http://www.nytimes.com/2015/12/18/world/asia/
beijing-issues-a-second-red-alert-on-pollution.html.

71. Hong Kong Special Administrative Region, Environmental Bureau, "A Clean Air
Plan for Hong Kong," 13.

72. Michael J. Enright, Edith E. Scott, and Ka-mun Chang, *Regional Powerhouse:
Greater Pearl River Delta and the Rise of China* (Singapore: John Wiley & Sons [Asia],

2005); Hong Kong Special Administrative Region, Environmental Protection Department, "Hong Kong and Guangdong Strengthen Co-Operation on Cleaner Production to Improve Regional Environmental Quality," October 29, 2015, http://www.info.gov.hk/gia/general/201510/29/P201510290629.htm.

73. Jintai Lin et al., "China's International Trade and Air Pollution in the United States," *PNAS* 111, no. 5 (February 4, 2014), 1736–1741, http://www.pnas.org/content/111/5/1736.full.pdf; Edward Wong, "China Exports Pollution to US, Study Finds," *New York Times*, January 20, 2014, http://www.nytimes.com/2014/01/21/world/asia/china-also-exports-pollution-to-western-us-study-finds.html; Tony Barboza, "China's Industry Exporting Air Pollution to U.S., Study Says," *Los Angeles Times*, January 20, 2014, http://www.latimes.com/science/sciencenow/la-sci-sn-china-exports-air-pollution-united-states-20140120-story.html.

74. "Ecology: Menace in the Skies," from the cover story "The Polluted Air," *Time*, January 27, 1967.

75. Intergovernmental Panel on Climate Change, *Climate Change 2014: Mitigation of Climate Change: Contribution of Working Group III to the Fifth Assessment Report of the Intergovernmental Panel on Climate Change* (Cambridge: Cambridge University Press, 2014), 46; Glen P. Peters et al., "Growth in Emission Transfers via International Trade from 1990 to 2008," *Proceedings of the National Academy of Sciences* 108, no. 21 (2011), 8903–8908, http://www.pnas.org/content/108/21/8903.full; Andreas Malm, "China as Chimney of the World: The Fossil Capital Hypothesis," *Organization and Environment* 25, no. 2 (2012), 146–177, http://oae.sagepub.com/content/25/2/146.abstract; Yan Yunfeng and Yang Laike, "China's Foreign Trade and Climate Change: A Case Study of CO_2 Emissions," *Energy Policy* 38, no. 1 (January 2010), 350–356.

Chapter 4

1. The Dongjiang (East River) is one of three tributaries (along with the North and West Rivers) that feed into the Pearl River estuary, which then flows from Guangzhou to the sea. The Pearl River (and its tributaries) is 2,215 kilometers long and drains approximately 450,000 square kilometers of land. The Pearl River Basin (including the nearly 40 million people living within the Dongjiang watershed) has an overall population of 141 million, or 12.1 percent of China's population, and makes up 16.7 percent of China's total water resources. Guangdong Province itself recorded the largest jump in population between 2000 and 2010 among all of China's provinces, at more than 104 million. See National Bureau of Statistics of China, "Communiqué of the National Bureau of Statistics of People's Republic of China on Major Figures of the 2010 Population Census" (no. 2), April 29, 2011, http://www.stats.gov.cn/english/NewsEvents/201104/t20110429_26450.html. See also World Bank, "Addressing China's Water Scarcity: Recommendations for Selected Water Resource Management Issues" (2009), 11, http://www-wds.worldbank.org/external/

default/WDSContentServer/WDSP/IB/2009/01/14/000333037_20090114011126/
Rendered/PDF/471110PUB0CHA0101OFFICIAL0USE0ONLY1.pdf.

2. The 2010 census placed Huizhou's population at 1.807 million and Heyuan's at 450 thousand. City Population. 2010 Census, "China: Guangdong," accessed July 31, 2016, http://www.citypopulation.de/China-Guangdong.html.

3. The industries that were encouraged to relocate included labor-intensive ones, such as those making apparel, toys, shoes, hardware, and packaging; those that were resource-intensive, such as those involving ceramics, furniture, recycled metal products, alloys, and die-casting of nonferrous metals; those that were capital-intensive, such as information technology; and agricultural operations. The mechanism for relocation included the construction of industrial relocation parks (IR parks), with as many as thirty-six recognized at the provincial level in just the first few years after the IR policy was adopted, although as of 2011 only twelve of these had wastewater treatment facilities. Liu Su and Berton Bian, "Liquid Assets II: Industrial Relocation in Guangdong Province: Avoid Repeating Mistakes," Civic Exchange, Hong Kong, January 10, 2012, 12, http://www.civic-exchange.org/en/publications/4292630.

4. While the name Heyuan literally translates as "origin of the river," it doesn't reference the origin of Dongjiang but is relatively close to the origins of three of its tributaries or subtributaries: Lianping He (連平河), Zhongxin He (忠信河), and Xinfeng rivers (新豐江). L. Ling, "Origin of Heyuan," *Heyuan Daily*, June 1, 2014, http://www.heyuan.cn/xw/20140530/96252.htm.

5. By 2013, the rapid construction of new wastewater treatment facilities had increased treatment to upward of 90 percent of Heyuan's wastewater. Bureau of Statistics of Heyuan, People's Republic of China, "Statistical Communiqué of Heyuan City on the 2013 Economic and Social Development,"March 28, 2014.

6. Su Liu et al., "Liquid Assets IIIB: Dongjiang Overloaded: A Photographic Report of the 2011 Dongjiang Expedition," Civic Exchange, Hong Kong, May 30, 2012, http://www.civic-exchange.org/en/publications/4292633.

7. Dick Roraback, "Up a Lazy River, Seeking the Source: Your Explorer Follows in Footsteps of Gaspar de Portola," *Los Angeles Times*, October 20, 1985. This was the first of a twenty-part series of articles based on Roraback's description of his own expedition from mouth to source, that ran intermittently between October 20, 1985, and January 30, 1986, with the articles poking fun of the idea that the LA River was any longer a river.

8. Joe Linton, *Down by the Los Angeles River: Friends of the Los Angeles River's Official Guide* (Los Angeles: Wilderness Press, 2005); Robert Gottlieb, *Reinventing Los Angeles: Nature and Community in the Global City* (Cambridge, MA: MIT Press, 2007), 135–172.

9. The kayak journey subsequently helped shape the decision of the head of the US Environmental Protection Agency, Lisa Jackson, to designate the entire LA River as

"traditional navigable waters," thereby allowing provisions of the federal Clean Water Act to be applied. United States Environmental Protection Agency, Pacific Southwest Media Center (Region 9), "EPA Takes Action to Strengthen Environmental and Public Health Protection for the L.A. River Basin," accessed July 30, 2016, http://www.epa.gov/region09/mediacenter/LA-river/index.html.

10. Joe Linton, "Kayaking the L.A. River: Day 1," *L.A. Creek Freak Blog*, July 26, 2008, http://lacreekfreak.wordpress.com/2008/07/26/kayaking-the-los-angeles-river-day-1/, and "Kayaking the L.A. River: Day 2," http://lacreekfreak.wordpress.com/2008/07/27/kayaking-the-los-angeles-river-day-2/.

11. Richard Bigger and James D. Kitchen, *How the Cities Grew: A Century of Municipal Independence and Expansionism in Metropolitan Los Angeles* (Los Angeles: Bureau of Governmental Research, University of California at Los Angeles, 1952), 5–27.

12. *The Metropolitan Water District of Southern California: History and First Annual Report, for the Period Ending June 30, 1938*, compiled and edited by Charles A. Bissell (Los Angeles: Metropolitan Water District of Southern California, 1939); Steven P. Erie, *Beyond Chinatown: The Metropolitan Water District and the Environment in Southern California* (Stanford: Stanford University Press, 2006).

13. The framework for continued expansion through annexation and a reliance on additional imported water was codified in a policy statement adopted in 1952 by the MWD board that came to be known as the Laguna Declaration; it is reprinted in Robert Gottlieb and Margaret FitzSimmons, *Thirst for Growth: Water Agencies as Hidden Government in California* (Tucson: University of Arizona Press, 1989), 15.

14. The Alaskan proposal was promoted by the former Secretary of Interior and Alaska Governor Walter Hickel, who printed up a poster with an image of the undersea aqueduct and these words: "Big projects define a civilization. So why war [over water sources]—why not big projects." Poster titled "Alaska-California Sub-Oceanic Fresh Water Transport System," in author's possession. Other proposals at the time were just as ambitious and expensive and impossible to imagine ever being implemented, and were presented with the argument that without these types of huge interbasin water transfer schemes a water crisis in parts of the country, including Southern California, was imminent. See, for example, Jim Wright, *The Coming Water Famine* (New York: Coward-McCann, 1966); see also Gottlieb and FitzSimmons, *Thirst for Growth*, 117.

15. A detailed account of the 1982 Peripheral Canal campaign can be found in Robert Gottlieb, *A Life of Its Own: The Politics and Power of Water* (San Diego: Harcourt Brace Jovanovich, 1988), 3–33.

16. The statement "people will have to drink water from the toilet bowl ..." was made by MWD Director Sam Rue, July 6, 1981; from the author's own notes and cited in Gottlieb and FitzSimmons, *Thirst for Growth*, 20. Coauthor Robert Gottlieb

served during this period, and ultimately from 1980 to 1987, on the Board of Directors of the Metropolitan Water District of Southern California.

17. Frederick Muir, "MWD Looking at Rate, Tax and Fee Hikes," *Los Angeles Times*, September 27, 1991.

18. Robert Gottlieb, "For the MWD, It Isn't Easy Being Green," *Los Angeles Times*, December 3, 1992. By the early 1990s, the debate indicated a recognition that a shift away from "unlimited water for unlimited growth" needed to take place, even as a more fundamental shift in Metropolitan's mission—from "not simply as suppliers but as managers of demand"—had yet to occur.

19. Mark Gold, "Keeping L.A.'s Taps Flowing," *Los Angeles Times*, July 15, 2011, http://articles.latimes.com/2011/jul/15/opinion/la-oe-gold-dwp-rate-hike-20110715.

20. San Diego County Water Authority, "Water Authority–Imperial Irrigation District Water Transfer," http://sdcwa.org/water-transfer; Bettina Boxall, "Seawater Desalination Plant Might Be Just a Drop in the Bucket," *Los Angeles Times*, February 17, 2013, http://articles.latimes.com/2013/feb/17/local/la-me-carlsbad-desalination-20130218; Metropolitan Water District of Southern California, "2015 Urban Water Management Plan," June 2016, http://www.mwdh2o.com/PDF_About_Your_Water/2.4.2_Regional_Urban_Water_Management_Plan.pdf.

21. Dan Rodrigo, "Lessons Learned from Southern California's Integrated Resources Plan," Metropolitan Water District of Southern California, 2011, 6, http://opensiuc.lib.siu.edu/cgi/viewcontent.cgi?article=1316&context=jcwre. See also Metropolitan Water District of Southern California, "2015 Urban Water Management Plan," June 2016, http://mwdh2o.com/Reports/2.4.1_Integrated_Resources_Plan.pdf. The development of the IRP, and its implication that Metropolitan should play a role in developing *local* sources was itself controversial at the time the IRP was developed in the 1990s, with some Metropolitan board members and agencies arguing that Metropolitan's role should be limited to only providing imported supplemental water supplies. See Dennis E. O'Connor, "The Governance of the Metropolitan Water District of Southern California: An Overview of the Issues," California Research Bureau, CRB-98-013, August 1998, pp. 31–32, California State Library, Sacramento, https://www.library.ca.gov/crb/98/13/98013.pdf.

22. Los Angeles Department of Water and Power, "Urban Water Management Plan 2010," 33–34, http://www.water.ca.gov/urbanwatermanagement/2010uwmps/Los%20Angeles%20Department%20of%20Water%20and%20Power/LADWP%20UWMP_2010_LowRes.pdf; Metropolitan Water District of Southern California, "Regional Urban Water Management Plan," I-20–22.

23. Metropolitan Water District of Southern California, "Metropolitan Board Approves Nation's Largest Conservation Program to Meet Unprecedented Consumer Demand in Drought's Fourth Year," news release, May 26, 2015; Executive

Department, State of California, "Executive Order B-29-15," April 1, 2015, http://gov.ca.gov/docs/4.1.15_Executive_Order.pdf.

24. The American Lawns organization, http://www.american-lawns.com/index.html, cited in Gottlieb, *Reinventing Los Angeles*, 31.

25. Comprehensive Yard Ordinance, adopted September 12, 1934, City of Los Angeles Archives and Records Center, Ordinance Number 74145, Box no 1534, File 2347.

26. By the turn of the twenty-first century, Los Angeles's fastest growing suburb, the city of Palmdale, had even instituted a rule that made it illegal for a resident not to maintain their grass, setting a height limit on weeds and proscribing the use of yards to park cars. Some US cities even mandated particular types of lawns that were more water-intensive and often required significant pesticide use to keep them from turning brown or becoming weed-infested. Salt Lake City, for example, established a zoning ordinance that mandated that all front lawns be covered with flat green grass, a result that required constant watering; this ordinance wasn't challenged until 2006. Melissa Sanford, "Salt Lake City Moving toward Less Thirsty Lawns," *New York Times*, August 25, 2006; Martha L. Willman, "'Lawn Police' May Be Coming to Palmdale," *Los Angeles Times*, May 6, 2001; see also Robert Messia, "Lawns as Artifacts: The Evolution of Social and Environmental Implications of Suburban Residential Land Use," in Matthew J. Lindstrom and Hugh Bartling, eds., *Suburban Sprawl: Culture, Theory and Politics* (London: Rowman and Littlefield, 2003), 74–75; Ted Steinberg, "Lawn and Landscape in World Context, 1945–2000," *Magazine of History* 19, no. 6 (November 2005), 62–68; and Ted Steinberg, "Lawn Mores," *Los Angeles Times*, March 18, 2006.

27. The 2014–2015 turf removal rebate program developed by Metropolitan and made available to all its member agencies, including the Los Angeles Department of Water and Power, totaled $450 million in rebates and incentives that removed 175 million square feet of turf. Although the vast majority of the requests for rebates came from residents, most of the funding went to businesses, especially golf courses, given the amount of turf they wanted to remove and the rebate program's lack of a limit on the amount of the rebate. Metropolitan Water District of Southern California, "Water Tomorrow: Integrated Water Resources Plan 2015 Update," Report no. 1518, January 2016, foreword, 1, http://www.mwdh2o.com/PDF_About_Your_Water/2015%20IRP%20Update%20Report%20(web).pdf; Matt Stevens and Javier Panzar, "Businesses Get Fat Rebates for Water-Saving Turf Replacement," *Los Angeles Times*, January 11, 2015, http://www.latimes.com/local/california/la-me-turf-rebate-20150112-story.html; Metropolitan Water District of Southern California, "Gearing Up for More Drought, Metropolitan Water Board Revises Supply Allocation Plan, Also Adds $40 Million to Rebate Program," press release, December 9, 2014. By 2015, as the mood of crisis about the drought deepened throughout the state, applications for rebates for turf removal skyrocketed and Metropolitan greatly expanded its program, identifying it as the largest in the country. Metropolitan Water District of

Southern California, "Metropolitan Board Approves Nation's Largest Conservation Program to Meet Unprecedented Consumer Demand in Drought's Fourth Year," news release, May 26, 2015; see also Los Angeles Department of Water and Power press release, November 3, 2014, http://www.ladwpnews.com/go/doc/1475/2404414/Residential-Cash-In-Your-Lawn-Rebate-now-3-75.

28. Governor Jerry Brown's emergency executive order, April 25, 2014, is available at Office of the Governor, State of California, http://gov.ca.gov/news.php?id=18496; see also drought regulations, http://openstates.org/ca/bills/20152016/AB1/. The model ordinance is available at https://cwc.ca.gov/Documents/2015/06_June/June2015_Agenda_Item_4_Attach_3_mwelo_revision_ordinance__memo_Final.pdf. On drought shaming, see Rose Hackman, "California Drought Shaming Takes on a Class Conscious Edge," *Guardian*, May 16, 2015, http://www.theguardian.com/us-news/2015/may/16/california-drought-shaming-takes-on-a-class-conscious-edge.

29. Ho Pui Yin, *Water for a Barren Rock: 150 Years of Water Supply in Hong Kong* (Hong Kong: Commercial Press, 2001), 47–48. The Los Angeles developments are described in David S. Torres-Rouff, "Water Use, Ethnic Conflict, and Infrastructure in Nineteenth-Century Los Angeles," *Pacific Historical Review* 75, no. 1 (2006), 119–121. Torres-Rouff also argues that the Spanish *zanja* system, which identified water as a common resource, came to be contested and replaced by the dominant Anglo approach toward water (including the sewerage system) that emphasized water as serving individual (and private) rather than collective needs.

30. At a meeting of the Legislative Council, the British Hong Kong Governor reported that storage had been reduced in the island and mainland reservoirs to only 7 percent and 15 percent of capacity, characterizing the yearlong 1929 drought as "much the most severe in the Colony's history." "Our water problem," the Governor proclaimed, "is the most pressing and most important of our domestic problems, and it is quite clear to all of us that an adequate solution to this problem is imperative." Hong Kong Legislative Council meeting September 5, 1929, "The Colony's Waterworks," 134, http://www.legco.gov.hk/1929/h290905.pdf.

31. Hong Kong Special Administrative Region, Water Supplies Department, "Plover Cove Reservoir," http://www.wsd.gov.hk/en/education/fun_of_fishing_in_hong_kong/brief_introduction_of_reservoirs/plover_cove_reservoir/; Olga Wong, "Seawater Flushing to Be Extended to More Hong Kong Toilets Next Year," *South China Morning Post*, April 3, 2014, http://www.scmp.com/news/hong-kong/article/1463333/seawater-flushing-be-extended-more-hong-kong-toilets-next-year.

32. Ho, *Water for a Barren Rock*, 182–183 and 192.

33. "Typhoon Ends Long Drought in Hong Kong," *Evening Independent*, May 28, 1964, https://news.google.com/newspapers?id=qaAvAAAAIBAJ&sjid=_FYDAAAAIBAJ&dq=storm+viola&pg=1310,4607194&hl=en.

34. Liu Su, "Hong Kong Water Governance under One Country, Two Systems," presentation at the Hong Kong Water Governance Conference, City University of Hong Kong, Hong Kong, June 6, 2014; see also Nelson Lee, "The Changing Nature of Border, Scale and the Production of Hong Kong's Water Supply System since 1959," *International Journal of Urban and Regional Research* 38, no. 3 (May 2014), 909–910.

35. Liu Su, "Liquid Assets IV: Hong Kong's Water Resources Management under 'One Country, Two Systems,'" Civic Exchange, Hong Kong, November 13, 2013, 26–30, http://www.civic-exchange.org/en/publications/4292629.

36. Liang Yang, Chunxiao Zhang, and Grace W. Ngaruiya, "Water Supply Risks and Urban Responses under a Changing Climate: A Case Study of Hong Kong," *Pacific Geographies* 39 (January/February 2013), 10, http://qa-pubman.mpdl.mpg.de/pubman/faces/viewItemFullPage.jsp?itemId=escidoc%3A2034671%3A2.

37. Prior to the passage of the 1972 legislation, the US Council on Environmental Quality, in its second annual report, indicated how widespread the pollution of surface waters due to various discharges had become. The report stated that "almost one-third of U.S. stream miles are characteristically polluted in the sense that they violate Federal water quality criteria," and it cited US EPA estimates that less than 10 percent of all watersheds "are unpolluted or even moderately polluted." Council on Environmental Quality, *Second Annual Report* (Washington, DC: Government Printing Office, 1971).

38. The "taste and odor" problem was considered particularly worrisome from a public relations viewpoint, which led Metropolitan in 1981 to establish the first "Flavor Panel Analysis (FPA)" among water agencies, a technique utilized in the food and beverage industries. See Water Research Foundation, "Advancing the Science of Water: WRF and Research on Taste and Odor in Drinking Water," 2014, http://www.waterrf.org/resources/StateOfTheScienceReports/TasteandOdorResearch.pdf; and Rob Hallwachs, "Water Scents," May 2008, *People.interactive* (Metropolitan's employee magazine), http://www1.mwdh2o.com/Peopleinteractive/archive_08/May_08/article01.html.

39. United States Environmental Protection Agency, Pacific Southwest Media Center, Region 9, "San Gabriel Valley (Area 4), City of Industry, Puente Valley: Description and History," accessed July 30, 2016, https://yosemite.epa.gov/r9/sfund/r9sfdocw.nsf/BySite/San%20Gabriel%20Valley%20(area%204)%20City%20Of%20Industry,%20Puente%20Valley?OpenDocument#descr.

40. United States Environmental Protection Agency, "Background on Drinking Water Standards in the Safe Drinking Water Act (SDWA)," updated December 23, 2015, https://www.epa.gov/dwstandardsregulations/background-drinking-water-standards-safe-drinking-water-act-sdwa.

41. Gottlieb and FitzSimmons, *Thirst for Growth*, 154–161.

42. The direct correlation between water quality impacts that reduced groundwater supplies and the consequent need to increase imported water became a major preoccupation of the Metropolitan Water District during the 1980s as the magnitude of the problem became apparent. This concern is spelled out by Metropolitan's General Manager Carl Boronkay in his April 29, 1985, memo to MET's member agency managers entitled "Study of the Quality of Groundwater Resources in Metropolitan's Service Area" that is in the author's possession.

43. For example, during the summer of 2014 when algae blooms in Lake Erie contaminated the drinking water supply of Toledo, Ohio, and caused the city's residents to be without water for four days, the occurrence heightened concerns about climate change. See Jane G. Lee, "Driven by Climate Change, Algae Blooms behind Ohio Water Scare Are New Normal," *National Geographic*, August 4, 2014, http://news .nationalgeographic.com/news/2014/08/140804-harmful-algal-bloom-lake-erie -climate-change-science/.

44. Government of Hong Kong, "Drinking Water Quality in Hong Kong," accessed November 16, 2014, http://www.gov.hk/en/residents/environment/water/drinking water.htm.

45. Hong Kong Special Administrative Region, Environmental Protection Department, "Twenty Years of River Water Quality Monitoring in Hong Kong, 1986–2005," 2006, http://www.epd.gov.hk/epd/misc/river_quality/1986-2005/eng/1_background _menu.htm; Hong Kong Special Administrative Region, Environmental Protection Department, "River Water Quality in Hong Kong in 2012," 2013, http://wqrc.epd .gov.hk/pdf/water-quality/annual-report/Report_2012_Eng_Combined.pdf.

46. Hong Kong Special Administrative Region, Water Supplies Department, "Dongjiang Raw Water," August 2014, http://www.wsd.gov.hk/en/water_resources/raw _water_sources/dongjiang_raw_water/index.html.

47. Wang Yunpeng et al., "Water Quality Change in Reservoirs of Shenzhen, China: Detection Using LANDST/TM Data," *Science of the Total Environment* 328 (2004), 195–206; K. C. Ho et al., "Chemical and Microbiological Qualities of the East River (Dongjiang) Water, with Particular Reference to Drinking Water Supply in Hong Kong," *Chemosphere* 52 (2003), 1441–1450; Fu Jiamo et al., "Persistent Organic Pollutants in Environment of the Pearl River Delta, China: An Overview," *Chemosphere* 52 (2003), 1411–1422. It's also interesting to note the parallels in air quality impacts due to the shift of specific industries from Hong Kong to Guangdong.

48. Ma Jun, *China's Water Crisis (Zhongguo shui weiji)* (Norwalk, CT: Eastbridge, 2004), ix.

49. The World Bank asserted in its report: "Based on standard definitions, North China is already a water-scarce region, and China as a whole will soon join the group of water-stressed countries." World Bank, "Addressing China's Water Scarcity: Recommendations for Selected Water Resource Management Issues," 2009, 1,

http://www-wds.worldbank.org/external/default/WDSContentServer/WDSP/IB/200
9/01/14/000333037_20090114011126/Rendered/PDF/471110PUB0CHA0101OFFICI
AL0USE0ONLY1.pdf.

50. Office of the South-to-North Water Diversion Commission of the State Council,
accessed November 16, 2014, www.nsbd.gov.cn/zx/gczz/201106/t20110630_188237
.html (Chinese only). The South-North Water Diversion project has been an object
of pride for the Chinese leadership due to its scale and ambition but has also been
subject to strong criticism for its environmental impact. Even as the first phase of
the project, the central canal route sending Yangtze River water through Hubei and
Henan provinces to the industrial areas in the North, got under way in late 2013,
officials began to encounter concerns that those two provinces were themselves
experiencing an extended drought that had affected the areas since the early 1990s.
The Deputy Director of the Yangtze River Water Resources Commission with responsibility for the project countered media reports that the Water Diversion project had
any relationship to the drought conditions, arguing that it would benefit rather
than harm the provinces. Yet the drought itself in Hubei and Henan underlined the
controversy over the most appropriate water supply strategies in a country where
drought cycles could last as long as twenty-five years. Mandy Zuo, "South-North
Water Diversion Project Not Cause of Hubei and Henan Droughts, Say Officials,"
South China Morning Post, October 25, 2014, http://www.scmp.com/news/china/
article/1580527/south-north-water-diversion-project-not-cause-hubei-and-henan
-droughts. See also Jon Barnett et al., "Transfer Project Cannot Meet China's Water
Needs," *Nature* 527, no. 7578 (November 18, 2015), http://www.nature.com/news/
sustainability-transfer-project-cannot-meet-china-s-water-needs-1.18792.

51. The different levels identifying the quality of a water source range from Grades
I–III, which are generally acceptable for human consumption, to Grades IV–V+,
which are acceptable for industrial and agricultural use, with Grade V+, the most
polluted, not available for any use. State Environmental Protection Administration
and General Administration of Quality Supervision, Inspection and Quarantine,
People's Republic of China, "Environmental Quality Standards for Surface Water,"
GB 3838-2002, April 28, 2002, http://english.mep.gov.cn/standards_reports/
standards/water_environment/quality_standard/200710/W020061027 50989667
2057.pdf. For an assessment of the Yangtze and Pearl River Basin water quality standards, see Ministry of Environmental Protection, People's Republic of China, "Report
on the State of the Environment in China 2013," May 27, 2014, http://www.mep
.gov.cn/zhxx/tpxw/201406/P020140604597469704833.pdf; China Water Risk,
"2013 State of Environment Report Review," *China Water Risk*, July 9, 2014, http://
chinawaterrisk.org/resources/analysis-reviews/2013-state-of-environment-report
-review.

52. Dow Chemical, Water and Process Solutions, "China's Thirst for Water," April
2011, accessed June 14, 2014, http://www.futurewecreate.com/water/includes/
DOW072_China%20White_Opt1_Rev1.pdf; Laura Ediger and Linda Hwang, "Water

Quality and Environmental Health in Southern China," July 2009, 4, from a BSR Forum, May 15, 2009, in Guangzhou, China, http://www.bsr.org/reports/BSR _Southern_China_Water_Quality_Environmental_Health_Forum.pdf.

53. "China's Underground Water Quality Worsens: Report," *Xinhua*, April 22, 2014, http://news.xinhuanet.com/english/china/2014-04/22/c_126421022.htm; see also Debra Tan, "The State of China's Agriculture," *China Water Risk*, April 9, 2014, http://chinawaterrisk.org/resources/analysis-reviews/the-state-of-chinas -agriculture/; Hu Feng, Debra Tan, and Inna Lazareva, "Eight Facts on China's Wastewater," *China Water Risk*, March 12, 2014, http://chinawaterrisk.org/resources/ analysis-reviews/8-facts-on-china-wastewater/.

54. State Council of the People's Republic of China, "Action Plan for Water Pollu-tion Prevention," April 2, 2015. This plan set a number of targets for improving water quality for rivers, groundwater, and aquatic environments, with strategies ranging from a pollution permit trade system to the gradual phase-out of the production and use of high-risk chemicals (e.g., endocrine disrupters). Premier Li Keqiang's March 5, 2014, speech before China's Twelfth National People's Congress, "Report on the Work of the Government," is available at http://online.wsj.com/ public/resources/documents/2014GovtWorkReport_Eng.pdf. See also Debra Tan, "The War on Water Pollution," *China Water Risk*, March 12, 2014, http:// chinawaterrisk.org/resources/analysis-reviews/the-war-on-water-pollution/; Ying Shen and Debra Tan, "Groundwater Crackdown—Hope Springs," *China Water Risk*, July 11, 2013, http://chinawaterrisk.org/resources/analysis-reviews/groundwater -crackdown-hope-springs/.

55. China Water Risk, "China's Water Crisis: Part 1—Introduction," March 2010, http://chinawaterrisk.org/wp-content/uploads/2011/06/Chinas-Water-Crisis-Part-1 .pdf, 7.

56. 2030 Water Resources Group, "Charting Our Water Future," 2010, http:// www.2030waterresourcesgroup.com/water_full/Charting_Our_Water_Future_Final .pdf, 59.

57. People's Government of Guangdong Province, "Guangdong Province Urban-Rural Sewage Treatment and Water Reuse Facilities Infrastructure under 12th Five-Year Guideline of People's Republic of China"; Ediger and Hwang, "Water Quality and Environmental Health in Southern China," 3–4.

58. United Nations Environment Programme, "The Songhua River Spill December 2005: Field Mission Report," http://www.unep.org/PDF/China_Songhua_River _Spill_draft_7_301205.pdf.

59. See, for example, Liu Hongqiao, "The Polluted Legacy of China's Largest Rice-Growing Province," *China Dialogue*, May 30, 2014, https://www.chinadialogue.net/ article/show/single/en/7008-The-polluted-legacy-of-China-s-largest-rice-growing -province.

60. Cited in "All Dried Up," *Economist*, October 12, 2013, http://www.economist
.com/news/china/21587813-northern-china-running-out-water-governments
-remedies-are-potentially-disastrous-all. Several years earlier, in 2007, while still
Water Minister, Wang Shucheng predicted that China's water problems could
become so severe that widespread rationing would need to be instituted over several
decades; however, he also cautioned that such a policy could run into difficulty
among the Beijing leadership and other wealthy residents who owned and maintained two or more apartments, among other reasons for resisting such an approach.
"Down to the Very Last Drop," *South China Morning Post*, April 18, 2007, http://
www.scmp.com/article/589365/down-very-last-drop.

61. Personal communications with Evan Griffith, February 1982.

62. See, for example, the central focus on Guangdong in Hong Kong's evaluation of
its overall water management objectives. Hong Kong Special Administrative Region,
Water Supplies Department, "Total Water Management in Hong Kong: Toward
Sustainable Use of Water Resources," 2008, http://www.wsd.gov.hk/filemanager/
en/share/pdf/TWM.pdf; and Paul Chan, "LCQ17: Total Water Management Strategy
and Related Measures," press release, Hong Kong Special Administrative Region
Government, April 17, 2013, http://www.info.gov.hk/gia/general/201304/17/
P201304170456.htm. See also Lee, "The Changing Nature of Border, Scale and the
Production of Hong Kong's Water Supply System since 1959," 917.

63. This divide between local, provincial, and central governments in China has
also been compounded by the split jurisdictions at the central government level
itself. This divide can be seen in the management of even a single river system,
where, as Ediger and Hwang describe, "the Ministry of Environmental Protection
has the mandate to protect surface water bodies and can issue pollution permits, the
Ministry of Construction manages municipal wastewater treatment, and the Ministry of Health sets drinking water standards." Ediger and Hwang, "Water Quality and
Environmental Health in Southern China," 5.

64. The City of Los Angeles has made some efforts in areas such as wastewater management, water recycling infrastructure, and even rainwater capture, but those
efforts have remained at the level of small pilot projects and yet require a complex
set of planning mechanisms due to the range of jurisdictions involved. See, for
example, City of Los Angeles, "Water IRP 5 Year Review: Final Documents," June
2012, http://www.lacity-irp.org/documents/FINAL_IRP_5_Year_Review_Document
.pdf; Jacques Leslie, "Los Angeles, City of Water," *New York Times*, December
6, 2014, http://www.nytimes.com/2014/12/07/opinion/sunday/los-angeles-city-of-
water.html.

65. Debra Tan, "Just What Is Bottled Water?," *China Water Risk*, March 11, 2011,
http://chinawaterrisk.org/opinions/just-what-is-bottled-water/; Mandy Lao Man-lei

and Carine Lai, "Reducing Plastic Waste in Hong Kong: Public Opinion Survey of Bottled Water Consumption and Attitudes towards Plastic Waste," Civic Exchange, Hong Kong, April 2015; Euromonitor International, "Bottled Water in Hong Kong," July 2014, http://www.euromonitor.com/bottled-water-in-hong-kong-china/report.

66. Natalie Ng Sze-Man, "Why Bottled Water Is NOT the Solution for China's Drinking Water Crisis?," *Globalization Monitor*, October 2013, http://globalmon.org.hk/sites/default/files/attachment/Why%20Bottled%20Water%20is%20not%20the%20solution%20for%20Water%20Crisis_SI.pdf water-crisis.

67. In an interesting survey of public attitudes toward voluntary and mandated conservation, a joint research team from Berkeley and Barcelona revealed that negative attitudes toward private operators were most pronounced when a controversial recent privatization of the utility had taken place. Researchers also identified a greater willingness to conserve in relation to public utilities that were more proactive and target-oriented in their conservation approaches. Giorgis Kallis et al., "Public versus Private: Does It Matter for Water Conservation? Insights from California," *Environmental Management* 45, no. 1 (January 2010), 177–191, http://www.ncbi.nlm.nih.gov/pmc/articles/PMC2815296/.

68. In one episode involving a joint public-private water supply company (Lanzhou Veolia Water Co., 45 percent owned by the global multinational Veolia China, a subsidiary of Veolia Environment, and 55 percent by the Lanzhou city government), a major water quality scandal took place when high levels of benzene were detected in the water supply, more than twenty times the national safety standard. The company failed to immediately alert the public, and when the information was released, the government responded by shutting down service in one area of the city and requesting that all of the 3.6 million residents in the city not drink tap water for the next twenty-four hours. Sui-Lee Wee, "Chinese Court Dismisses Water Pollution Lawsuit," *Reuters*, April 15, 2014, http://www.reuters.com/article/2014/04/15/us-china-water-veolia-idUSBREA3E05P20140415; see also Lijin Zhong, Arthur P. J. Mol, and Tao Fu, "Public-Private Partnerships in China's Urban Water Sector," *Environmental Management*, June 2008, 863–877, http://www.ncbi.nlm.nih.gov/pmc/articles/PMC2359833/; Chris Dodd, "France's Suez Hoping to Clean Up in China," *Finance Asia*, February 25, 2014, http://www.financeasia.com/News/373254,frances-suez-hoping-to-clean-up-in-china.aspx; Au Loong Yu and Wei Yan-Zhu, "The Privatization of Water Supply in China," *Globalization Monitor*. December 8, 2006, http://www.globalmon.org.hk/content/privatization-water-supply-china-0.

69. See Robert Gottlieb and Don Villarejo, "Urban, Rural Unite for a New Water Ethic," *Los Angeles Times*, April 17, 1989, http://articles.latimes.com/1989-04-17/local/me-1877_1_water-users-water-industry-california-water-agencies; David Beckman, "The Threats to Our Drinking Water," *New York Times*, August 7, 2014, http://www.nytimes.com/2014/08/07/opinion/the-threats-to-our-drinking-water.html.

Chapter 5

1. The discussion of the Tyson-LAUSD school menu debacle is partly based on personal observation of one of the authors of this book, who helped found the Los Angeles Food Policy Council and served on its Leadership Board. It is also based on personal communications with David Binkle, then director of Food Services at LAUSD, and LAUSD school board member Steve Zimmer.

2. See Christopher Leonard, *The Meat Racket: The Secret Takeover of America's Food Business* (New York: Simon and Schuster, 2014); Steve Striffler, *Chicken: The Dangerous Transformation of America's Favorite Food* (New Haven: Yale University Press, 2005). Tyson also has a long history of health and safety and wage violations at its various plants. That record is available at the Occupational Health and Safety Administration Integrated Management Information System (IMIS) at https://www.osha.gov/pls/imis/establishment.html, and at the US Department of Labor's Data Enforcement website at http://ogesdw.dol.gov/views/search.php. The Tyson information has been compiled by the Los Angeles Food Policy Council as part of its assessment of suppliers for its Good Food Purchasing Policy.

3. Tyson Foods, International Operations, http://www.tysonfoods.com/Working-Smart/Around-the-World/International-Operations.aspx.

4. International Centre for Trade and Sustainable Development, "China Sets Anti-Dumping Penalties on US Poultry Imports," *Bridges* 14, no. 5 (February 10, 2010), http://www.ictsd.org/bridges-news/bridges/news/china-sets-anti-dumping-penalties-on-us-poultry-imports; US Department of Agriculture, "Chickens, Turkeys, and Eggs: Annual and Cumulative Year-to-Date US Trade—All Years and Countries," May 11, 2014, http://www.ers.usda.gov/datafiles/Livestock__Meat_International_Trade_Data/Annual_and_Cumulative_YeartoDate_US_Livestock_and_Meat_Trade_by_Country/BroilerTurkey_YearlyFull.xls; "US Poultry, Egg Exports Set Records in 2011," *World Poultry*, February 15, 2012, http://www.worldpoultry.net/Broilers/Markets--Trade/2012/2/US-poultry-egg-exports-set-records-in-2011-WP009994W/; Kim Souza, "US to Accept Some China Poultry Imports," *The City Wire*, September 4, 2013, http://talkbusiness.net/2013/09/u-s-to-accept-some-chinese-poultry-imports/#.UuvIyH7TnIU.

5. Tyson Foods Fiscal 2015 Fact Book, "Sales by Distribution Channel," http://ir.tyson.com/investor-relations/investor-overview/tyson-factbook/.

6. Personal communication with David Binkle, January 30, 2014; Teresa Watanabe, "L.A. Schools' Healthful Lunch Menu Panned by Students," *Los Angeles Times*, December 17, 2011, http://articles.latimes.com/2011/dec/17/local/la-me-food-lausd-20111218.

7. Comments by LAUSD school board member Steve Zimmer at Occidental College, February 11, 2015; see also Alexa Delwiche and Joann Lo, "Los Angeles' Good Food

Purchasing Policy: Worker, Farmer and Nutrition Advocates Meet ... and Agree!,"
Progressive Planning (Fall 2013), 24–28. In 2015, the success of the Good Food Purchasing Policy in Los Angeles led its organizers to establish a national Center for
Good Food Purchasing (http://www.cfgfp.org/) to help set up similar procurement
programs in school districts, hospitals, government offices, museums, and other
institutions across the country. Personal communication with Alexa Delwiche, June
4, 2015.

8. Rachel Surls and Judith Gerber, *From Cows to Concrete: The Rise and Fall of Farming
in Los Angeles* (Los Angeles: Angel City Press, 2016).

9. Juan Crespí, *A Description of Distant Roads: Original Journals of the First Expedition
into California, 1769–1770*, ed. and trans. Alan K. Brown (San Diego: San Diego State
University Press, 2001), 337.

10. William Deverell, *Whitewashed Adobe: The Rise of Los Angeles and the Remaking of
Its Mexican Past* (Berkeley: University of California Press, 2005).

11. Sucheng Chan, *This Bittersweet Soil: The Chinese in California Agriculture, 1860–
1910* (Berkeley: University of California Press, 1986), 115, 159. See also Matt Garcia,
*A World of Its Own: Race, Labor, and Citrus in the Making of Greater Los Angeles, 1900–
1970* (Chapel Hill: University of North Carolina Press, 2001), 54–55; Steven Stoll,
The Fruits of Natural Advantage: Making the Industrial Countryside in California (Berkeley: University of California Press, 1998). Stoll argues that the Chinese taught the
fruit growers strategies for pruning, packing, and "in general how to manage a clean
garden" (228, n. 35).

12. Garcia, *A World of Its Own*, 22–23; Carey McWilliams, *Southern California: An
Island on the Land* (Santa Barbara, CA: Peregrine Smith, 1973), 217; Ronald Tobey
and Charles Wetherell, "The Citrus Industry and the Revolution of Corporate Capitalism in Southern California, 1887–1944," *California History* 74, no. 1 (Spring 1995),
9; Edward D. Melillo, "The First Green Revolution: Debt Peonage and the Making of
the Nitrogen Fertilizer Trade, 1840–1930," *American Historical Review* 117, no. 4
(October 2012), 1028–1060. Melillo argues that the trade in nitrate and guano fertilizer constituted a "first green revolution" that also laid the groundwork, with its
Pacific nexus of inputs, labor, trade, and corporate consolidation, for the subsequent
or second green revolution and the global food-system changes of the 1950s through
1970s. See also H. Vincent Moses, "G. Harold Powell and the Corporate Consolidation of the Modern Citrus Enterprise, 1904–1922," *Business History Review* 69, no. 2
(1995), 119–155.

13. Daniel Cletus, *A History of California Farmworkers, 1870–1941* (Ithaca: Cornell
University Press, 1981); Don Mitchell, "The Scales of Justice: Localist Ideology,
Large-Scale Production, and Agricultural Labor's Geography of Resistance in 1930s
California," in Andrew Herod, ed., *Organizing the Landscape: Geographical Perspectives
on Labor Unionism* (Minneapolis: University of Minnesota Press, 1998); Carey

McWilliams, *Factories in the Field: The Story of Migratory Farm Labor in California* (Santa Barbara, CA: Peregrine Publishers, 1971).

14. Air quality impacts, particularly on leafy vegetables, were increasingly noted as early as the mid to late 1940s, with one estimate of the loss of value in LA agriculture, during a single week of a severe decline in air quality, set at a quarter of a million dollars (or the equivalent of $3.3 million in 2016 dollars). Ronald Schiller, "The Los Angeles Smog," *National Municipal Review* (December 1955), 560.

15. Surls and Gerber, *From Cows to Concrete.*

16. Po-keung Hui, "Comprador Capitalism and Middleman Capitalism," in Takwing Ngo, ed., *Hong Kong's History: State and Society under Colonial Rule* (London: Routledge, 1999), 36.

17. Hong Kong Heritage Museum, "Hong Kong's Food Culture," December 2, 1999, http://www.heritagemuseum.gov.hk/documents/2199315/2199693/Hong_Kong _Food_Culture-E.pdf, accessed August 2, 2016.

18. Chris Li, "Hong Kong Exporter Guide, 2013," Global Agricultural Information Network Report no. HK 1317, US Department of Agriculture, Foreign Agricultural Service, May 1, 2013, http://gain.fas.usda.gov/Recent%20GAIN%20Publications/ Exporter%20Guide_Hong%20Kong_Hong%20Kong_4-30-2013.pdf.

19. Chan Shun-hing, "Understanding Anew the Value of an Everyday Life with Its Roots in *Nong*," in Partnerships for Community Development, *Touching the Heart, Taking Root: CSA in Hong Kong, Taiwan and Mainland China* (Hong Kong: Partnerships for Community Development, 2015), 13, http://www.pcd.org.hk/en/content/ touching-heart-taking-root-csa-hong-kong-taiwan-and-mainland-china.

20. Hong Kong Special Administrative Region, Agriculture, Fisheries, and Conservation Department, "Agriculture in HK," March 10, 2016, http://www.afcd.gov.hk/ english/agriculture/agr_hk/agr_hk.html.

21. Hong Kong Legislative Council, Panel on Food Safety and Environmental Hygiene, "Prevention and Control of Avian Influenza," November 19, 2013, LC Paper no. CB(2)277/13–14(03), http://www.legco.gov.hk/yr13-14/english/panels/ fseh/papers/fe1119cb2-277-3-e.pdf; Chris Li, "Hong Kong: Retail Foods. Annual 2015," Global Agricultural Information Network Report no. HK 1532, US Department of Agriculture, Foreign Agricultural Service, http://gain.fas.usda.gov/ Recent%20GAIN%20Publications/Retail%20Foods_Hong%20Kong_Hong%20 Kong_12-9-2015.pdf.

22. Chris Li and Annie Lai, "Hong Kong Exporter Guide, 2015," Global Agricultural Information Network (GAIN) Report no. HK 1528, prepared for US Department of Agriculture, Foreign Agricultural Service, November 10, 2015, 2, http://gain.fas.usda .gov/Recent%20GAIN%20Publications/Exporter%20Guide_Hong%20Kong

_Hong%20Kong_11-10-2015.pdf; Li, "Hong Kong: Retail Foods. Annual 2015," 14–22.

23. Chang-won Lee et al., "H5N2 Avian Influenza Outbreak in Texas in 2004: The First Highly Pathogenic Strain in the United States in 20 Years?," *Journal of Virology* 79, no. 17 (September 2005), 11412–21, http://www.ncbi.nlm.nih.gov/pmc/articles/ PMC1193578/?tool=pubmed; Li, "Hong Kong: Retail Foods. Annual 2015," 14–15, 19.

24. Angus Lam, "Farmers' Market: Manifesting the Spirit of Everyday Life," in Partnerships for Community Development, *Touching the Heart, Taking Root*, 41.

25. Hong Kong Special Administrative Region, Agriculture, Fisheries, and Conservation Department, "Organic Farming in Hong Kong," 2014, http://www.afcd.gov.hk/ textonly/english/agriculture/agr_orgfarm/agr_orgfarm.html; see also Tracy Wong, "The Organic Food Market: U.S. and Hong Kong," Occidental College, 2013, http:// www.oxy.edu/sites/default/files/assets/UEP/Summer_Research/Tracy%20 research%20Organic.pdf; "Sustainable Development of Local Agriculture," press release, Hong Kong Legislative Council, January 21, 2015, http://www.info.gov.hk/ gia/general/201501/21/P201501210568.htm.

26. Chow Sung Ming, "Sharing Hong Kong: From Social and Solidarity Economy to Sharing Economy, and from Fair Trade to Community Supported Agriculture," paper submitted to the Asian Solidarity Economy Council, 5th RIPESS International Meeting, Manila, Philippines, October 15–18, 2013, http://www.google.com/url?sa =t&rct=j&q=&esrc=s&source=web&cd=1&ved=0CCQQFjAA&url=http%3A%2F%2F www.ripess.org%2Fwp-content%2Fuploads%2F2013%2F06%2FChow-Sung-Ming -Sharing-Hong-Kong-1.doc&ei=Q3U-VfLHEILlsAWWioGgAg&usg=AFQjCNGs7 -wBBcqFFT-HDVIJSNpbz3geXQ&sig2=jhgcB_t2gcomLfIgQIGQkw.

27. "Seelig's Chain Is Now Safeway," *Los Angeles Times*, March 15, 1925, B8; Marc Levinson, *The Great A&P and the Struggle for Small Business in America* (New York: Hill and Wang, 2011), 128.

28. Marion Bruce, "Concentration-Relationship in Food Retailing," in Leonard W. Weiss, ed., *Concentration and Price* (Cambridge, MA: MIT Press, 1989), 183–194; Alden Manchester, "The Transformation of U.S. Food Marketing," in Lyle P. Schertz and Lynn M. Daft, eds., *Food and Agricultural Markets: The Quiet Revolution*, USDA Economic Research Service, National Planning Association, 1994. On the relationship of neighborhood markets to a sense of place, see Richard Longstreth, *The Drive-In, the Supermarket, and the Transformation of Commercial Space in Los Angeles, 1914–1941* (Cambridge, MA: MIT Press, 1999), 111.

29. Amanda Shaffer, "The Persistence of L.A.'s Grocery Gap: The Need for a New Food Policy and Approach to Market Development," Urban & Environmental Policy Institute, Occidental College, 2002, http://scholar.oxy.edu/cgi/viewcontent .cgi?article=1395&context=uep_faculty; Samina Raja et al., "Beyond Food Deserts:

Measuring and Mapping Racial Disparities in Neighborhood Food Environments,"
Journal of Planning Education and Research 27, no. 4 (2008), 469–482; Jane Black,
"Food Deserts versus Food Swamps: The USDA Weighs In," *Washington Post*,
June 25, 2009, http://voices.washingtonpost.com/all-we-can-eat/food-politics/food
-deserts-vs-swamps-the-usd.html. Several organizations based in low-income com-
munities have begun to use the term "food apartheid" to highlight the intent as well
as the outcome of supermarket abandonment. Comments by Marqueece Harris
Dawson, Los Angeles City Councilmember, at LA Food Policy Council Food Gala,
June 11, 2015.

30. Walmart was not the first to explore a "supercenter" (or what European retailers
called a "hypermarket") format in the United States. The French global retailer
Carrefour sought to enter the US market in 1988 with its hypermarket approach,
opening a megastore in Philadelphia, with a 330,000-square-foot footprint, 60
checkouts, roller skating grounds, and wide corridors. But Carrefour was unable to
secure a role in the highly competitive grocery retail environment and pulled out of
the United States in 1993. Scott Flander, "Au Revoir, Carrefour Carrefour [sic] to
U.S.: Au Revoir!: Two Area Stores to Close Soon," *Philadelphia Daily News*, September
7, 1993, http://articles.philly.com/1993-09-07/news/25984620_1_stores-low-prices
-philadelphia-area.

31. Tim Lang and Michael Heasman, *Food Wars: The Global Battle for Mouths, Minds
and Markets* (London: Earthscan, 2004), 139; Bobby J. Martens, Frank Dooley, and
Sounghun Kim, "The Effect of Entry by Wal-Mart Supercenters on Retail Grocery
Concentration," paper presented at the 2006 America Agricultural Economics Association Annual Meeting, Long Beach, California, http://ageconsearch.umn.edu/
bitstream/21101/1/sp06ma03.pdf; Nelson Lichtenstein, *The Retail Revolution: How
Wal-Mart Created a Brave New World of Business* (New York: Metropolitan Books,
2009).

32. Walmart is not the only big-box store to explore the grocery business. One of its
rivals, Target, also expanded its grocery line, in part as a strategy to help drive customers to its stores. In 2014, Target's selection of its new CEO, Brian Conwell, who
had prior experience at Sam's Club (a Walmart chain) and Pepsico, highlighted its
interest in grocery expansion. Kavita Kumar, "Target Hints at Grocery Makeover
after Reporting Stronger Sales," *Minneapolis Star Tribune*, February 25, 2015.

33. Amanda Shaffer et al., "Shopping for a Market: Evaluating Tesco's Entry into Los
Angeles and the United States," Occidental College, Urban & Environmental Policy
Institute, August 1, 2007, http://clkrep.lacity.org/onlinedocs/2010/10-1537_misc
_plum_12-8-10.pdf; Evelyn Iratini, "Retail Giant Cries Unfair: Wal-Mart Chief's
Remarks That a British Rival Might Be Too Big Raises Critics' Eyebrows," *Los Angeles
Times*, September 5, 2005; Jessica Garrison et al., "Wal-Mart to Push Southland
Agenda: Retail Giant Downplays Inglewood Defeat, Vows to Continue Its Drive in
the Region; Opponents Say Battle Could Be Repeated in Other Cities," *Los Angeles*

Times, April 8, 2004; Sara Lin and Monte Morin, "Voters in Inglewood Turn Away Wal-Mart," *Los Angeles Times*, April 7, 2004.

34. Tiffany Hsu, "Fresh & Easy Fail: Tesco Exits US after Profit Tanks 96%," *Los Angeles Times*, April 17, 2013, http://articles.latimes.com/2013/apr/17/business/la-fi-mo -fresh-easy-tesco-us-20130417.

35. Richard Guzman, "As Wal-Mart Opening Nears, Division Remains in China-town," *DT News*, August 26, 2013, http://www.ladowntownnews.com/news/as-wal -mart-opening-nears-division-remains-in-chinatown/article_8e9ce710-0c4b-11e3 -adc5-0019bb2963f4.html; Denver Nicks, "Walmart Food Drive for Employees Raises Questions about Wages," *Time*, November 18, 2013, http://business.time .com/2013/11/18/walmart-seeks-food-donations-to-help-needy-employees/; Saman-tha Masunaga and Ivan Penn, "Wal-Mart to Shut 269 Stores," *Los Angeles Times*, January 16, 2016.

36. Jesse Azrilian et al., "Creating Healthy Corner Stores: An Analysis of Factors Necessary for Effective Corner Store Conversions," Los Angeles Food Policy Coun-cil, May 2012, http://goodfoodla.org/wp-content/uploads/2013/06/Pages-from -Creating-Healthy-Corner-Stores-Report-prepared-for-LAFPC.11.pdf.

37. Carrefour did seek entry in Hong Kong in 1996 and operated four large super-markets there until September 2000 when it closed down, citing an inability to access large-enough sites to accommodate its large-format store operation and lack of cooperation from a number of its suppliers which undermined Carrefour's low-cost pricing strategy. Victoria Button, Antoine So, and Chow Chung-Yan, "Supermarket Giants 'Will Continue Unchallenged,'" *South China Morning Post*, June 21, 2001, http://www.scmp.com/article/350532/supermarket-giants-will-continue -unchallenged.

38. Hong Kong Consumer Council, "Grocery Market Study: Market Power of Super-market Chains under Scrutiny," December 19, 2013, https://www.consumer.org .hk/sites/consumer/files/competition_issues/20131219/GMSReport20131219.pdf.

39. The phrase "captive base of customers," referencing the 2003 Consumer Coun-cil report, is from Mark Williams, "The Supermarket Sector in China and Hong Kong: A Tale of Two Systems," *Competition Law Review* 3, no. 2 (March 2007), 263; Hong Kong Legislative Council on Economic Services, "Consumer Council's Report on Wet Markets versus Supermarkets: Competition in the Retailing Sector," LC Paper no. CB(1)1017/03–04(03), February 23, 2004, http://www.legco.gov.hk/yr03 -04/english/panels/es/papers/es0223cb1-1017-3e.pdf.

40. Oswald Chan, "Mighty Li Takes Stock of Empire," *China Daily USA*, January 15–17, 2016; Hutchison Whampoa Limited, "About HWL," http://www.hutchison-whampoa.com/en/about/overview.php; see also Alice Poon, *Land and the Ruling Class in Hong Kong* (Hong Kong: Enrich Professional Publishing, 2010), 29–32; Pru-dence Ho, "Li Ka-shing's Hutchison Stops Shopping Parknshop," *Wall Street Journal*,

October 19, 2013, http://www.wsj.com/articles/SB1000142405270230438410457914
3342398492638; Vinicy Chan and Eleni Himaras, "Li Ka-shing Watson Spinoff May Bring Biggest Asia IPO in Three Years," *Bloomberg*, October 21, 2013, http://www.bloomberg.com/news/2013-10-18/hutchison-scraps-parknshop-private-sale-after-review.html.

41. Dairy Farm International Holdings, "Our Company," http://www.dairyfarm group.com/en-US/Our-Company/Our-History.

42. For example, Simon Wong Ka-wo, the President of the Federation of Restaurants and Related Trades stated that one of the two supermarket duopolies, ParknShop, charged up to HK$300,000 to place a new product on shelves. Amy Nip, "Grocery Giants ParknShop, Wellcome Accused of Pressuring Suppliers," *South China Morning Post*, December 20, 2013, http://www.scmp.com/news/hong-kong/article/1386263/grocery-giants-parknshop-wellcome-accused-pressuring-suppliers?page=all; Hong Kong Consumer Council, "Grocery Market Study: Market Power of Supermarket Chains under Scrutiny," December 19, 2013, http://www.consumer.org.hk/competition_issues/grocery/GMSReport20131219.pdf.

43. Kathleen Megan Blake, "Ordinary Food Places in a Global City: Hong Kong," *Streetnotes* 21 (2013), 1–12, http://escholarship.org/uc/item/11d2j987#page-1. Historically, the hawkers were identified with both low-income vendors and low-income neighborhoods. Since the 1970s when licensing systems became more stringent, the government established enforcement policies against the vendors that have become a continuing flashpoint and were the occasion for the fishball riots of February 2016. See Jennifer Ngo et al., "Closing Time: How Hong Kong's Hawkers Face a Struggle to Survive," *South China Morning Post*, November 23, 2014, http://www.scmp.com/news/hong-kong/article/1646845/closing-time-how-hong-kongs-hawkers-face-struggle-survive.

44. Linda Ashman et al., "Seeds of Change: Strategies for Food Security for the Inner City," UCLA Department of Urban Planning and the Interfaith Hunger Coalition, 1993; Robert Gottlieb, *Environmentalism Unbound: Exploring New Pathways for Change* (Cambridge, MA: MIT Press, 2001), 184–185.

45. Wendelin Slusser et al., "A School Salad Bar Increases Frequency of Fruit and Vegetable Consumption among Children Living in Low-Income Households," *Public Health Nutrition* (December 2007), 1490–1496, http://journals.cambridge.org/action/displayAbstract?fromPage=online&aid=1427932; Robert Gottlieb and Anupama Joshi, *Food Justice* (Cambridge, MA: MIT Press, 2013), 68–69.

46. Cynthia L. Ogden et al., "Prevalence of Childhood and Adult Obesity in the United States, 2011–2012," *Journal of the American Medical Association* 311, no. 8 (2014), 806–814, http://jama.jamanetwork.com/article.aspx?articleid=1832542; Let's Move, www.letsmove.gov/.

47. United States Bureau of Labor Statistics, "Consumer Expenditures for the Los Angeles Area, 2013–2014," November 10, 2015, http://www.bls.gov/regions/west/news-release/ConsumerExpenditures_LosAngeles.htm.

48. See, for example, Michael Moss, *Salt, Sugar, Fat: How the Food Giants Hooked Us*, rev. ed. (New York: Random House, 2014).

49. UCLA Center for Health Policy Research, "Child and Teen, 2011-2012 Health Profiles: Los Angeles County," UCLA Center for Health Policy Research, March 2014, http://healthpolicy.ucla.edu/health-profiles/Child_Teen/Documents/2011-2012/Regions/LosAngelesCounty.pdf; Francine Kaufman, *Diabesity: A Doctor and Her Patients on the Front Lines of the Obesity-Diabetes Epidemic* (New York: Bantam, 2006).

50. Mary Kaye Schilling, "The Silver Tongue: Jonathan Gold," *Town and Country*, August 6, 2014, http://www.townandcountrymag.com/leisure/dining/a1646/jonathan-gold-la-times/.

51. Mark Vallianatos and Elizabeth Medrano, "The Transformation of the School Food Environment in Los Angeles: The Link between Grass Roots Organizing and Policy Implementation," July 2009, Center for Food and Justice/Healthy School Food Coalition, Urban & Environmental Policy Institute, http://scholar.oxy.edu/cgi/viewcontent.cgi?article=1431&context=uep_faculty.

52. Centre for Food Safety, "Hong Kong Population-Based Food Consumption Survey: 2005–2007, Final Report," Chinese University of Hong Kong Food and Environmental Hygiene Department, April 2010, http://www.cfs.gov.hk/english/programme/programme_firm/files/FCS_final_report.pdf, x–xi. Other figures about obese or overweight populations in Hong Kong have varied, but they are still high numbers, ranging from 36 percent to the study's 47.1 percent assessment. These also may reflect differences even within the World Health Organization about ways to measure BMI in adult Asian populations. See "Health Facts of Hong Kong," 2013, http://www.dh.gov.hk/english/statistics/statistics_hs/files/Health_Statistics_pamphlet_E.pdf, and the World Health Organization, "Appropriate Body-Mass Index for Asian Populations and Its Implications for Policy and Intervention Strategies," http://www.who.int/nutrition/publications/bmi_asia_strategies.pdf.

53. See, for example, Leo F. Goodstadt, *Poverty in the Midst of Affluence: How Hong Kong Mismanaged Its Prosperity* (Hong Kong: Hong Kong University Press, 2013).

54. Oxfam Hong Kong, "Survey on the Impact of Soaring Food Prices on Poor Families in Hong Kong," August 2011, http://www.oxfam.org.hk/filemgr/1630/FoodSurveyReportAug2011_revised.pdf. See also Feeding Hong Kong, "Hunger Stats," http://feedinghk.org/hunger-stats/; Goodstadt, *Poverty in the Midst of Affluence*, 73.

55. Hong Kong Special Administrative Region, Department of Health, "Action Plan to Promote Healthy Diet and Physical Activity Participation in Hong Kong," 2013,

http://www.change4health.gov.hk/filemanager/common/image/strategic
_framework/action_plan/action_plan_e.pdf; Hong Kong Special Administrative
Region, Centre for Health Protection, "Assessment of Dietary Patterns in Primary
Schools 2008," July 2009, http://www.chp.gov.hk/files/pdf/report_part1_english
.pdf. The Hong Kong government, recognizing the impact of this dietary shift,
established in 2015 an international panel designed to identify strategies to reduce
sodium and sugar in residents' diets. See Hong Kong Special Administrative Region,
Centre for Food Safety, "Reduction of Dietary Sodium and Sugar," last modified May
7, 2016, http://www.cfs.gov.hk/english/programme/programme_rdss/programme
_rdss.html.

56. He Zhixiong, Zhang Lanying, and Wen Tiejun, "New Rural Regeneration in
Contemporary China," in Partnerships for Community Development, *Touching the
Heart, Taking Root*, 111.

57. G. William Skinner, "Vegetable Supply and Marketing in Chinese Cities," *China
Quarterly*, no. 76 (December 1978); Graeme Lang and Bo Miao, "Food Security for
China's Cities," *International Planning Studies* 18, no. 1 (2013).

58. Dorothy J. Solinger, *Contesting Citizenship in Urban China: Peasant Migrants, the
State, and the Logic of the Market* (Berkeley: University of California Press, 1999); Jikun
Huang, Xiaobing Wang, and Huanguang Qiu, "Small-Scale Farmers in China in the
Face of Modernization and Globalization," May 2012, International Institute for
Environment and Development (IIED/Hivos), http://pubs.iied.org/16515IIED.html,
35; Kam Wing Chan, "The Global Financial Crisis and Migrant Workers in China:
'There Is No Future as a Labourer; Returning to the Village Has No Meaning,'" *Inter-
national Journal of Urban and Regional Research* 34, no. 3 (September 2010); Yan Hai-
rong, *New Masters, New Servants: Migration, Development, and Women Workers in
China* (Durham: Duke University Press, 2008).

59. China National Bureau of Statistics, "Basic Statistics on National Population
Census: 1953, 1964, 1982, 1990, 2000, 2010," 2013, http://www.stats.gov.cn/tjsj/
ndsj/2013/html/Z0308E.HTM; Zhengzhou Yang, "Demographic Changes in China's
Farmers: The Future of Farming in China," *Asian Social Science* 9, no. 7 (2013), 136–
143.

60. You-tien Hsing, *The Great Urban Transformation: Politics of Land and Property in
China* (New York: Oxford University Press, 2010), 75; Philip C. C. Huang, Gao Yuan,
and Yusheng Peng, "Capitalization without Proletarianization in China's Agricultural Development," *Modern China* 38, no. 2 (2012), 159; Kam Wing Chan, "Crossing the 50 Percent Population Rubicon: Can China Urbanize to Prosperity?,"
Eurasian Geography and Economics 53, no. 1 (2012); Yang, "Demographic Changes
in China's Farmers," 140; Aijuan Chen and Steffanie Scott, "Rural Development
Strategies and Government Roles in The Development of Farmers' Cooperatives
in China," *Journal of Agriculture, Food Systems, and Community Development* 4,

no. 4 (2014), 35–55, http://www.agdevjournal.com/volume-4-issue-4/471-farmers -cooperatives-china.html.

61. Philip C. C. Huang, "China's New-Age Small Farms and Their Vertical Integration: Agribusiness or Coops?," *Modern China* 37, no. 2 (2011); see also Qian Forrest Zhang and John A. Donaldson, "The Rise of Agrarian Capitalism with Chinese Characteristics: Agricultural Modernization, Agribusiness and Collective Land Rights," *China Journal*, no. 60 (July 2008), 26–27; Huang, Yuan, and Peng, "Capitalization without Proletarianization in China's Agricultural Development"; Yongshun Cai, *Collective Resistance in China: Why Popular Protests Succeed or Fail* (Stanford: Stanford University Press, 2010), 54; Richard Sanders, *Prospects for Sustainable Development in the Chinese Countryside: The Political Economy of Chinese Ecological Agriculture* (Aldershot, UK: Ashgate Publishing, 2000); He Guangwei, "In China's Heartland: A Toxic Trail Leads from Factories to Food," *Yale Environment 360*, July 7, 2014, http://e360 .yale.edu/feature/chinas_toxic_trail_leads_from_factories_to_food/2784/; Edward Wong, "One-Fifth of China's Farmland Is Polluted, State Study Finds," *New York Times*, April 17, 2014, http://www.nytimes.com/2014/04/18/world/asia/one-fifth-of-chinas-farmland-is-polluted-state-report-finds.html.

62. Huang, Yuan, and Peng, "Capitalization without Proletarianization in China's Agricultural Development," 166.

63. Emelie K. Peine, "Trading on Pork and Beans: Agribusiness and the Construction of the Brazil-China-Soy-Pork Commodity Complex," in H. S. James Jr., ed., *The Ethics and Economics of Agrifood Competition* (New York: Springer, 2013); see also Mindi Schneider, "Feeding China's Pigs: Implications for the Environment, China's Smallholder Farmers and Food Security," Institute for Trade and Agricultural Policy, Minneapolis, May 2011.

64. Pi Chendong with Zhang Rou and Sarah Horowitz, "Fair or Fowl? Industrialization of Poultry Production in China," Institute for Agriculture and Trade Policy, Minneapolis, February 2014, http://www.iatp.org/documents/fair-or-fowl -industrialization-of-poultry-production-in-china.

65. As of 2016, China had yet to export any processed or slaughtered chicken products to the US, although the 2013 equivalence ruling of USDA remained intact. US Department of Agriculture, Food Safety and Inspection Service, "Frequently Asked Questions: Equivalence of China's Poultry Processing and Slaughter Inspection Systems," March 4, 2016, http://www.fsis.usda.gov/wps/portal/fsis/newsroom/news -releases-statements-transcripts/news-release-archives-by-year/archive/2016/faq -china-030416. See also US Department of Agriculture, Food Safety and Inspection Service, "Frequently Asked Questions: Equivalence of China's Poultry Processing System," September 26, 2013, http://www.fsis.usda.gov/wps/portal/fsis/newsroom/ news-releases-statements-transcripts/news-release-archives-by-year/archive/2013/ faq-china-08302013; Brian Wingfield and Shruti Date Singh, "Chicken Processed in China Triggers U.S. Food Safety Protests," *Bloomberg*, September 26, 2013,

http://www.bloomberg.com/news/2013-09-27/chicken-processed-in-china-triggers
-u-s-food-safety-protests.html.

66. The newly named WH Group also changed its logo to signify (as a global player) "four streams representing the Earth's four major oceans, separating the space into five parts to represent the world's continents," as its website put it (http://www .shuanghui-international.com/). Although most press accounts described the merger as Shuanghui's (WH Group) effort to obtain a secure product and operations that could counter food safety concerns, Mark Bittman pointed out that Smithfield's own environmental and health-related track record was problematic, including its use of CAFOs, antibiotics, and energy- and water-intensive operations. The deal, Bittman argued, essentially transferred "the environmental damage of large-scale pork production from China to the United States without even guaranteeing us pork with as few chemicals as that shipped to China." Bittman, "On Becoming China's Farm Team," *New York Times*, November 5, 2013, http://www.nytimes.com/2013/ 11/06/opinion/bittman-on-becoming-chinas-farm-team.html. And see also Tom Philpott, "Are We Becoming China's Factory Farm?," *Mother Jones*, November-December 2013, http://www.motherjones.com/media/2014/03/china-factory-farm -america-pork?google_editors_picks=true.

67. Dinghuan Hu et al., "The Emergence of Supermarkets with Chinese Characteristics: Challenges and Opportunities for China's Agricultural Development," *Development Policy Review* 22, no. 5 (2004), 558–560.

68. Williams, "The Supermarket Sector in China and Hong Kong," 254.

69. Fung Business Intelligence Centre, "China Retail: Hypermarkets and Supermarkets," Kowloon, Hong Kong, February 2015, http://www.funggroup.com/eng/ knowledge/research/china_dis_issue125.pdf; Louise Herring, Daniel Hui, Paul Morgan, and Caroline Tufft, "Inside China's Hypermarkets: Past and Prospects," *McKinsey & Company*, May 2012, https://www.mckinseyonmarketingandsales.com/ sites/default/files/pdf/Inside_China_Hypermarket.pdf; Brad Gilmour and Fred Gale, "A Maturing Food Retail Sector: Wider Channels for Food Imports?," in "China's Food and Agriculture: Issues for the 21st Century," US Department of Agriculture, Economic Research Service, Agriculture Information Bulletin no. AIB-775, April 2002, 14–16.

70. In an unusual turn for Walmart, a local unit of the government-sanctioned All China Federation of Trade Unions (ACFTU) led demonstrations in 2014 against the slated closure of one of Walmart's stores in Changde, in Hunan province. The demonstrations occurred despite ACFTU's long-standing role of being amenable to the company's objectives, including helping Walmart secure several of its previous store locations. The unprecedented union action in Changde potentially represented the increased bargaining power of workers, not just on wage issues but also in contesting Walmart's ability to make decisions without taking into account the impact on its workforce. See Tom Mitchell and Barney Jopson, "Official China Union Raises

Stakes in Walmart Closure Programme," *Financial Times*, March 23, 2014, https://next.ft.com/content/2038fd78-b262-11e3-b891-00144feabdc0; Barney Jopson, "Walmart Slows International Expansion," *Financial Times*, August 16, 2012, http://www.ft.com/intl/cms/s/0/876f65a4-e79d-11e1-86bf-00144feab49a.html?siteedition =intl#axzz2x1H3V6Sm; Matthew Kirdahy, "Walmart Strikes Pay Deal with Chinese Union," *Forbes*, July 25, 2008, http://www.forbes.com/2008/07/25/walmart-china -unions-lead-cx_mk_0725acftu.html. Walmart also attempted to demonstrate its commitment to a sustainability agenda in China (similar to its high-visibility environmental claims in the United States) by seeking to get its suppliers to reduce energy use and chemical inputs. While initially receiving some favorable press, its auditing system was challenged by critics and was connected by the press to the ability of those suppliers to hide, adjust, or make phony assertions about their environmental claims. See Andy Kroll, "Are Walmart's Chinese Factories as Bad as Apple's?," *Mother Jones*, March–April 2012, http://www.motherjones.com/ print/161316; Michelle Chen, "Walmart Empire Clashes with China," *The Progressive*, April 2014, http://www.progressive.org/walmart-empire-clashes-with-china; Orville Schell, "How Walmart Is Changing China," *Atlantic*, December 2011, http://www.theatlantic.com/magazine/print/2011/12/how-walmart-is-changing -china/8709. See also Wang Zhuoqiong, "Wal-Mart to Open 60 New Stores by 2017," *China Daily*, April 8–10, 2016.

71. Laurie Burkitt and Peter Evans, "Tesco, China Resources in Talks about Joint Supermarket Venture in China," *Wall Street Journal*, August 9, 2013, http://online .wsj.com/news/articles/SB10001424127887323977304579001752585974952; Chad Bray and Neil Gough, "Tesco Plans Venture with Leading Chinese Grocery Chain," *New York Times*, October 1, 2013, http://dealbook.nytimes.com/2013/10/01/tesco -to-pay-550-million-in-supermarket-tie-up-in-china/?_php=true&_type=blogs.

72. Haiyan Liu et al., "Household Composition and Food Away from Home Expenditures in Urban China," paper presented at the Agricultural and Applied Economics Association (formerly the American Agricultural Economics Association), August 12–14, 2012, Seattle, Washington, http://ageconsearch.umn.edu/ bitstream/131057/3/Household%20Composition%20and%20Food%20Away%20 From%20Home%20Expenditures%20in%20Urban%20China.pdf. The increased consumption levels are from the *Statistical Yearbook of China*, 1994 and 1995, cited by Richard Smith, "Creative Destruction: Capitalist Development and China's Environment," *New Left Review* I-222 (March–April 1997), http://newleftreview.org/I/222/ richard-smith-creative-destruction-capitalist-development-and-china-s -environment.

73. Huang, Yuan, and Peng, "Capitalization without Proletarianization in China's Agricultural Development," 151; Barry Popkin and Shufa Du, "Dynamics of the Nutrition Transition toward the Animal Foods Sector in China and Its Implications: A Worried Perspective," *Journal of Nutrition* 133 (November 2003), 3898S–3906S; World Bank Office, Beijing, "China Economic Update. Special Topic: Changing

Food Consumption Patterns in China: Implications for Domestic Supply and International Trade," June 2014, http://www.worldbank.org/content/dam/Worldbank/document/EAP/China/China_Economic_Update_June2014.pdf.

74. Don Weinland, "Sale of Yum Brands China Franchise Stalls," *Financial Times*, July 26, 2016, https://next.ft.com/content/b9a6bd32-531f-11e6-befd-2fc0c26b3c60. In recent years both KFC and McDonald's also experienced a sequence of food scares (e.g., their use of excessive amounts of antibiotics and sale of expired meat) that led to depressed sales and only reinforced the more general concern over food safety that impacted all major producers, suppliers, restaurants, and stores as well as other fast food chains such as Burger King and Starbucks. See Brianna Sacks, "Fast Food Titans Sorry over China Scandal," *Los Angeles Times*, July 22, 2014; Tom Huddleston Jr., "Chinese Meat Scandal Hurts Sales at Fast-Food Giant Yum Brands," *Fortune*, October 7, 2014, http://fortune.com/2014/10/07/chinese-meat-scandal-hurts-sales -at-fast-food-giant-yum-brands/.

75. The *Lancet* article also reported that globally more than 32 percent of all men and women age twenty or older were overweight or obese, and also identified the rapid increase in weight gain. Marie Ng et al., "Global, Regional, and National Prevalence of Overweight and Obesity in Children and Adults during 1980–2013: A Systematic Analysis for the Global Burden of Disease Study, 2013," *Lancet* 384, no. 9945 (August 30, 2014), http://www.thelancet.com/pdfs/journals/lancet/PIIS0140 -6736(14)60460-8.pdf; Jeanette Wang, "62m People in China Considered Obese," *South China Morning Post*, May 30, 2014, http://www.scmp.com/lifestyle/health/article/1521011/21-billion-people-are-obese-china-and-us-among-worlds-fattest -new; Paul French and Matthew Crabbe, *Fat China: How Expanding Waistlines Are Changing a Nation* (London: Anthem Press, 2010); Sun Xioachen and Lei Lei, "Obesity Rate on the Rise," *China Daily*, August 6, 2013, http://usa.chinadaily.com.cn/china/2013-08/06/content_16872878.htm; Yangfeng Wu, "Overweight and Obesity in China," *British Medical Journal* 333 (2006), 362–363, http://archive.oxha .org/knowledge/publications/china_overweightandobesity_bmjed_aug06.pdf; Tom Levitt, "China Facing Bigger Dietary Health Crisis than the U.S. (Interview with Barry Popkin)," *China Dialogue*, April 7, 2014, https://www.chinadialogue.net/article/show/single/en/6880-China-facing-bigger-dietary-health-crisis-than-the-US. In this interview, Popkin, professor of global nutrition at the University of North Carolina and coordinator of the China health and nutrition survey, commented that "China's speed of change in snacking, beverage consumption, [and] meat and oil intake has exceeded every other part of the world in speed of change."

76. Yu Xu et al., "Prevalence and Control of Diabetes in Chinese Adults," *Journal of the American Medical Association* 310, no. 9 (2013), 948–958, http://jamanetwork .com/journals/jama/fullarticle/1734701; "China 'Catastrophe' Hits 114 Million as Diabetes Spreads," *Bloomberg*, September 3, 2013, http://www.bloomberg.com/news/2013-09-03/china-catastrophe-hits-114-million-as-diabetes-spreads.html; Jing

Wu and Catherine Vasilikas-Scaramozza, "Type 2 Diabetes in China," *Decision Resources*, August 2013, http://decisionresources.com/Products-and-Services/Report ?r=emermd0113.

77. See Yunxiang Yan, "Food Safety and Social Risk in China 2011," *Journal of Asian Studies* 71, no. 3 (August 2012), 705–729. While the food safety problems in China have received prominent attention in the US media, far less attention has been paid to widespread food safety problems in the United States, where as many as 5.5 million people are made ill annually due to foodborne illnesses, many of them linked to unsafe workplace conditions that also impact food-processing and restaurant workers as well as farmworkers. Yet US food workers are largely exempt from any regulatory oversight and are even prevented from raising concerns due to a number of "ag-gag" laws in several food-producing states, which criminalize any video or audio recording of those production facilities. See Jacob E. Gerson and Benjamin I. Sachs, "Protect Those Who Protect Our Food," *New York Times*, November 12, 2014, http://www.nytimes.com/2014/11/13/opinion/protect-those-who-protect-our-food.html.

78. As Chengdu writer Sascha Matuszak has noted, urban desire for a farm-based ecotourism has led to the development of rural bed-and-breakfast hotels and restaurants known as *nongjiale* (happy farmhouses), which were given a major boost when then President Hu Jintao visited one of the original nongjiales as part of his promotion of the new ecological "Socialist Countryside" slogan. Sascha Matuszak, "China's Organic Food Cooperatives Must Overcome Trust Deficit," *Tea Leaf Nation*, April 23, 2013, http://www.rivers.org.cn/forum/201308/210.html.

79. National Health and Family Planning Commission of the PRC, "China's National Program for Food and Nutrition (2014–2020)," press release, May 16, 2014, http://www.chinadaily.com.cn/m/chinahealth/2014-05/16/content_17514010.htm; Barbara Demick, "In China, What You Eat Tells Who You Are," *Los Angeles Times*, September 16, 2011, http://articles.latimes.com/2011/sep/16/world/la-fg-china -elite-farm-20110917.

80. Elizabeth Henderson, "CSA in Taiwan and China—a Tantalizing Glimpse," *Chelsea Green Publishing*, January 16, 2012, http://chelseagreen.com/blogs/ elizabethhenderson/2012/01/16/csa-in-taiwan-and-china-%E2%80%93-a -tantalizing-glimpse/. Some of the 2012 Hong Kong CSA conference proceedings are available in Partnerships for Community Development, *Touching the Heart, Taking Root*; see also Zhang Chun, "China's Declining Crop Diversity Threatens Its Food Sovereignty," *China Dialogue*, October 4, 2014, https://www.chinadialogue.net/ article/show/single/en/6872-China-s-declining-crop-diversity-threatens-its-food -sovereignty.

81. Anupama Joshi, Moira Beery, and Marion Kalb, *Going Local: Paths to Success for Farm to School Programs*, Urban & Environmental Policy Institute and the Community Food Security Coalition, December 2006, http://www.farmtoschool.org/

Resources/Going_Local_Paths_to_Success.pdf; and the websites for the National Farm to School Network, www.farmtoschool.org, and Farm to Preschool, www .farmtopreschool.org, both accessed July 20, 2016.

82. Elizabeth Bowman, "Growing Los Angeles' Urban Agriculture Policy," March 2012, Antioch Urban Sustainability Program, Antioch University, Los Angeles; Los Angeles Food Policy Council, Working Group on Urban Agriculture, http:// goodfoodla.org/policymaking/working-groups-2/urban-agriculture/; Laura Benjamin, "Growing a Movement: Community Gardens in Los Angeles County," Urban and Environmental Policy Department, Occidental College, May 2008, http://www .oxy.edu/sites/default/files/assets/UEP/Comps/2008/benjaminCommunityGardens .pdf.

83. A campaign for effective policy mechanisms for urban agriculture culminated in the 2014 passage of the California Urban Agriculture Incentive Zones Act (AB 551), which allowed urban landowners to receive tax incentives for putting land into agricultural use and was adopted in Los Angeles in 2016. Available at University of California Division of Agriculture and Natural Resources (2016), http://ucanr.edu/sites/ UrbanAg/Laws_Zoning_and_Regulations/The_Urban_Agriculture_Incentive_Zones _Act_AB551/; see also Los Angeles Food Policy Council, "AB 551: A Powerful New Tool for Urban Agriculture," October 2014, http://goodfoodla.org/wp-content/ uploads/2014/10/AB551-Fact-Sheet-2-pager-020122014.pdf; Rebecca Kessler, "Urban Gardening: Managing the Risk of Contaminated Soil," *Environmental Health Perspectives* 121, nos. 11–12 (November-December 2013), A327–A333, http://ehp.niehs.nih .gov/121-a326/.

84. On the Good Food Procurement Policy program, see Delwiche and Lo, "Los Angeles' Good Food Purchasing Policy."

85. Now called "China Food Safety," the webpage has the following explanation for its existence: "A crisis in confidence in China's food industry emerged after melamine was found in domestically produced baby formula in 2008. The scandal sickened 300,000 babies and resulted in six premature deaths. Other stories of fake eggs, diseased pork, recycled oil, mislabelled meat and more have only led to more calls for industry reform." See "China Food Safety," *South China Morning Post*, 2016, http://www.scmp.com/topics/china-food-safety.

86. The 2008 melamine scandal (powdered milk laced with the toxic chemical melamine), which caused 6 babies to die and another 860 to be hospitalized, was seen as a turning point in food safety concerns, as continuing problems were identified among all the major companies involved in the powdered milk business. As *China Dialogue* correspondent Nan Xu put it, "the reputation of a whole food category was destroyed overnight." Nan Xu, "A Decade of Food Safety in China," *China Dialogue*, August 6, 2012, https://www.chinadialogue.net/article/show/single/ en/5083-A-decade-of-food-safety-in-China. On the Hong Kong limit of two purchases of baby milk formula, see Samuel Chan, "No Plan to Lift Two-Can Milk

Formula Limit at the Border, Says Hong Kong Health Minister," *South China Morning Post*, September 8, 2015, http://www.scmp.com/news/hong-kong/health -environment/article/1856056/hong-kong-health-minister-says-no-plan-ease-ban.

87. Samuel Fromartz, *Organic Inc.: Natural Foods and How They Grew* (Boston: Houghton Mifflin Harcourt, 2007).

88. Hong Kong Special Administrative Region, Agriculture, Fisheries, and Conservation Department, "Organic Farming in Hong Kong"; see also Produce Green Foundation website, accessed July 20, 2016, http://www.producegreen.org.hk/eng/index_e .htm.

89. Hong Kong Special Administrative Region, Agriculture, Fisheries, and Conservation Department, "Organic Farming in Hong Kong"; Mary Hui, "In Organic-Hungry Hong Kong, Corn as High as an Elevator's Climb," *New York Times*, October 3, 2012, http://www.nytimes.com/2012/10/04/world/asia/fearing-tainted-imports-hong -kong-squeezes-in-farms.html?_r=0.

90. Prior to the establishment of the KFBG, the Kadoories had created the Kadoorie Agricultural Aid Association in 1951 in order to focus on the development of land in the New Territories that was made available to peasants migrating to Hong Kong after the end of the Civil War and the triumph of the Chinese Communists. The original source of land was 3.5 acres of steep hillsides donated by the government that the peasant refugees helped to reclaim, with the Kadoories also providing inputs such as machinery, insecticides, poultry, and pigs for the pigsties that had been built from the stones the peasants had removed from the land as part of their effort to reclaim the land for agriculture. Aside from the electric utility they owned, the Kadoories also became majority stockholders in such Hong Kong trademarks as the Peak Tram and the Star Ferry. Matthias Messmer, *Jewish Wayfarers in Modern China* (Lanham, MD: Lexington Books, 2012), 9–10. See also Hong Kong Heritage, *Kadoorie Agricultural Aid Association: A Documentary* (1982), https://www.youtube .com/watch?v=WgbiDMBwVf4; "Lawrence Kadoorie Is Dead: A Leader in Hong Kong's Growth," *New York Times*, August 26, 1993, http://www.nytimes.com/1993/ 08/26/obituaries/lawrence-kadoorie-94-is-dead-a-leader-in-hong-kong-g-growth .html.

91. Kadoorie Farm and Botanic Garden, "Sustainable Living," 2016, http://www .kfbg.org/eng/sustainable-living-intro.aspx; personal communication with Idy Wong and May Cheng, Kadoorie Farm and Botanic Garden, Hong Kong, May 15, 2014; personal communication with Angus Lam, Partnerships for Community Development, Hong Kong, May 28, 2014; see also Partnerships for Community Development, "Ecological Agriculture," 2016, http://pcd.org.hk/en/foci/ecological -agriculture.

92. Sanders, *Prospects for Sustainable Development in the Chinese Countryside*; Paul Thiers, "From Grassroots Movement to State-Coordinated Market Strategy: The

Transformation of Organic Agriculture in China," *Environment and Planning C: Government and Policy* 20 (2002), 357–373; John Paull, "China's Organic Revolution," in S. Bhaskaran and Suchitra Mohanty, eds., *Marketing of Organic Products: Global Experiences* (Hyderabad: Icfai University Press, 2008), http://orgprints.org/14846/1/14846 .pdf; Richard Sanders and Xingji Xiao, "The Sustainability of Organic Agriculture in Developing Countries: Lessons from China," *International Journal of Environmental, Economic and Social Sustainability* 6, no. 6 (2010), 233–243.

93. In an interesting illustration of a "niche" market, the *Southern Weekly*, a respected newspaper in Guangzhou, published an article (subsequently removed from the newspaper's website) about an organic farm called the "Beijing Customs Administration Vegetable Base and Country Club" whose products were reserved for special consumers, such as government officials, elite athletes, foreign diplomats, and other elite constituencies. As the *Los Angeles Times* reported, in following up on the story, what had been a practice in the Maoist period of a food supply reserved for leadership cadres had become "a food system for the elite" due to environmental and health-related concerns. Demick, "In China, What You Eat Tells Who You Are."

94. Jordan Calinoff, "China's New Organic Industry," *Global Post*, February 19, 2009, http://www.globalpost.com/dispatch/china-and-its-neighbors/090217/chinas -new-organic-industry.

95. Chemical Inspection and Regulatory Service, "The Strictest 'Food Safety Law" in China," April 2015, http://www.cirs-reach.com/news/The_Strictest_Food_Safety _Law_in_China.html. See also Joe Balzano, "Lingering Food Safety Regulatory Issues for China in 2016," *Forbes Asia*, January 10, 2016, http://www.forbes.com/sites/ johnbalzano/2016/01/10/lingering-food-safety-regulatory-issues-for-china-in -2016/#4885fa0e5f5a; US Department of Agriculture Foreign Agricultural Service, "China's Food Safety Law (2015)," Global Agricultural Information Network Report CH 15016, May 18, 2015, http://gain.fas.usda.gov/Recent%20GAIN%20 Publications/Amended%20Food%20Safety%20Law%20of%20China_Beijing _China%20-%20Peoples%20Republic%20of_5-18-2015.pdf. See also Bian Yongmin, "The Challenges for Food Safety in China," *China Perspectives* 53 (May-June 2004), https://chinaperspectives.revues.org/819.

96. Steffanie Scott et al., "Contradictions in State- and Civil Society–Driven Developments in China's Ecological Agriculture Society," *Food Policy* 45 (April 2014), 158–166; Shi Yan et al., "Safe Food, Green Food, Good Food: Chinese Community Supported Agriculture and the Rising Middle Class," *International Journal of Agricultural Sustainability* 9, no. 4 (2011), 551–558, http://www.tandfonline.com/doi/pdf/1 0.1080/14735903.2011.619327; Simiao Lui, personal communication with Shi Yan, August 2013. The "ethical eaters" comment was made during a presentation by Judith Redmond, Community Alliance with Family Farmers, Occidental College, December 1, 2004. See also Judith Redmond and Thomas Nelson, "Defining a Food

Ethic: Common Values for the Sustainable Food and Farm Movement," Full Belly Farm, Guinda, CA, January 28, 2001.

97. Personal communication with Li Kai Kuen and May Cheng, and participation at the Tai Po CSA, May 15, 2014; Li Kai Kuen, "Taipo Community Supported Agriculture: Notes to Members," 2014.

98. Wen Tiejun, "Deconstructing Modernization," *Chinese Sociology and Anthropology* 39, no. 4 (Summer 2007), 10, 20–21, http://chinastudygroup.net/wp-content/uploads/2010/12/Wen_2007_Deconstructing_Modernization.pdf; Wen Tiejun, "China's Century-Long Quest for Industrialization," *China Study Group*, November 18, 2003, http://chinastudygroup.net/2003/11/chinas-century-long-quest-for-industrialization/ (translation from *Dushu*, no. 3, March 2001); see also Wen Tiejun, Lau Kinchi, Cheng Cunwang, He Huili, and Qiu Jiansheng, "Ecological Civilization, Indigenous Culture, and Rural Reconstruction in China," *Monthly Review* 63, no. 9 (February 2012), http://monthlyreview.org/2012/02/01/ecological-civilization-indigenous-culture-and-rural-reconstruction-in-china; Pan Jia'en and Du Jie, "Alternative Responses to 'The Modern Dream': The Sources and Contradictions of Rural Reconstruction in China," *Inter-Asia Cultural Studies* 12, no. 3 (2011), 454–464; Alexander F. Day, *The Peasant in Postsocialist China: History, Politics, and Capitalism* (Cambridge: Cambridge University Press, 2013).

Chapter 6

1. Henri Lefebvre, *The Urban Revolution* (Minneapolis: University of Minnesota Press, 2003), 18.

2. Designing Hong Kong, "Lost in TST," *YouTube*, April 25, 2013, https://www.youtube.com/watch?v=ThFiS_Nn0Q4&feature=youtube.

3. Gloria J. Jeff, General Manager, Department of Transportation, City of Los Angeles, "Subject of Pedestrians" (CF 06-0762), interdepartmental memo to the Los Angeles City Council, May 4, 2006.

4. Deborah Murphy, "Sunland Incident with Mavis Coyle, Pedestrian Safety Forum," memo to Los Angeles City Council member Wendy Greuel and Laurie Pollner, May 2, 2006.

5. Dana Bartholomew, "Officer Cites 82-Year-Old Woman for Being Too Slow to Negotiate Busy Street," *Los Angeles Daily News*, April 10, 2006; Damien Newton, "Mayvis Coyle Redux," *Streetsblog*, February 13, 2008, http://la.streetsblog.org/2008/02/13/mayvis-coyle-redux/; Steve Lopez, "Guilty of 'Crossing while Elderly,'" *Los Angeles Times*, April 15, 2006, http://articles.latimes.com/2006/apr/15/local/me-lopez15; Amanda Covarrubias and Cynthia H. Cho, "She Has the Whole World at Her Not-So-Fleet Feet," *Los Angeles Times*, April 14, 2006, http://articles.latimes.com/2006/apr/14/local/me-mayvis14.

6. Hong Kong Special Administrative Region, Transport Department, "Hong Kong Transport: 40 Years," November 2008, http://www.td.gov.hk/filemanager/en/publication/td-booklet-final-251108.pdf.

7. Transport Advisory Committee, "Report on Study of Road Traffic Congestion in Hong Kong," December 2014, http://www.thb.gov.hk/eng/boards/transport/land/Full_Eng_C_cover.pdf; Trading Economics, "Vehicles (per km of Road) in Hong Kong," 2015, http://www.tradingeconomics.com/hong-kong/vehicles-per-km-of-road-wb-data.html; Neil Padukone, "The Unique Genius of Hong Kong's Transportation System," *Atlantic*, September 10, 2013, http://www.theatlantic.com/china/archive/2013/09/the-unique-genius-of-hong-kongs-public-transportation-system/279528/.

8. David Johnson, *Star Ferry: The Story of a Hong Kong Icon* (Auckland, New Zealand: Remarkable View Ltd. for the Star Ferry Co. Ltd., 1998); Wai Chi Sham, "The History of Hong Kong and Yaumati Ferry Company Limited, 1923 to the 1970s," master's thesis, Lingnan University, 2007, http://commons.ln.edu.hk/cgi/viewcontent.cgi?article=1001&context=his_etd.

9. Hong Kong Special Administrative Region, Transport Department, "Hong Kong: The Facts—Transport," 2015, http://www.gov.hk/en/about/abouthk/factsheets/docs/transport.pdf.

10. Hong Kong Tramways, "Our Story," 2015, https://www.hktramways.com/en/our-story/; Joy Chen, "A Brief History of Hong Kong Trams," blog, 2015, https://joyxiaonan.wordpress.com/.

11. Olivia Chung, "Hong Kong Trams Change Driver," *Asia Times*, April 10, 2009, http://www.atimes.com/atimes/China_Business/KD10Cb02.html; Justine Lau, "Veolia Climbs Aboard to Run HK Tram Group," *Financial Times*, April 7, 2009, https://next.ft.com/content/134cc4a2-238b-11de-996a-00144feabdc0.

12. C. K. Leung, "The Growth of Public Passenger Transport," in D. J. Dwyer, ed., *Asian Urbanization: A Hong Kong Casebook* (Hong Kong: Hong Kong University Press, 1971), 137–154.

13. Russell Flannery, "An Informal Look at Cheng Yu-tung's Dynasty," *Forbes Asia*, January 9, 2014, http://www.forbes.com/sites/russellflannery/2014/01/09/an-informal-look-at-cheng-yu-tungs-dynasty/; Chai Hua, "New Concept Mall Draws the Crowds to Qianhai," *China Daily Asia*, December 8, 2015, http://www.chinadailyasia.com/hknews/2015-12/08/content_15355288.html.

14. Hong Kong Special Administrative Region, Transport Department, "Hong Kong: The Facts—Transport."

15. Freeman, Fox, Wilbur Smith and Associates, *Hong Kong Mass Transport Study: Report Prepared for the Hong Kong Government* (Hong Kong: Hong Kong Government,

1967); Hong Kong Mass Transit Infocenter, "The History," July 2004, http://www .hkmtr.net/Past_History.htm.

16. Siman Tang and Hong K. Lo, "The Impact of Public Transport Policy on the Viability and Sustainability of Mass Railway Transit: The Hong Kong Experience," *Transportation Research Part A* 42 (2008), 569; Kowloon-Canton Railway Corporation, "One Hundred Years of Railway Operations in Hong Kong," 2015, http://www .kcrc.com/download/common/about-kcrc/history/KCRC018_Booklet.pdf.

17. Bo-sin Tang et al., "Study of the Integrated Rail-Property Development Model in Hong Kong," Hong Kong Polytechnic University, November 2004, http://www .reconnectingamerica.org/assets/Uploads/mtrstudyrpmodel2004.pdf; Rikki Yeung, *Moving Millions: The Commercial Success and Political Controversies of Hong Kong's Railways* (Hong Kong: Hong Kong University Press, 2008). Yeung points out that the public-private joint partnerships established with major property companies, such as Sun Hung Kai Properties controlled by the Kwok family (who owned the KMB bus system), were seen as a boon by private property owners and their pro-business political allies during housing boom periods when profits were huge; however, these same people then complained during housing bust periods when they took losses while the quasi-public rail entities could "unfairly" take advantage of its public subsidies and support that reduced their losses (87–91).

18. S. C. Yeung, "What Hong Kong People Should Learn from the High Speed Rail Link Saga," *EJinsight*, May 11, 2015, http://www.ejinsight.com/20150511-what -hk-people-should-learn-from-high-speed-rail-link-saga/; Hong Kong Special Administrative Region, Highways Department, "Hong Kong Section of the Guangzhou-Shenzhen-Hong Kong Express Rail Link," 2011, http://www.hyd.gov.hk/xrl/eng/ home/.

19. Hong Kong Special Administrative Region, Transport and Housing Bureau, "Railway Development Strategy 2014," September 2014, 6, http://www.thb.gov.hk/ eng/psp/publications/transport/publications/rds2014.pdf.

20. Hong Kong Special Administrative Region, Transport Department, "2015 Annual Transport Digest: Registration and Licensing of Vehicles and Drivers," 2015, http://www.td.gov.hk/mini_site/atd/2015/en/section3_2.html; Hong Kong Special Administrative Region, Transport Department, "The Annual Traffic Census: 2014," August 2015, 43, http://www.td.gov.hk/filemanager/en/content_4726/annual%20 traffic%20census%202014.pdf; P. F. Leeds, "Evolution of Urban Transport," in H. T. Dimitrou and A. H. S. Cook, eds., *Land-Use/Transport Planning in Hong Kong: The End of an Era: A Review of Principles and Practices* (Aldershot: Ashgate, 1998), quoted in Barrie Shelton, Justyna Karakiewicz, and Thomas Kvan, *The Making of Hong Kong: From Vertical to Volumetric* (New York: Routledge, 2011), 56.

21. T. D. Denham, "California's Great Cycle-Way," *Good Roads Magazine*, November 1901; Cecilia Rasmussen, "Bikeway Was Ahead of Its Time," *Los Angeles Times*,

November 29, 1998, http://articles.latimes.com/1998/nov/29/local/me-48885. In their history of the Schwinn Bike Company, Judith Crown and Glenn Coleman note that the bicycle allowed a woman to determine "where she wants to go and what she plans to do when she gets there, regardless of a male companion or lack of one." Judith Crown and Glen Coleman, *No Hands: The Rise and Fall of the Schwinn Bicycle Company, an American Institution* (New York: Henry Holt, 1996), 20.

22. William B. Friedricks, *Henry E. Huntington and the Creation of Southern California* (Columbus: Ohio State University Press, 1992). "Wherever the Huntington railways extend their lines," the *Los Angeles Times* proclaimed about the real estate syndicates, "the Huntington electric and gas companies are preparing to furnish fuel and light"—a reference to the role of the Huntington-controlled electric utility that became Southern California Edison. The *Times* quotation is cited in Robert Gottlieb and Irene Wolt, *Thinking Big: The Story of the Los Angeles Times, Its Publishers, and Their Influence on Southern California* (New York: G. P. Putnam's Sons, 1977), 145.

23. George W. Hilton and John F. Due, *The Electric Interurban Railways in America* (Stanford: Stanford University Press, 1960), 406; Spencer Crump, *Ride the Big Red Cars: How Trolleys Helped Build Southern California* (Los Angeles: Trans-Anglo Press, 1970).

24. "City-Owned Street Cars?," *Los Angeles Times*, February 10, 1927.

25. Martin Wachs, "Autos, Transit, and the Sprawl of Los Angeles: The 1920s," *Journal of the American Planning Association* 50, no. 3 (1984), 297–310, http://www .tandfonline.com/doi/abs/10.1080/01944368408976597?journalCode=rjpa20.

26. Jeremiah Axelrod, *Inventing Autopia: Dreams and Visions of the Modern Metropolis in Jazz Age Los Angeles* (Berkeley: University of California Press, 2009); Bruce Henstell, *Sunshine and Wealth: Los Angeles in the Twenties and Thirties* (San Francisco: Chronicle Books, 1984). The *Los Angeles Times* weighed in as well on the street infrastructure debate, urging that the traffic problem be solved quickly and "correctly" if the city was going to be able "to continue to grow and prosper": see "Two Strides Forward," *Los Angeles Times*, October 9, 1924.

27. Greg Hise and William Deverell, *Eden by Design: The 1930 Olmsted-Bartholomew Plan for the Los Angeles Basin* (Berkeley: University of California Press, 2000).

28. Anastasia Loukaitou-Sideris and Robert Gottlieb, "Putting the Pleasure Back in the Drive: Reclaiming Urban Parkways for the Twenty-first Century," *Access* 22 (Spring 2003), http://scholar.oxy.edu/cgi/viewcontent.cgi?article=1427&context =uep_faculty; David W. Jones, *California's Freeway Era in Historical Perspective* (Berkeley: Institute of Transportation Studies, University of California at Berkeley, 1989), 169–175.

29. Brian Taylor, "When Finance Leads Planning: Urban Planning, Highway Planning and Metropolitan Freeways in California," *Journal of Planning Education and Research* 20, no. 2 (2000), 196–214, http://jpe.sagepub.com/content/20/2/196.short.

30. Kenneth Jackson, in his classic study of suburbanization, characterized the "automobile suburbs" as including the overall pattern of settlement and the length and direction from home to work, arguing that "the automobile had a greater spatial and social impact on cities than any technological innovation since the development of the wheel." Kenneth Jackson, *Crabgrass Frontier: The Suburbanization of the United States* (New York: Oxford University Press, 1983), 188. See also John Chynoweth Burnham, "The Gasoline Tax and the Automobile Revolution," *Mississippi Valley Historical Review* 48, no. 3 (December 1961), 435–469.

31. Martin Webster, "Transportation: A Civic Problem," *Engineering and Science,* December 1949, 11–15; Martin Wachs, "The Evolution of Transportation Policy in Los Angeles: Images of Past Policy and Future Prospects," in Allen J. Scott and Edward Soja, eds., *The City: Los Angeles and Urban Theory at the End of the Twentieth Century* (Berkeley: University of California Press, 1996), 132–134.

32. Jones, *California's Freeway Era in Historical Perspective,* 235. The "substituting rubber for rail" comment is in a letter from Henry Workman Keller to Fletcher Bowron, June 5, 1937, in the Henry Workman Keller Papers, University of California at Los Angeles. On the nostalgia front, see the documentary of the last ten days in 1961 of the Big Red Car ("Ride the Last of the Big Red Cars"), *Candlelight Stories,* http://www.candlelightstories.com/2012/06/25/ride-the-last-of-the-big-red-cars -1961-los-angeles-streetcar-documentary/. The first successfully revived streetcar system, the Blue Line, whose service commenced three decades later in 1990, essentially followed the same route from Los Angeles to Long Beach.

33. Scott Hamilton Dewey, *Don't Breathe the Air: Air Pollution and U.S. Environmental Politics, 1945–1970* (College Station: Texas A&M University Press, 2000), 58–59; David Brodsly, *L.A. Freeway: An Appreciative Essay* (Berkeley: University of California Press, 1981).

34. Eric Avila, *The Folklore of the Freeway: Race and Revolt in the Modernist City* (Minneapolis: University of Minnesota Press, 2014), 138–139; George J. Sanchez, "'What's Good for Boyle Heights Is Good for the Jews': Creating Multiculturalism on the Eastside During the 1950s," *American Quarterly* 56 (2004), 633–661; Robert E. G. Harris, "Have the Freeways Failed Los Angeles?," *Frontier,* January 1957, 7–11.

35. For example, the main thrust of a 1988 symposium at UCLA entitled "The Car and the City" that brought together leading transportation planners assumed that auto use was inevitable, if not preferable, in Los Angeles and that the American love affair with the automobile had become "a well-established marriage." Martin Wachs and Margaret Crawford, eds., *The Car and the City: The Automobile, the Built Environment, and Daily Life* (Ann Arbor: University of Michigan Press, 1992), 56. In arguing in favor of the dominance of the car in Los Angeles, Scott Bottles wrote, in an essay from that same volume, that the change from streetcars to the automobile represented a "democratic mass movement," resulting in a "new kind of public transportation ... [where] the public [i.e., the government] provided the streets, and the

individual brought along his or her own car." Bottles, "Mass Politics and the Adoption of the Automobile in Los Angeles" (202).

36. The 1974 effort to fund a rail revival would have relied on a sales tax and was criticized at the time by UCLA Professor Peter Marcuse as potentially contributing to additional sprawl. A follow-up initiative in 1976 to jump-start an extensive rail system, which also failed, was similarly opposed by UCLA Professor Martin Wachs for not providing for bus systems, which he considered far more appropriate for a spread-out region. Martin Wachs, "Are Trains the Ticket for L.A.? No: System Would Fail Those Who Need It Most," *Los Angeles Times*, May 23, 1976; Ethan Elkind, *Railtown: The Fight for the Los Angeles Metro Rail and the Future of the City* (Berkeley: University of California Press, 2014), 17.

37. The consent decree is available at http://www.clearinghouse.net/chDocs/public/PB-CA-0038-0002.pdf.

38. Los Angeles Metropolitan Transportation Authority, "Metro Customer Survey Results," accessed August 2, 2016, http://media.metro.net/projects_studies/research/images/infographics/metro_infographic_02.pdf.

39. After a new California state law was passed that allowed undocumented immigrants to obtain driver's licenses, more than half of people who applied for licenses in the first six months—443,000 of the 883,000 issued—were undocumented, many of them from Southern California. Officials predicted that as many as 1.5 million could be issued within three years. Jennifer Medina, "California Effort to Issue Driver's Licenses to Immigrants Receives Surge of Applications," *New York Times*, August 8, 2015, http://www.nytimes.com/2015/08/09/us/california-effort-to-issue-drivers -licenses-to-immigrants-receives-surge-of-applicants.html.

40. Christopher Weber, "Carmageddon 2: 405 Freeway Ahead of Schedule: Closure Deemed a Success," *Huffington Post*, October 1, 2012; Christopher Hawthorne, "Fast Lane to Gridlock," *Los Angeles Times*, July 12, 2011, http://articles.latimes.com/2011/jul/12/entertainment/la-et-carmageddon-20110712; California Department of Motor Vehicles, "Estimated Vehicles Registered by County, for the Period of January 1 through December 31, 2014," https://www.dmv.ca.gov/portal/wcm/connect/add5eb07-c676-40b4-98b5-8011b059260a/est_fees_pd_by_county.pdf?MOD =AJPERES.

41. Jon Christensen and Mark Gold, "Welcome to Christopher Hawthorne's Third Los Angeles," *L.A. Observed*, February 8, 2015, http://www.laobserved.com/intell/2015/02/welcome_to_christopher_hawthor.php.

42. Transport Advisory Committee, "Report on Study of Road Traffic Congestion in Hong Kong"; Hong Kong Special Administrative Region, Transport Department, "Section 3: Registration and Licensing of Vehicles and Drivers" and "Section 6: Vehicle Parking," in 2015 Annual Transport Digest, September 16, 2015, http://www.td.gov.hk/mini_site/atd/2015/tc/index.html.

43. Becky P. Y. Loo and Alice S. Y. Chow, "Changing Urban Form in Hong Kong: What Are the Challenges on Sustainable Transportation?," *International Journal of Sustainable Transportation* 2, no. 3 (2008), 185–186, http://www.tandfonline.com/doi/pdf/10.1080/15568310701517331#.VInbVH7TnIU.

44. Transport Advisory Committee, "Report on Study of Road Traffic Congestion in Hong Kong," 6.

45. Hong Kong Special Administrative Region, Environmental Protection Department, "An Overview of Noise Pollution and Control in Hong Kong," 2015, http://www.epd.gov.hk/epd/english/environmentinhk/noise/noise_maincontent.html; Kin-che Lam et al., "Annoyance Response to Mixed Transportation Noise in Hong Kong," *Applied Acoustics* 70 (2009), 1–10.

46. Hong Kong Special Administrative Region, Transport Department, "Road Traffic Casualties by Age, 2004–2014," 2015, figure 1.1, http://www.td.gov.hk/filemanager/en/content_4717/f1.1.pdf.

47. California Department of Transportation, "2010 California Public Road Data," http://www.dot.ca.gov/hq/tsip/hpms/hpmslibrary/hpmspdf/2010PRD.pdf; see also California Department of Transportation (Caltrans) Office of System and Freight Planning, "Freight Planning Fact Sheet: Caltrans District 7," June 2012, http://dot.ca.gov/hq/tpp/offices/ogm/district_freight_fact_sheets/updated_092412/District_7_GM_Fact_Sheet_061312.pdf.

48. Los Angeles Department of City Planning, "Mobility Plan 2035: An Element of the General Plan," January 20, 2016, 40, http://planning.lacity.org/documents/policy/mobilityplnmemo.pdf; Aaron Mendelson, "Watch a Decade of Growth in L.A.'s Bike Infrastructure," KPCC, April 10, 2015, http://www.scpr.org/news/2015/04/10/50849/watch-a-decade-of-growth-in-la-s-bike-infrastructu/. Since nearly all bike lanes are limited to a sliver of the roadway, the proportion of road miles to bike miles is even greater than the numbers suggest.

49. Adam Millard-Ball, "Phantom Trips: Overestimating the Traffic Impacts of New Development," *Journal of Transport and Land Use* 8, no. 1 (2015), 31–49; see also Adam Millard-Ball, "Phantom Trips," *Access*, Fall 2014, http://www.accessmagazine.org/articles/fall-2014/phantom-trips/.

50. Donald Shoup, *The High Cost of Free Parking* (Chicago: American Planning Association, 2011); Donald C. Shoup, "The Trouble with Minimum Parking Requirements," *Transportation Research Part A* 33 (1999), 549–574, http://shoup.bol.ucla.edu/Trouble.pdf.

51. Peter D. Norton, *Fighting Traffic: The Dawn of the Motor Age in the American City* (Cambridge, MA: MIT Press, 2008), 220–222.

52. Catherine Saillant, "For Downtown LA's Pedestrians, Citations Send a 'Don't Walk' Signal," *Los Angeles Times*, April 24, 2015, http://www.latimes.com/local/

california/la-me-walkability-downtown-20150412-story.html#page=1; Laura J. Nelson, Armand Emamdjomeh, and Joseph Serna, "Crossing Examination," *Los Angeles Times*, July 12, 2015, http://www.pressreader.com/usa/los-angeles-times/20150712/281479275090678/TextView.

53. "Loyd Sigmon, 95, Creator of California Traffic Alerts," *New York Times*, June 7, 2004, http://www.nytimes.com/2004/06/07/us/loyd-sigmon-95-creator-of-california-traffic-alerts.html; California Department of Transportation, "What Are Sig-Alerts?," 2016, http://www.dot.ca.gov/hq/paffairs/faq/faq18.htm; Robert Gottlieb, *Reinventing Los Angeles: Nature and Community in the Global City* (Cambridge, MA: MIT Press, 2007), 174.

54. Eric Eidlin, "What Density Doesn't Tell Us about Sprawl," *Access*, Fall 2010, http://www.accessmagazine.org/articles/fall-2010/density-doesnt-tell-us-sprawl/.

55. See Vanessa Carter, Manuel Pastor, and Madeline Wander, "An Agenda for Equity: A Framework for Building a Just Transportation System in Los Angeles County," November 2013, USC Program for Environmental and Regional Equity, https://dornsife.usc.edu/assets/sites/242/docs/Agenda_Equity_Full_Report_Web02.pdf; Gregory M. Rowangould, "A Census of the US Near-Roadway Population: Public Health and Environmental Justice Considerations," *Transportation Research Part D* 25 (2013), 59–67.

56. E. M. Bjorklund, "The Danwei: Socio-Spatial Characteristics of Work Units in China's Urban Society," *Economic Geography* 62, no. 1 (January 1986), 19–29.

57. Pengjun Zhao, "The Impact of the Built Environment on Bicycle Commuting: Evidence from Beijing," *Urban Studies* 51, no. 5 (April 2014), 1019–1037, http://usj.sagepub.com/content/early/2013/07/12/0042098013494423.

58. Walter Hook, "The Sound of China's Bicycle Industry? One Hand Clapping," *Sustainable Transport*, Fall 2002, 20–22; Annemarie de Boom, Richard Walker, and Rob Goldup, "Shanghai: The Greatest Cycling City in the World?," *World Transport Policy and Practice* 7, no. 3 (2001), 55.

59. Robert Dreyfuss, "China: A 'Kingdom of Bicycles' No Longer," National Public Radio, November 25, 2009, http://www.npr.org/templates/story/story.php?storyId=120811453.

60. Jonathan Watts, "China Restores Bike Lanes Lost to Car Boom," *Guardian*, June 16, 2006, http://www.theguardian.com/world/2006/jun/16/china.transport. The most notable of the bike-share programs was in Hangzhou, which launched its program in 2008 through a membership service called Hangzhou Public Bicycle. More than 60,000 bicycles and 2,500 bike stations were made available, with estimates of more than 250,000 users. Susan A. Shaheen et al., "China's Hangzhou Public Bicycle: Understanding Early Adoption and Behavioral Response to Bikesharing," *Transportation Research Record: Journal of the Transportation Research Board*, no. 2247 (Washing-

ton, DC: Transportation Research Board of the National Academies, 2011), 33, http://www.tsrc.berkeley.edu/sites/default/files/China's%20Hangzhou%20 Public%20Bicycle%20(article)%20-%20Shaheen.pdf; see also Tom Miller, *China's Urban Billion: The Story Behind the Biggest Migration in Human History* (London: Zed Books, 2012), 138.

61. "Bicycle Manufacturing in China: Market Research Report," *IBIS World*, May 2016, http://www.ibisworld.com/industry/china/bicycle-manufacturing.html; Ken Roberts, "Imports of Bicycles Topping $1 Billion Annually," *Miami Herald*, February 1, 2015, http://www.miamiherald.com/news/business/biz-monday/article8855768 .html; Grace S. Ruan, "Profile of the Chinese Bicycle Market," Special Report, Bike Market/CBES 2013–2014 (2014), 58–61, http://biketaiwan.com/resource/ article/6/157/article-03.pdf; Timothy Aeppel, "Bringing Manufacturing to the U.S. after a World Tour," *Wall Street Journal*, July 17, 2014, http://www.wsj.com/articles/ bringing-manufacturing-to-the-u-s-after-a-world-tour-1405629164.

62. Kelly Sims Gallagher, *China Shifts Gears: Automakers, Oil, Pollution, and Development* (Cambridge, MA: MIT Press, 2006), 22; Rachel Tang, "The Rise of China's Auto Industry and Its Impact on the U.S. Motor Vehicle Industry," Congressional Research Service, November 16, 2009, https://www.fas.org/sgp/crs/row/R40924.pdf.

63. Feng Suwei and Li Qiang, "Car Ownership Control in China's Mega Cities: Shanghai, Beijing, and Guangzhou," *Journeys*, 2013, http://www.lta.gov.sg/ ltaacademy/doc/13Sep040-Feng_CarOwnershipControl.pdf.

64. Wu Yan, "Shenzhen Explains Abrupt Limit on Car Purchase," *China Daily*, December 30, 2015, http://www.chinadaily.com.cn/china/2014-12/30/content _19203728.htm; Colum Murphy, "Southern Chinese City of Shenzhen to Place Restrictions on Car Purchases," *Wall Street Journal*, December 30, 2014, http://www .wsj.com/articles/southern-chinese-city-of-shenzhen-to-place-restrictions-on-car -purchases-1419918102; An Baijie, "Putting the Brakes on Car Owners," *Asia Weekly*, February 6, 2015, http://epaper.chinadailyasia.com/asia-weekly/article-3962.html.

65. Liu Jie, "Official Data Confirm China as World's Biggest Auto Producer, Consumer, Challenges Remain," *Xinhua*, January 11, 2010, http://www.china-embassy .org/eng/xw/t650869.htm; Jerry Hirsch, "Sales Gridlock?," *Los Angeles Times*, September 3, 2015, http://www.pressreader.com/usa/los-angeles-times/20150903/281887 297074397/TextView.

66. Levi Tillemann, "China's Electric Car Boom: Should Tesla Motors Worry?," *Fortune*, February 19, 2015, http://fortune.com/2015/02/19/chinas-electric-car-boom -should-tesla-motors-worry/; Kelly Sims Gallagher, *The Globalization of Clean Energy Technology: Lessons from China* (Cambridge, MA: MIT Press, 2014).

67. Hans-Ulrich Riedel, "Chinese Metro Boom Shows No Sign of Abating," *International Railway Journal*, November 19, 2014, http://www.railjournal.com/index. php/metros/chinese-metro-boom-shows-no-sign-of-abating.html.

68. Xin Dingding et al., "Subway Costs Feared to Go Off the Rails," *China Daily*, July 31, 2012, http://usa.chinadaily.com.cn/china/2012-07/31/content_15633448.htm; Steven Smith, "Why China's Subway Boom Went Bust," *CityLab*, September 10, 2012, http://www.citylab.com/commute/2012/09/why-chinas-subway-boom-went -bust/3207/.

69. Martin de Jong et al., "Introducing Public-Private Partnerships for Metropolitan Subways in China: What Is the Evidence?," *Journal of Transport Geography* 18 (2013), 308; "Shenzhen's Pre-Universiade Eviction of 80,000 'High Risk People' Sparks Controversy," *Xinhua*, April 13, 2011, http://news.xinhuanet.com/english2010/ china/2011-04/13/c_13826681.htm.

70. Robert Cervero and Jennifer Day, "Suburbanization and Transit-oriented Development in China," *Transport Policy* 15, no. 5 (2008), 315–323, http://www .worldtransitresearch.info/research/1401/.

71. John Scales, a World Bank transport specialist, argued that China was in fact the only country in history that had ever attempted such a rapid and enormous increase in its rail network. Elizabeth Fischer, "China's High Speed Rail Revolution," *Railway-technology.com*, November 21, 2012, http://www.railway-technology.com/features/ feature124824/.

72. Angus Grigg, "Chinese High Speed Rail Should Confine the XPT to History," *Financial Review*, June 30, 2015, http://www.afr.com/business/infrastructure/rail/ chinese-high-speed-rail-should-confine-the-xpt-to-history-20150629-gi11pb.

73. "China Focus: Full Speed Ahead for High Speed Rail Expansion," *Xinhua*, January 28, 2015, http://news.xinhuanet.com/english/indepth/2015-01/28/c_133953786 .htm.

74. Jiao Jingjuan et al., "Impacts on Accessibility of China's Present and Future HSR Network," *Journal of Transport Geography* 40 (October 2014), 123; Zhao Pengjun and Phillipa Howden-Chapman, "Social Inequalities in Mobility: The Impact of the *Hukou* System on Migrants' Job Accessibility and Commuting Costs in Beijing," *International Development Planning Review* 32, nos. 3–4 (2010), 363–384.

75. Anastasia Loukaitou-Sideris and Robert Gottlieb, "The Day That People Filled the Freeway: Re-Envisioning the Arroyo Seco Parkway and the Urban Environment in Los Angeles," *DISP Journal* (Zurich) 159, no. 4 (2004).

76. ICLEI Local Governments for Sustainability, "Shenzhen Guangming New District Expressed Interest to Develop Closer Partnership with ICLEI," December 3, 2015, http://eastasia.iclei.org/newsdetails/article/shenzhen-guangming-new-district -expressed-interest-to-develop-closer-partnership-with-iclei.html.

77. Shaojie Tang et al., "Exploration and Practice of Implementable Planning of Low-Carbon New Town," 45th ISOCARO Congress, Porto, Portugal, October 18–22, 2009, http://www.isocarp.net/Data/case_studies/1406.pdf; Energy Smart

Communities Initiative: Knowledge Sharing Platform, "Guangming New District," April 11, 2014, http://esci-ksp.org/project/guangming-new-district/.

78. Yang Liuqing, "Landfill in Shenzhen Disaster Was Cited as Risk to Industrial Park," *Caixin*, December 21, 2015, http://english.caixin.com/2015-12-21/100891075 .html; Chris Buckley and Austin Ramzy, "Before Shenzhen Landslide, Many Saw Warning Signs as Debris Spread," *New York Times*, December 22, 2015, http://www .nytimes.com/2015/12/23/world/asia/landslide-shenzhen-china.html.

79. Tang and Lo, "The Impact of Public Transport Policy on the Viability and Sustainability of Mass Railway Transit"; Hong Kong Special Administrative Region, Transport Department, "Fees and Charges," http://www.td.gov.hk/en/public _services/fees_and_charges/.

80. Hong Kong Special Administrative Region, Transport Department, "Guidelines for Registration and Importation of Motor Vehicles," 2015, http://www.td.gov.hk/ en/public_services/licences_and_permits/vehicle_first_registration/guidelines_for _importation_and_registration_of_mot/index.html; Hong Kong Special Administrative Region, Environmental Protection Department, "Tax Incentives for Environment-Friendly Petrol Private Cars (Terminated on April 1)," 2015, http://www.epd .gov.hk/epd/english/environmentinhk/air/prob_solutions/environment_friendly _private_cars.html#1.

81. Transport and Housing Bureau, HKSAR Government, "LCQ7: Electronic Road Pricing," press release, Hong Kong Special Administrative Region Government, June 22, 2016, http://www.info.gov.hk/gia/general/201606/22/P201606220362.htm.

82. Enrique Rivero, "L.A.'s CicLAvia Significantly Improves Air Quality in Host Neighborhoods, UCLA Study Finds," *UCLA Newsroom*, October 19, 2015, http:// newsroom.ucla.edu/releases/l-a-s-ciclavia-significantly-improves-air-quality-in-host -neighborhoods-ucla-study-finds.

83. Janette Sadik-Khan, "New York's Streets? Not So Mean Anymore," TED talk, New York, September 2013, https://www.ted.com/talks/janette_sadik_khan_new _york_s_streets_not_so_mean_any_more?language=en.

84. The Vision Zero concept was first introduced in Sweden in the 1970s as a way to address the high number of traffic fatalities and "base every transportation design, construction and enforcement decision around a basic premise: 'will it help reduce Sweden's total traffic deaths to zero?'" The term "Vision Zero" was formally adopted in Sweden in 1997 as part of its transportation planning; thanks to the success of the plan, the concept began to be introduced (including the use of the name) in several cities around the world. Alternative transportation advocates for a more walkable and bikeable Los Angeles began to tout this approach but without much initial success. During Mayor Eric Garcetti's administration, it was picked up by the new leadership at the Department of Transportation and effectively incorporated into the language of Garcetti's various plans rolled out in 2014 and 2015. Damien Newton,

"Vision Zero or Zero Vision: LA Needs to Change the Way It Thinks about Safety," *Streetsblog LA*, March 24, 2014, http://la.streetsblog.org/2014/03/24/vision-zero-or -zero-vision-l-a-needs-to-change-the-way-it-thinks-about-safety/.

85. Los Angeles Department of Transportation, "Great Streets for Los Angeles: Strategic Plan," 2014, http://www.smartgrowthamerica.org/documents/cs/impl/ ca-losangeles-dot-strategicplan2014.pdf; Damien Newton and Joe Linton, "LA DOT's Bold New Strategic Vision: Eliminate Traffic Deaths by 2025," *Streetsblog LA*, September 30, 2014, http://la.streetsblog.org/2014/09/30/ladots-bold-new-strategic-vision -eliminate-l-a-traffic-deaths-by-2025/; Los Angeles Department of City Planning, "Mobility Plan 2035: An Element of the General Plan."

86. Bike and pedestrian advocates, while welcoming the Mobility 2035 Plan, remained concerned about its implementation and urged further community mobilization to turn the vision into "broader mobility choices." Joe Linton, "L.A. City Council Linton Approves New Mobility Plan, Including Vision Zero," *Streetsblog LA*, August 11, 2015, http://la.streetsblog.org/2015/08/11/l-a-city-council-approves-new -mobility-plan-vision-zero/. See also Fix the City, "Why Fix the City Opposes MP2035—The Immobility Plan," August 12, 2015, http://fixthecity.org/?p=156. Garcetti's comment is in Ian Lovett, "A Los Angeles Plan to Reshape the Streetscape Sets Off Fears of Gridlock," *New York Times*, September 7, 2015, http://www.nytimes .com/2015/09/08/us/a-los-angeles-plan-to-reshape-the-streetscape-sets-off-fears-of -gridlock.html; see also Sam Schwartz, "Why Mobility Is Overrated," *Los Angeles Times*, September 2, 2015, http://www.pressreader.com/usa/los-angeles-times/ 20150902/281775627922693/TextView.

87. *L.A. Streetsblog* editor and long-time alternative transportation advocate Joe Linton commented that while vehicle miles traveled had been reduced, "very few transportation agencies and traffic engineers have incorporated such trends in their predictions of future car traffic. Engineers are still predicting that upticks, 1990s-style or 1970s-style, are right around the corner." Joe Linton, "Some Perspective on How Angelenos Are Driving Less," *L.A. Streetsblog*, June 22, 2015, http://la.streetsblog .org/2015/06/22/some-perspective-on-how-angelenos-are-driving-less/;   "Vehicle Registrations (Estimated Fee Paid) Los Angeles County, as of December 31, 2014," *Los Angeles Almanac*, accessed July 25, 2016, http://www.laalmanac.com/transport/ tr02.htm.

Chapter 7

This chapter included major contributions from Carine Lai of Civic Exchange and Mark Vallianatos of the Urban & Environmental Policy Institute.

1. BBC, "Hong Kong: Tear Gas and Clashes at Democracy Protest," September 28, 2014, United Kingdom, http://www.bbc.com/news/world-asia-china-29398962.

2. The term "collective strolls" has also been used to identify this type of social-media-generated action. Sam Geall, ed., *China and the Environment: The Green Revolution* (London: Zed Books, 2013), 165; Jeffrey Hou, ed., *Insurgent Public Space: Guerrilla Urbanism and the Remaking of Contemporary Cities* (London: Routledge, 2010), 2.

3. Hong Kong Federation of Students, "Insist Our Demands Persist Our Occupation," October 18, 2014, Hong Kong, http://www.hkfs.org.hk/2014/10/19/insist-our-demands-persist-our-occupation/?lang=en.

4. Anna Almendrala, "Occupy L.A. Protest Underway at City Hall," October 1, 2011, http://www.huffingtonpost.com/2011/10/01/occupy-la-protest-_n_990439.html.

5. "HKSAR Government Headquarters / Rocco Design Architects," *Archdaily*, February 28, 2014, http://www.archdaily.com/481237/hksar-government-headquarters-rocco-design-architects/.

6. Damien Newton, "Bikes, Bikeshare, and the Occupy L.A. Movement," October 14, 2011, http://la.streetsblog.org/2011/10/14/66301/.

7. Hong Kong Special Administrative District, Transport Department, "The Annual Traffic Census—2013," Traffic and Transport Survey Division Publication No. 14CAB1, June 2014, http://www.td.gov.hk/filemanager/en/content_4677/annual%20traffic%20census%202013.pdf.

8. See for example Joyce Lau, "Art Spawned by Protest: Now to Make It Live On," *New York Times*, November 16, 2014, http://www.nytimes.com/2014/11/15/world/asia/rescuing-protest-artwork-from-hong-kongs-streets.html.

9. Julian Ng and Brian Yu, "The Politics of Public Space," *Varsity*, April 2015, http://varsity.com.cuhk.edu.hk/index.php/2015/04/politics-of-public-space/.

10. Saskia Sassen, "The Global Street: Making the Political," *Globalizations* 8, no. 5 (2011), 574.

11. Hong Kong Special Administrative District, Information Services Department, "Hong Kong: The Facts—Population," April 2015, http://www.gov.hk/en/about/abouthk/factsheets/docs/population.pdf.

12. Hong Kong Special Administrative District, Planning Department, "Land Utilization in Hong Kong 2014," May 15, 2015, http://www.pland.gov.hk/pland_en/info_serv/statistic/landu.html; see also "Urban Age Cities Compared: Where People Live," *LSE Cities* (London School of Economics and Political Science), November 2011, https://lsecities.net/media/objects/articles/urban-age-cities-compared/en-gb/.

13. Hong Kong Special Administrative District, Census and Statistics Department, "Table 005: Statistics on Domestic Households," 2015, http://www.censtatd.gov.hk/hkstat/sub/sp150.jsp?tableID=005&ID=0&productType=8; Hong Kong Special Administrative District, Ratings and Valuations Department, "Private Domestic—

Stock and Vacancy by Class" and "Hong Kong Property Review 2015: Tables Appendix Plans," 2015, http://www.rvd.gov.hk/doc/en/hkpr15/05.pdf.

14. O. Wong, "The Uphill Battle for Our Green Havens—Hong Kong's Country Parks," *South China Morning Post*, September 20, 2015, http://www.scmp.com/news/hong-kong/article/1313224/uphill-battle-our-green-havens-hong-kongs-country-parks.

15. The lower figure, 2.6 square meters per person, comes from the Hong Kong Planning Department in 2008. This figure is for open space that is "countable" toward targets laid out in the Hong Kong Planning Standards and Guidelines. The definition is problematic for our purposes, as it includes some private open space but excludes at least half of the large urban parks. See Hong Kong Audit Commission, "Audit 60," chapter 4: "Development and Management of Parks and Gardens," March 28, 2013, http://www.aud.gov.hk/pdf_e/e60ch04.pdf. The higher figure comes from the Hong Kong Planning Department, which reported that there were about 2.5 square kilometers of "parks, stadiums, playgrounds and recreational facilities" in 2014. Divided between 7.24 million inhabitants, this works out to about 3.5 square meters per person. See Hong Kong Special Administrative District, Planning Department, "Land Utilization in Hong Kong 2014," May 15, 2015, http://www.pland.gov.hk/pland_en/info_serv/statistic/landu.html.

16. P. Harnik, A. Martin, and T. O'Grady, "2014 City Park Facts," The Trust for Public Land, February 2014, https://www.tpl.org/sites/default/files/files_upload/2014_CityParkFacts.pdf.

17. Barrie Shelton, Justyna Karakiewicz, and Thomas Kvan, *The Making of Hong Kong: From Vertical to Volumetric* (New York: Routledge, 2011), 5–6. The British later acquired the Kowloon Peninsula after the Second Opium War in 1860, and signed a ninety-nine-year lease on the New Territories in 1898.

18. R. Nissim, *Land Administration and Practice in Hong Kong* (Hong Kong: Hong Kong University Press, 2008), 3–9; Hong Kong Legislative Council Secretariat, "Fact Sheet: Major Sources of Government Revenue," FS19/12–13, August 7, 2013, http://www.legco.gov.hk/yr12-13/english/sec/library/1213fs19_20130807-e.pdf; W. Chu, "Law, Environment and Governance: Harbour Reclamation in Hong Kong, a Case Study," speech given at University College of London Faculty of Law, *YouTube*, March 27, 2012, https://www.youtube.com/watch?v=CZQ-ozP9avY; Christine Loh, "Government and Business Alliance: Hong Kong's Functional Constituencies," in Christine Loh and Civic Exchange, *Functional Constituencies: A Unique Feature of the Hong Kong Legislative Council* (Hong Kong: Hong Kong University Press, 2006), 23–24.

19. Darren Man Wai Cheung, "Land Supply and Land-Use Planning of Public Open Space in Hong Kong," PhD thesis, University of Hong Kong, 2015, 81–82, 96, and 75–76, http://hub.hku.hk/bitstream/10722/209500/2/FullText.pdf?accept=1.

20. C. Chu, "Combating Nuisance: Sanitation, Regulation and the Politics of Property in Colonial Hong Kong," in Robert Peckham and David M. Pomfret, eds., *Imperial Contagions: Medicine, Hygiene and Cultures of Planning in Asia* (Hong Kong: Hong Kong University Press, 2013), 17–36; Shelton, Karakiewicz, and Kvan, *The Making of Hong Kong*, 38.

21. W. M. E. Lo, "Open Space Planning in Metropolitan Hong Kong: From Theory to Practice," MSc thesis, University of Hong Kong, 1986, 23–25; B. S. Tang and S. W. Wong, "A Longitudinal Study of Open Space Zoning and Development in Hong Kong," *Landscape and Urban Planning* 87 (2008), 258–268.

22. Shelton, Karakiewicz, and Kvan, *The Making of Hong Kong*, 77–83.

23. Hong Kong Special Administrative District, Information Services Department, "Fact Sheet: New Towns, New Development Areas and Urban Developments," December 2014, http://www.gov.hk/en/about/abouthk/factsheets/docs/towns%26urban_developments.pdf; Shelton, Karakiewicz, and Kvan, *The Making of Hong Kong*, 68–77.

24. Tang and Wong, "A Longitudinal Study of Open Space Zoning and Development in Hong Kong"; Urban Council, *Parks and Playgrounds in Hong Kong* (Hong Kong: Hong Kong Government Printer, 1969); Lo, "Open Space Planning in Metropolitan Hong Kong," 45; Cheung, "Land Supply and Land-Use Planning of Public Open Space in Hong Kong," 113–114.

25. Lo, "Open Space Planning in Metropolitan Hong Kong," 20–22.

26. Hong Kong Special Administrative District, Planning Department, "Hong Kong Planning Standards and Guidelines," chapter 4: "Recreation, Open Space and Greening," October 2015, http://www.pland.gov.hk/pland_en/tech_doc/hkpsg/full/ch4/ch4_text.htm.

27. Hong Kong Audit Commission, "Development and Management of Parks and Gardens," March 28, 2013, http://www.aud.gov.hk/pdf_e/e60ch04.pdf.

28. In early 2015, the government announced that it had rezoned about 150 ha of green belt, government, institution/community, and open space land scattered over about seventy sites for residential use. Legislative Council Panel on Development, "Increasing Land Supply," CB(1)407/14–15(01), Hong Kong Development Bureau, January 2, 2015, http://www.legco.gov.hk/yr14-15/english/panels/dev/papers/devcb1-407-1-e.pdf. Cedric Sam, Fanny W. Y. Fung, and Olga Wong, "Where Will the Government Rezone for New Housing Sites?," *South China Morning Post*, April 9, 2015, http://www.scmp.com/news/hong-kong/article/1760200/where-will-government-rezone-new-housing-sites.

29. Yvonne Liu, "Rezoning Hurdles Slow Residential Land Release," *South China Morning Post*, September 30, 2014, http://www.scmp.com/property/hong-kong-china/article/1604556/rezoning-hurdles-slow-residential-land-release.

30. Jeremy Rosenberg, "Laws that Shaped L.A.: Why Los Angeles Isn't a Beach Town," "Departures," KCET, January 9, 2012, http://www.kcet.org/socal/departures/columns/laws-that-shaped-la/the-laws-that-shaped-la-laws-of-the-indies.html.

31. Anne V. Howell, *City Planners and Planning in Los Angeles (1781–1998)* (Los Angeles: Department of City Planni ng, 1998).

32. Kathy A. Kolnick, "Order before Zoning: Land Use Regulation in Los Angeles, 1880–1915," PhD dissertation, University of Southern California, May 2008, http://digitallibrary.usc.edu/cdm/ref/collection/p15799coll127/id/61051.

33. "The Sixteen-Hundred-Foot Limit," *Los Angeles Times*, October 2, 1898, B4.

34. "Factories to Build, Where?," *Los Angeles Times*, October 3, 1909, I9.

35. Quoted in Marc A. Weiss, *The Rise of the Community Builders: The American Real Estate Industry and Urban Land Planning* (New York: Columbia University Press, 1987), 10.

36. Richard W. Longstreth, *City Center to Regional Mall: Architecture, the Automobile, and Retailing in Los Angeles, 1920–1950* (Cambridge, MA: MIT Press, 1997).

37. Andrew Whittemore, "How the Federal Government Zoned America: The Federal Housing Administration and Zoning," *Journal of Urban History* 39, no. 4 (July 2013), 620–642; Stefanos Polyzoides et al., *Courtyard Housing in Los Angeles* (New York: Princeton Architectural Press, 1996).

38. Fred W. Viehe, "Black Gold Suburbs: The Influence of the Extractive Industry on the Suburbanization of Los Angeles, 1890–1930," *Journal of Urban History* 8 (November 1981), 3–26; Greg Hise, *Magnetic Los Angeles: Planning the Twentieth-Century Metropolis* (Baltimore: Johns Hopkins University Press, 1999); Wade Graham, "Blueprinting the Regional City: The Urban and Environmental Legacies of the Air Industry in Southern California," in Peter J. Westwick, ed., *Blue Sky Metropolis: Aerospace and Southern California* (Berkeley: University of California Press, 2012).

39. Robert Gottlieb and Irene Wolt, *Thinking Big: The Story of the Los Angeles Times, Its Publishers, and Their Influence on Southern California* (New York: Putnam, 1977), 263.

40. Becky Nicolaides, *My Blue Heaven: Life and Politics in the Working-Class Suburbs of Los Angeles, 1920–1965* (Chicago: University of Chicago Press, 2002); Joe Saltzman and Barbara Saltzman, "Proposition 14: Appeal to Prejudice," *Frontier*, October 1964, 5–8.

41. Greg Morrow, "The Homeowner Revolution, Democracy Land Use and the Los Angeles Slow-Growth Movement, 1965–1992," PhD dissertation, University of California, Los Angeles, 2013, https://escholarship.org/uc/item/6k64g20f.

42. James Rojas, "The Enacted Environment: The Creation of 'Place' by Mexicans and Mexican-Americans in East Los Angeles," master's thesis, MIT, 1991; Wei Li, *Ethnoburb: The New Ethnic Community in Urban America* (Honolulu: University of Hawai'i Press, 2009).

43. Cheung, "Land Supply and Land-Use Planning of Public Open Space in Hong Kong," 79–102.

44. Ibid., 107–116.

45. Hong Kong Planning Standards and Guidelines, 2015, http://www.pland.gov .hk/pland_en/tech_doc/hkpsg/full/.

46. Ibid.

47. Jun Wang and Stephen S. Y. Lau, "Hierarchical Production of Privacy: Gating in Compact Living in Hong Kong," *Current Urban Studies* 1, no. 2 (2013), 11–18.

48. L. W. C. Lai, "Discriminatory Zoning in Colonial Hong Kong: A Review of the Post-war Literature and Some Further Evidence for an Economic Theory of Discrimination," *Property Management* 29, no. 1 (2011), 50–86, http://hub.hku.hk/bitstream/ 10722/134452/1/Content.pdf?accept=1.

49. Wang and Lau, "Hierarchical Production of Privacy," 11.

50. Shelton, Karakiewicz, and Kvan, *The Making of Hong Kong*, 114–116.

51. Wang and Lau, "Hierarchical Production of Privacy," 17–18; Shelton, Karakiewicz, and Kvan, *The Making of Hong Kong*, 114–115.

52. Hong Kong Special Administrative District, Development Bureau, "Provision of Public Facilities in Private Developments," paper presented for discussion to the Legislative Council Panel on Development, CB(1)319/08–09(03), December 8, 2008.

53. See Diana Lee, "Democrats Enter Fray in Times Square Rent Row," *The Standard*, March 6, 2008; Nick Gentle, "Mall Sued over Public Space Rents," *South China Morning Post*, June 18, 2008.

54. Civic Party, "Deputation on Public Open Space—Legislative Council Panel on Development," accessed July 26, 2016, http://www.civicparty.hk/cp/media/pdf/ 090216_dev0216cb1-761-1-e.pdf.

55. Hong Kong Special Administrative District, Development Bureau, "Design and Public Open Space in Private Developments: Design and Management Guidelines," accessed August 8, 2016. http://www.devb.gov.hk/filemanager/en/content_582/ guidelines_english.pdf.

56. Hong Kong Special Administrative District, Development Bureau, "Public Open Space in Private Developments: Design and Management Guidelines," 2011, https:// www.devb.gov.hk/filemanager/en/content_582/Guidelines_English.pdf.

57. Hong Kong Special Administrative District, Audit Commission, "Provision of Public Open Space in Private Developments," October 30, 2014, http://www.aud .gov.hk/pdf_e/e63ch07.pdf.

58. Ibid.

59. Edward J. Blakely and Mary Gail Snyder, "Divided We Fall: Gated and Walled Communities in the United States," in Nan Ellin, ed., *Architecture of Fear* (Princeton: Princeton Architectural Press, 1997), 85–100, http://www.asu.edu/courses/aph294/ total-readings/blakely%20--%20dividedwefall.pdf; Mike Davis, *City of Quartz: Excavating the Future in Los Angeles* (London: Verso, 1990), 223–263.

60. Merlin Chowkwanyun and Jordan Segall, "How an Exclusive Los Angeles Suburb Lost Its Whiteness," *City Lab*, August 27, 2012, http://www.citylab .com/politics/2012/08/how-exclusive-los-angeles-suburb-lost-its-whiteness/3046/; Sam Watters, "It Was Huntington's Master Plan," *Los Angeles Times*, September 6, 2008, http://articles.latimes.com/2008/sep/06/home/hm-lostla6; City of San Marino, "Guide to the Usage of Lacy Park," https://ca-sanmarino.civicplus.com/ DocumentCenter/Home/View/776.

61. Annie Lubinsky, "Authors Unearth History of Rancho Palos Verdes, Palos Verdes Estates," *Daily Breeze*, January 17, 2013, http://www.dailybreeze.com/20130117/ authors-unearth-history-of-rancho-palos-verdes-palos-verdes-estates; Sam Generre, "The Gated City of Rolling Hills," *South Bay Daily Breeze*, July 14, 2010, http://blogs .dailybreeze.com/history/2010/07/14/rolling-hills/; A. E. Hanson, "Rolling Hills: The Early History," published by the City of Rolling Hills, 1978, http://www.rolling-hills .org/DocumentCenter/Home/View/243.

62. Renaud Le Goix, "Gated Communities: Sprawl in Southern California and Social Segregation," *Housing Studies* 20, no. 2 (2005), 323–343, https://halshs .archives-ouvertes.fr/halshs-00004576/document; Tanya Mohn, "America's Most Exclusive Gated Communities," *Forbes*, July 3, 2012, http://www.forbes.com/sites/ tanyamohn/2012/07/03/americas-most-exclusive-gated-communities/. The Wikipedia list of "notable people" at Hidden Hills, accessed July 26, 2016, is at *Wikipedia*, https://en.wikipedia.org/wiki/Hidden_Hills,_California.

63. "Leisure World Celebrates 50th Today," *Orange County Register*, June 23, 2012, http://www.ocregister.com/articles/world-360367-leisure-residents.html. Del Webb's Sun City development is discussed in Peter Wiley and Robert Gottlieb, *Empires in the Sun: The Rise of the New American West* (Tucson: University of Arizona Press, 1985), 184–185.

64. Davis, *City of Quartz*, 223–228; Le Goix's statement is cited by Setha Low, "How Private Interests Take Over Public Space: Zoning, Taxes, and the Incorporation of Gated Communities," in Setha Low and Neil Smith, eds., *The Politics of Public Space* (New York: Routledge, 2006), 89.

65. Anastasia Loukaitou-Sideris, "Designing the Inaccessible Plaza," in Marc Angelil, ed., *On Architecture, the City, and Technology* (Stoneham, MA: Butterworth Publishing, 1991), 142; Paolo Cosulich-Schwartz, "Spatial Injustice in Los Angeles: An Evaluation of Downtown L.A.'s Privately Owned Public Space," Occidental College, 2009, https://www.oxy.edu/sites/default/files/assets/UEP/Comps/2009/cosulich%20 master%20doc_final.pdf.

66. M. L. Chung and K. W. Chow, *Pictures of Hong Kong and Its People, 1950s–1970s* (Hong Kong: Commercial Press, 1997).

67. Advisory Council on Food and Environmental Hygiene, "Review on Hawker Licensing Policy," January 15, 2009, http://www.fhb.gov.hk/download/committees/ board/doc/2009/paper20090115_92.pdf; Urban Council, "A Consultative Document on Hawker and Market Policies: Being the Report of a Working Party of the Urban Council Markets and Street Traders Select Committee to Review Hawker and Related Policies," 1985; Urban Council, "The Report and Recommendations of the Urban Council's Working Party to Review Hawker and Related Policies," 1987.

68. Hong Kong Special Administrative District, Food and Environmental Hygiene Department, "Pleasant Environment," February 29, 2016, http://www.fehd.gov.hk/ english/pleasant_environment/hawker/overview.html.

69. Hong Kong Tourism Board, "Street Markets and Shopping Streets," http://www .discoverhongkong.com/eng/shop/where-to-shop/street-markets-and-shopping -streets/index.jsp, and "Local Dining Places," http://www.discoverhongkong.com/ eng/dine-drink/what-to-eat/local-flavours/local-dining-places.jsp, accessed July 26, 2016.

70. Shelton, Karakiewicz, and Kvan, *The Making of Hong Kong*, 138–139.

71. Candy Chan, "Mong Kok's Pedestrian Zone to Operate Only at Weekends," *South China Morning Post*, November 22, 2013, http://www.scmp.com/news/hong -kong/article/1362246/mong-koks-pedestrian-zone-operate-only-weekends.

72. Prior to 1999, the District Councils (formerly called District Boards) were a layer of local government below the Urban and Regional Councils. They had few formal powers and played a mostly advisory role. After the return to Chinese sovereignty, the Tung Chee Hwa administration abolished the Urban and Regional Councils, leaving the District Councils as the only remaining tier of local government.

73. "Living at Density: Voices of Hong Kong Residents," *LSE Cities*, accessed July 26, 2016, https://lsecities.net/media/objects/articles/living-at-density/en-gb/.

74. Anastasia Loukaitou-Sideris and Renia Ehrenfeucht, *Sidewalks: Conflict and Negotiation over Public Space* (Cambridge, MA: MIT Press, 2009).

75. Ashleigh E. Brilliant, "Some Aspects of Mass Motorization in Southern California, 1919–1929," *Southern California Quarterly* 47, no. 2 (1965), 191–208; Peter D.

Norton, *Fighting Traffic: The Dawn of the Motor Age in the American City* (Cambridge, MA: MIT Press, 2011), 159–168.

76. City of Los Angeles, "Mobility Plan 2035: An Element of the General Plan," August 2015, http://planning.lacity.org/documents/policy/mobilityplnmemo.pdf.

77. Joe Linton, "Muddled L.A. Sidewalk Repair Hearing Inconclusive on How to Proceed," *Streetsblog Los Angeles*, November 17, 2015, http://la.streetsblog .org/2015/11/17/muddled-l-a-sidewalk-repair-hearing-inconclusive-on-how-to -proceed/; "Re: City of Los Angeles Sidewalk/Path of Travel Repair Program," *Investing in Place*, November 9, 2015, http://investinginplace.org/2015/11/09/lasidewalks -2/.

78. Rojas, "The Enacted Environment." See also June M. Tester et al., "An Analysis of Public Health Policy and Legal Issues Related to Mobile Food Vending," *American Journal of Public Health* 100, no. 11 (November 2010), 2038–2046, http://www.ncbi .nlm.nih.gov/pmc/articles/PMC2951932/.

79. Gustavo Arellano, *Taco USA: How Mexican Food Conquered America* (New York: Scribner, 2012), 55.

80. Los Angeles Municipal Code IV. 42 (m). Added by Ord. No. 169,319, Eff. 2/18/94.

81. Los Angeles Street Vendors Campaign, accessed July 26, 2016, http:// streetvendorcampaign.blogspot.com/p/partners.html; Lucia Torres, "Families Precariously Make a Living from Street Vending; City Remains Ambivalent," KCET, November 11, 2015, http://www.kcet.org/socal/departures/families-make-a-living -from-illicit-street-vending-city-remains-ambivalent.html?utm_source=tumblr &utm_medium=social&utm_campaign=departures.

82. Los Angeles Homeless Services Authority, "2015 Greater Los Angeles Homeless Count: 2015 Results," May 11, 2015, http://documents.lahsa.org/Planning/ homelesscount/2015/HC2015CommissionPresentation.pdf; "Residential Neighborhoods as Unregulated Campgrounds/Homelessness and Constitutional Rights/Legal Options," City of Los Angeles Council file No. 14-1057, https://cityclerk.lacity.org/ lacityclerkconnect/index.cfm?fa=ccfi.viewrecord&cfnumber=14-1057; Gary Blasi, "Homeless When El Nino Hits," *Los Angeles Times*, November 23, 2015, http://www .pressreader.com/usa/los-angeles-times/20151123/281646779047519/TextView.

83. Benedict W. Wheeler et al., "Does Living by the Coast Improve Health and Wellbeing?," *Health and Place* 18, no. 5 (September 2012), 1198–1201.

84. The Industrial History of Hong Kong Group, "The Development of Containerization at the Port of Hong Kong," 2014, http://industrialhistoryhk.org/ development-containerization-port-hong-kong/.

85. S. Lee, "Trading in Nostalgia," *South China Morning Post*, October 1, 2002, http://www.scmp.com/article/393036/trading-nostalgia.

86. "Built in Hong Kong: Hollywood Road," *Hong Kong Heritage Series*, June 1, 2012, http://www.heritage.gov.hk/hollywoodroad/en/story.html.

87. Patsy Moy, "District Office to Take Over Night Market," *South China Morning Post*, December 3, 2003, http://www.scmp.com/article/436776/district-office-take-over-night-market.

88. Cap 531, Protection of the Harbour Ordinance, Section 3, December 3, 1999, http://www.legislation.gov.hk/blis_pdf.nsf/6799165D2FEE3FA94825755E0033E532/A6F680241E02ADBD482575EF00152C69?OpenDocument&bt=0. The Protection of the Harbour Ordinance also had the distinction of being the only private member's bill to have been passed in Hong Kong's legislative council; it was passed only three days before the return to Chinese sovereignty. Changes in voting rules by the incoming Chinese administration have made it more difficult for private members' bills to pass, giving the government an effective veto.

89. Chu, "Law, Environment and Governance"; Harbour Business Forum, "Hong Kong Harbour," 2010, http://www.harbourbusinessforum.com/en-us/hkharbour.

90. Civic Exchange, "Civil Society Declares Victory," press release, December 17, 2005, http://www.civic-exchange.org/en/publications/164987260.

91. Designing Hong Kong, "New Rezoning Request for the Central Waterfront," media briefing, May 29, 2007, http://www.harbourdistrict.com.hk/enews/20070531/PR_Rezoning_Central.pdf.

92. Rachel Jaqueline, "Vicious Cycle," *Post Magazine*, May 31, 2015.

93. Hong Kong Special Administrative District, Development Bureau, "Energizing Kowloon East Office," August 3, 2015, http://www.ekeo.gov.hk/en/award/ekeo/index.html.

94. Jenny Price, quoted in Karen Jao, "Lingering Thoughts on the Los Angeles River: A Q&A with Jenny Price," *KCET*, May 29, 2013, http://www.kcet.org/socal/departures/lariver/confluence/river-notes/lingering-thoughts-on-the-los-angeles-river-a-qa-with-jenny-price.html.

95. Martha Groves, "Malibu App Celebrates Beach Access—Right Next to David Geffen's House," *Los Angeles Times*, June 7, 2014, http://www.latimes.com/local/lanow/la-me-ln-malibu-beach-app-20140607-story.html; Robert Garcia, Erica Flores Baltodano, and Edward Mazzarella, "Free the Beach! Public Access, Equal Justice, and the California Coast," Policy Report, Center for Law in the Public Interest and the Surfrider Foundation, 2005, http://www.surfrider.org/images/uploads/publications/Free_the_Beach.pdf.

96. Robin Abcarian, "Long Overdue: Malibu Elitists Who Impede Public Access Now Face Fines," *Los Angeles Times*, June 23, 2014, http://www.latimes.com/local/abcarian/la-me-ra-malibu-public-access-fines-20140623-column.html#page=1.

97. Douglas Flamming, *Bound for Freedom: Black Los Angeles in Jim Cow America* (Berkeley: University of California Press, 2005), 273–275.

98. You-tien Hsing, *The Great Urban Transformation: Politics of Land and Property in China* (New York: Oxford University Press, 2010), 21.

99. Kam Wing Chan, "Crossing the 50 Percent Population Rubicon: Can China Urbanize to Prosperity?," *Eurasian Geography and Economics* 53, no. 1 (2012), 68; Li Zhang, *Strangers in the City: Reconfigurations of Space, Power, and Social Networks within China's Floating Population* (Stanford: Stanford University Press, 2002).

100. Li Zhang, "Migrant Enclaves and Impacts of Redevelopment Policy in Chinese Cities," in Laurence J. C. Ma and Fulong Wu, eds., *Restructuring the Chinese City: Changing Society, Economy and Space* (London: Routledge, 2005), 255.

101. Mimi Lau, "Guangzhou Faces Population Crisis," *South China Morning Post*, March 25, 2011, http://www.scmp.com/article/741938/guangzhou-faces-population-crisis.

102. Nick Smith, "City in the Village: Huanggang and China's Urban Renewal," in Stefan Al, ed., *Villages in the City: A Guide to South China's Informal Settlements* (Hong Kong: Hong Kong University Press, 2014), 29–41; Helen Siu, "Grounding Displacement: Uncivil Urban Spaces in Postreform South China," *American Anthropologist* 34, no. 2 (2007), 329–350; Jonathan Bach, "'They Come In Peasants and Leave Citizens': Urban Villages in the Making of Shenzhen, China," *Cultural Anthropology*, August 2010, 421–458, http://onlinelibrary.wiley.com/doi/10.1111/j.1548-1360.2010.01066.x/abstract.

103. Margaret Crawford and Jiong Wu, "The Beginning of the End: The Destruction of Guangzhou's Urban Villages," in Al, *Villages in the* City, 25; Ye Liu et al., "Growth of Rural Migrant Enclaves in Guangzhou, China: Agency, Everyday Practice, and Social Mobility," *Urban Studies* 52, no. 16 (December 2015), 3086–3105; see also Fulong Wu, "Neighborhood Attachment, Social Participation, and Willingness to Stay in China's Low-Income Communities," *Urban Affairs Review* 48, no. 4 (2012), 547–570.

104. Him Chung, "The Planning of 'Villages-in-the-City' in Shenzhen, China: The Significance of the New State-Led Approach," *International Planning Studies* 14, no. 3 (2009), http://www.tandfonline.com/doi/abs/10.1080/13563470903450606; Pu Hao et al., "Spatial Analyses of the Urban Village Development Process in Shenzhen, China," *International Journal of Urban and Regional Research* 37, no. 6 (November 2013), 2177–2197.

105. Pu Miao, "Deserted Streets in a Jammed Town: The Gated Community in Chinese Cities and Its Solution," *Journal of Urban Design* 8, no. 1 (2003), 45–66.

106. James J. Wang and Qian Liu, "Understanding the Parking Supply Mechanism in China: A Case Study of Shenzhen," *Journal of Transport Geography* 40 (October 2014), 82, http://www.sciencedirect.com/science/article/pii/S0966692314000866; Choon-Piew Pow and Lily Kong, "Marketing the Chinese Dream Home: Gated Communities and Representations of the Good Life in (Post-) Socialist Shanghai," *Urban Geography* 28, no. 2 (2007), 129–159, http://www.tandfonline.com/doi/abs/10.2747/0272-3638.28.2.129; Fulong Wu, "Gated and Packaged Suburbia: Packaging and Branding Chinese Suburban Residential Development," *Cities* 27, no. 5 (October 2010), 385–396. Wu gives the example of one gated community that named itself "Orange County" after that Southern California suburb.

107. Mary Ann O'Donnell, "Maillen Hotel and Apartments in Shenzhen, China," *Architectural Review*, August 28, 2012, http://www.architectural-review.com/buildings/maillen-hotel-and-apartments-in-shenzhen-china-by-c/8634622.article.

108. Hsing, *The Great Urban Transformation*, 2–3.

109. "How Innovation Helps Large Cities Like Shenzhen Face the Challenges of Urbanization" (interview with Huang Weiwen), *Opinion Internationale*, March 24, 2014, http://www.opinion-internationale.com/en/2014/03/24/how-innovation-helps-large-cities-like-shenzhen-face-the-challenges-of-urbanization_23792.html; Stanley C. T. Yip, "Planning for Eco-cities in China: Visions, Approaches and Challenges," 44th ISOCARP Congress, 2008, http://www.isocarp.net/data/case_studies/1162.pdf; Jonathan Kaiman, "China's 'Eco-cities': Empty of Hospitals, Shopping Centres, and People," *Guardian*, April 14, 2014, http://www.theguardian.com/cities/2014/apr/14/china-tianjin-eco-city-empty-hospitals-people.

110. Bai Xuemei et al., "Society: Realizing China's Urban Dream," *International Weekly Journal of Science* 509, no. 7499 (May 7, 2014), http://www.nature.com/news/society-realizing-china-s-urban-dream-1.15151; Xiaodong Ming, "Infrastructure's Central Role in China's 'New Urbanization,'" McKinsey & Co., May 2014, http://www.mckinsey.com/global-themes/urbanization/infrastructures-central-role-in-china-new-urbanization.

111. Gottlieb and Irene Wolt, *Thinking Big*, 106.

112. Yannie Chan, "Yes. We Can Have a Walkable Central," *HK Magazine*, November 20, 2014, http://hk-magazine.com/article/12894/yes-we-can-have-walkable-central; Hong Kong Institute of Planners et al., "Proposed Tram and Pedestrian Precinct in Des Voeux Road Central (Summary Report)," Civic Exchange, Hong Kong, April 2014, http://www.civic-exchange.org/en/publications/166723584.

113. Christopher Hawthorne, "'Latino Urbanism' Influences a Los Angeles in Flux," *Los Angeles Times*, December 6, 2014, http://www.latimes.com/entertainment/arts/la-et-cm-latino-immigration-architecture-20141206-story.html.

Chapter 8

1. "50 People You'll Be Talking About in 2016: Angelo Logan: The Mechanic-Turned-Activist," *Grist*, 2016, http://grist.org/grist-50/profile/angelo-logan/; Margot Roosevelt, "A New Crop of Eco-warriors Take to Their Own Streets," *Los Angeles Times*, September 24, 2009, http://www.latimes.com/local/la-me-air-pollution24 -2009sep24-story.html#page=2. The discussion about Angelo Logan and Christine Loh in this chapter draws on the more than fifteen-year history of connection and engagement that the authors have had with them.

2. Christine Loh, "Time for an Open Mind and Cool Head," *South China Morning Post*, May 5, 1997, http://www.scmp.com/article/194690/time-open-mind-and-cool -head.

3. Loh surprised political observers when she decided to not run again for her LegCo seat, but she felt that the Hong Kong administration had essentially obstructed the Legislature's work. Civic Exchange, instead, became a place independent of that political process; through its focus on the environment, it could expand the civic and democratic discourse and identify concrete opportunities for change. See, for example, Christine Loh and Civic Exchange, *Getting Heard: A Handbook for Hong Kong Citizens* (Hong Kong: Hong Kong University Press, 2002); Christine Loh and Civic Exchange, eds., *Building Democracy: Creating Good Government for Hong Kong* (Hong Kong: Hong Kong University Press, 2003).

4. The *Time* article lauded Loh's "big-picture environmental thinking and policy suggestions." Liam FitzPatrick, "Heroes of the Environment: Christine Loh," *Time*, October 17, 2007, http://content.time.com/time/specials/2007/article/ 0,28804,1663317_1663320_1669919,00.html. See Christine Loh, *Underground Front: The Chinese Communist Party in Hong Kong* (Hong Kong: Hong Kong University Press, 2010); Christine Loh and Civic Exchange, eds., *At the Epicenter: Hong Kong and the SARS Outbreak* (Hong Kong: Hong Kong University Press, 2004).

5. Cheung Chi-fai and Lo Wei, "C. Y. Leung Names Christine Loh New Environ-mental Undersecretary," *South China Morning Post*, September 13, 2012, http://www .scmp.com/news/hong-kong/article/1035319/c-y-leung-names-christine-loh-new -environmental-undersecretary; Mark L. Clifford, *The Greening of Asia: The Business Case for Solving Asia's Environmental Emergency* (New York: Columbia Business School Publishing, Columbia University Press, 2015), 88.

6. Hamish McDonald, "Crouching Tiger, Hidden Power," *Sydney Morning Herald*, March 26, 2005, http://www.smh.com.au/news/World/Crouching-tiger-hidden -power/2005/03/25/1111692627792.html?oneclick=true.

7. Wang Hui, "Son of the Jinsha River: In Memory of Xiao Liangzhong," in Hui, *The End of the Revolution: China and the Limits of Modernity* (London: Verso, 2009), 173–190.

8. Liu Jianqiang, "Defending Tiger Leaping Gorge," in Sam Geall, ed., *China and the Environment: The Green Revolution* (London: Zed Books, 2013), 203.

9. The reference to the LA Dodger stadium is from coauthor Robert Gottlieb and is in Jocelyn Y. Stewart, "Conference Celebrates the Value of Coalitions," *Los Angeles Times*, October 4, 1998, http://articles.latimes.com/1998/oct/04/local/me-29219; see also Harold Meyerson, "Less Than the Sum of Its Parts," *LA Weekly*, October 2–8, 1998, 27.

10. For the SARS connection to the 2003 demonstration, see Christine Loh's newsletter to her "subscribers and friends," issued shortly before the demonstration in June, at http://www.article23.org.hk/english/newsupdate/june03/0622Christine .htm. Hong Kong's "city of protests" reputation has been underlined by the frequency and volume of protests—as many as 50,000 protests in the sixteen years after the Special Administrative Region was established in 1997 through mid-2013, the year before Occupy Central. Daniel Garrett, *Counter-hegemonic Resistance in China's Hong Kong: Visualizing Protest in the City* (Singapore: Springer, 2015), 2; see also Ngok Ma, "Social Movement, Civil Society, and Democratic Development in Hong Kong," paper presented at the Conference on Emerging Social Movements in China, University of Hong Kong, March 23, 2005.

11. Joy Y. Zhang and Michael Barr, "Recasting Subjectivity through the Lenses: New Forms of Environmental Mobilization in China," *Environmental Politics* 22, no. 5 (2013), 849–865, http://www.tandfonline.com/doi/abs/10.1080/09644016.2013.8 17761.

12. Shu-Yun Ma, "The Chinese Discourse on Civil Society," *China Quarterly* 137 (March 1994), 180–193, http://unpan1.un.org/intradoc/groups/public/documents/ un-dpadm/unpan041405.pdf.

13. Daniel A. Farber, "Pollution Markets and Social Equity: Analyzing the Fairness of Cap and Trade," *Ecology Law Quarterly* 39, no. 1 (2012), 1–56, http://scholarship.law .berkeley.edu/cgi/viewcontent.cgi?article=3047&context=facpubs.

14. Rachel Morello-Frosch, "Environmental Policies Must Tackle Social Inequities," *Environmental Health News*, June 21, 2012, http://www.environmentalhealthnews .org/ehs/news/2012/pollution-poverty-people-of-color-op-ed-morello-frosch; James K. Boyce and Manuel Pastor, "Clearing the Air: Incorporating Air Quality and Environmental Justice into Climate Policy," *Climatic Change* 120, no. 4 (October 2013), 801–814, http://www.peri.umass.edu/fileadmin/pdf/other_publication _types/magazine__journal_articles/Boyce__Pastor_-_Climatic_Change_2013.pdf.

15. One important setback was the firing of the more environment-friendly executive director of the South Coast Air Quality Management District when the Air District's Board of Directors shifted to a more conservative, industry-influenced body in 2016. However, environmental groups, in league with a sympathetic state legislator, immediately mobilized to expand the board to reestablish it as more

environment-friendly. Steve Lopez, "Power Grab Topples Another Defender of California's Environment," *Los Angeles Times*, March 4, 2016, http://www.latimes.com/local/california/la-me-lopez-district-20160306-column.html.

16. Lam Wai-man, *Understanding the Political Culture of Hong Kong: The Paradox of Activism and De-politicization* (London: Routledge, 2004); Ngok Ma, "Social Movements and State-Society Relationship in Hong Kong," in Khun Eng Kuah-Pearce and Gilles Guiheux, eds., *Social Movements in China and Hong Kong: The Expansion of Protest Space* (Amsterdam: Amsterdam University Press, 2009), 45–64. Margaret Thatcher's recollection of Deng's words is cited in her private papers at the Margaret Thatcher Foundation, "Release of MT's Private Files for 1982—China and Hong Kong," accessed August 1, 2016, http://www.margaretthatcher.org/archive/1982cac5.asp.

17. Herbert S. Yee and Wong Yiu-chung, "Hong Kong: The Politics of the Daya Bay Nuclear Plant Debate," *International Affairs* 63, no. 4 (Autumn 1987), 617–630; Joseph A. Reaves, "Nuclear Plant Jolts Hong Kong out of Apathy," *Chicago Tribune*, August 17, 1986, http://articles.chicagotribune.com/1986-08-17/news/8603010439_1_hong-kong-daya-bay-nuclear-plant-build; Hsin-Huang Michael Hsiao et al., "The Making of Anti-Nuclear Movements in East Asia: State-Movements Relationships in Policy Outcomes," in Yok-shiu F. Lee and Alvin Y. So, eds., *Asia's Environmental Movements: Comparative Perspectives* (London: Routledge, 1999), 264.

18. Two of the nuclear plants, the Taishan facilities, each with a capacity of 1.75 gigawatts, were located 130 kilometers from Hong Kong. These plants have experienced considerable delays, with concerns raised by the French nuclear developer about insufficient information coming from China; in addition, a handful of environmental groups, led by Greenpeace and the research think tank Professional Commons, called the development "the most dangerous nuclear plant on Hong Kong's doorstep." See Steven Chen, "Chinese Nuclear Reactors 'Did Not Receive Latest Safety Tests before Installation,'" *South China Morning Post*, April 10, 2015, http://www.scmp.com/news/china/article/1763315/taishan-nuclear-reactors-did-not-receive-most-updated-safety-tests; Tara Patel and Benjamin Haas, "China Regulators 'Overwhelmed' as Reactors Built at Pace," *Bloomberg Business*, June 19, 2014, http://www.bloomberg.com/news/articles/2014-06-18/french-nuclear-regulator-says-china-cooperation-lacking; Ernest Kao, "Green Groups Fear 'Most Dangerous Nuclear Power Plant on Hong Kong's Doorstep,'" *South China Morning Post*, September 5, 2013, http://www.scmp.com/news/hong-kong/article/1303433/green-groups-fear-most-dangerous-nuclear-power-plant-hong-kongs.

19. As one example of the less confrontational approach by the mainstream environmental groups, Linda Siddell, the executive director of the Hong Kong Friends of the Earth organization (which had been expelled from the international Friends of the Earth parent organization for its acceptance of corporate funding), argued in a 1991 letter to the UK *New Scientist* that "times are changing and so too are industrial

attitudes. We have therefore sought to establish a relationship with industry which is suited to the present." Siddell's letter is cited in Sharon Beder, "Activism versus Negotiation: Strategies for the Environmental Movement," *Social Alternatives* 10, no. 4 (December 1991), 53–56, http://www.uow.edu.au/~sharonb/activism.html#fn13; see also Daniel Garrett, "Visualizing Protest Culture in China's Hong Kong: Recent Tensions over Integration," *Visual Communication*, February 2013, 55–70; Stephen Wing-Kai Chiu and Tai Lok Lui, *The Dynamics of Social Movements in Hong Kong* (Hong Kong: Hong Kong University Press, 2000).

20. "Regulations on the Registration and Management of Social Organizations," Order No. 250 of the State Council of the People's Republic of China, October 25, 1998, reprinted and translated by the Congressional-Executive Commission on China, http://www.cecc.gov/resources/legal-provisions/regulations-on-the-registration-and-management-of-social-organizations.

21. Feng Jie and Wang Tao, "Officials Struggling to Respond to China's Year of Environmental Protests," *China Dialogue*, December 6, 2012, https://www .chinadialogue.net/article/show/single/en/5438-Officials-struggling-to-respond-to -China-s-year-of-environment-protests-; Bao Maohong, "Environmental NGOs in Transforming China," *Nature and Culture* 4, no. 1 (Spring 2009), 1–16.

22. The 2007 protests at Xiamen, which introduced the idea of the "strolling protest" and included a number of academic and political figures, led to the plant being located in an industrial zone in Zhangzhou, ninety kilometers from Xiamen, after a concerted public relations offensive by Zhangzhou officials to have the plant built there. But in 2013 and again in 2015, major explosions occurred at the relocated plant. Although government officials labeled them as accidents, the explosions only reinforced people's concerns about PX specifically and about chemical industry management and safety issues in general. Liu Qin, "Government Assurances on Petchem Plants Vaporized after Latest Explosion," *China Dialogue*, April 8, 2015. See also Liu Qin, "Shanghai Residents Throng Streets in 'Unprecedented' Anti-PX Protest," *China Dialogue*, July 2, 2015, https://www.chinadialogue.net/article/show/single/en/8009-Shanghai-residents-throng-streets-in-unprecedented-anti-PX -protest; Kingsyhon Lee and Ming-Sho Ho, "The Maoming Anti-PX Protest of 2014," *China Perspectives* 3 (2014), 33–39, http://chinaperspectives.revues.org/6537?file=1+; Amruthra Gayathri, "Chinese Protests over Chemical Factory Reflect Government Mistrust," *International Business Times*, October 29, 2012, http://www.ibtimes.com/ chinese-protests-over-chemical-factory-reflect-government-mistrust-855415.

23. Peter Ho and Richard Louis Edmonds, *China's Embedded Activism: Opportunities and Constraints of a Social Movement* (London: Routledge, 2008). Ho and Edmonds quote Liao Xiaoyi, founder of the Beijing-based Global Village, from an interview in the *Beijing Review* in August 2000: "We guide the public instead of blaming them and help the government instead of complaining about it. This, perhaps, is the 'female mildness' referred to by the media. I don't appreciate extremist methods.

I'm engaged in environmental protection and I don't want to use it for political aims. This is my way, and my principle too." See also Peter Ho, "Greening without Conflict: Environmentalism, NGOs and Civil Society in China," *Development and Change* 32 (2001), 893–921; Anthony J. Spires, Lin Tao, and Kin-man Chan, "Societal Support for China's Grassroots NGOs: Evidence from Yunnan, Guangdong, and Beijing," *China Journal* 71 (January 2014), 65–90.

24. Institute of Public and Environmental Affairs (China), "China Water, Air, and Solid Waste Pollution Maps," accessed August 3, 2016, http://www.ipe.org.cn/en/pollution/index.aspx.

25. Congressional-Executive Commission on China, 2014 Annual Report, Section on the Environment, http://www.cecc.gov/publications/annual-reports/2014-annual-report.

26. Stanley Lubman, "China Asserts More Control over Foreign and Domestic NGOs," *Wall Street Journal*, June 16, 2015, http://blogs.wsj.com/chinarealtime/2015/06/16/china-asserts-more-control-over-foreign-and-domestic-ngos/.

27. Mark Purcell, "Excavating Lefebvre: The Right to the City and Its Urban Politics of the Inhabitants," *GeoJournal* 58 (2002), 99–108, http://faculty.washington.edu/mpurcell/geojournal.pdf.

28. Andre Gorz, "Reform and Revolution," *Socialist Register* 5 (1968), 1–34, http://socialistregister.com/index.php/srv/article/view/5272/2173#.V6kdivkrK00.

Bibliography

Chapter 1

Andreas, Joel. *Rise of the Red Engineers: The Cultural Revolution and the Origins of the New Class*. Stanford: Stanford University Press, 2009.

Barber, Benjamin R. *If Mayors Ruled the World: Dysfunctional Nations, Rising Cities*. New Haven: Yale University Press, 2013.

Barboza, David. "In China, Projects to Make Great Wall Feel Small." *New York Times*, January 12, 2015. http://www.nytimes.com/2015/01/13/business/international/in-china-projects-to-make-great-wall-feel-small-.html.

Bjorklund, E. M. "The Danwei: Socio-spatial Characteristics of Work Units in China's Urban Society." *Economic Geography* 62 (1) (January 1986): 19–29.

Brown, Stephen. "The Economic Impact of SARS." In *At the Epicentre: Hong Kong and the SARS Outbreak*, ed. Christine Loh and Civic Exchange, 179–193. Hong Kong: Hong Kong University Press, 2004.

Buijs, Steef, Wendy Tan, and Devisari Tunas, eds. *Megacities: Exploring a Sustainable Future*. Rotterdam: 010 Publishers, 2010.

Cann, Cynthia W., Michael C. Cann, and Gao Shangquan. "China's Road to Sustainable Development: An Overview." Chapter 1 in *China's Environment and the Challenge of Sustainable Development*, ed. Kristen A. Day. Armonk, NY: M. E. Sharpe, 2005.

Castells, Manuel, Lee Goh, and R. Yin-Wang Kwok. *The Shek Kip Mei Syndrome: Economic Development and Public Housing in Hong Kong and Singapore*. London: Pion, 1990.

Chan, Kam Wing. "The Global Financial Crisis and Migrant Workers in China: 'There Is No Future as a Labourer; Returning to the Village Has No Meaning.'" *International Journal of Urban and Regional Research* 34 (3) (September 2010): 659–677.

Chan, M. K., ed. *Precarious Balance: Hong Kong between China and Britain, 1842–1992.* Armonk, NY: M. E. Sharpe, 1994.

Chan, M. K., and A. Y. So, eds. *Crisis and Transformation in China's Hong Kong.* Armonk, NY: M. E. Sharpe, 2002.

Choy, Tim. *Ecologies of Comparison: An Ethnography of Endangerment in Hong Kong.* Durham: Duke University Press, 2011.

Davis, Mike. "The Case for Letting Malibu Burn." *Environmental History Review* 19 (2) (Summer 1995): 1–36.

Day, Alexander F. *The Peasant in Postsocialist China: History, Politics, and Capitalism.* Cambridge: Cambridge University Press, 2013.

Deverell, William. *Whitewashed Adobe: The Rise of Los Angeles and the Remaking of Its Mexican Past.* Berkeley: University of California Press, 2005.

Economy, Elizabeth. "The Great Leap Backward." *Foreign Affairs* 86 (September–October 2007): 5.

Enright, Michael J., Edith E. Scott, and Ka-mun Chang. *Regional Powerhouse: Greater Pearl River Delta and the Rise of China.* Singapore: John Wiley & Sons (Asia), 2005.

Friedmann, John. "The World City Hypothesis." *Development and Change* 17 (1986): 69–83.

Friedmann, John, and Goetz Wolff. "World City Formation: An Agenda for Research and Action." *International Journal of Urban and Regional Research* 6 (3) (1982): 309–344.

Geall, S., ed. *China and the Environment: The Green Revolution.* London: Zed Books, 2013.

Goodstadt, Leo F. *Poverty in the Midst of Affluence: How Hong Kong Mismanaged Its Prosperity.* Hong Kong: Hong Kong University Press, 2013.

Gottlieb, Robert. *Reinventing Los Angeles: Nature and Community in the Global City.* Cambridge, MA: MIT Press, 2007.

Gottlieb, Robert, and Irene Wolt. *Thinking Big: The Story of the Los Angeles Times, Its Publishers, and Their Influence on Southern California.* New York: G. P. Putnam's Sons, 1977.

Green, Emily. "Green Dreams." *Los Angeles Times*, October 31, 2004.

Haila, Anne. "Real Estate in Global Cities: Singapore and Hong Kong as Property States." *Urban Studies* (Edinburgh) 37 (12) (November 2000): 2241–2256.

Hairong, Yan. *New Masters, New Servants: Migration, Development, and Women Workers in China.* Durham: Duke University Press, 2008.

Hairong, Yan, and Yiyuan Chen. "Debating the Rural Cooperative Movement in China, the Past and the Present." *Journal of Peasant Studies* 40 (6) (2013): 955–981.

Hall, Peter. "Megacities, World Cities, Global Cities." In Steef Buijs, Wendy Tan, and Devisari Tunas, eds., *Megacities: Exploring a Sustainable Future*, 35–46. Rotterdam: 010 Publishers, 2010.

Hall, Peter. *The World Cities*. New York: McGraw-Hill, 1966.

Harney, Alexandra. *The China Price: The True Cost of Chinese Competitive Advantage*. New York: Penguin Press, 2008.

Harvey, David. "Possible Urban Worlds." In Steef Buijs, Wendy Tan, and Devisari Tunas, eds., *Megacities: Exploring a Sustainable Future*, 167–179. Rotterdam: 010 Publishers, 2010.

Ho, Peter. *Institutions in Transition: Land Ownership, Property Rights and Social Conflict in China*. New York: Oxford University Press, 2005.

Hu, Fox, and Michelle Yun. "Hong Kong Poverty Line Shows Wealth Gap with One in Five Poor." *BloombergBusinessWeek*, September 29, 2013. http://www.bloomberg.com/news/articles/2013-09-29/hong-kong-poverty-line-shows-wealth-gap-with-one-in-five-poor.

Hui, Wang. *China's New Order: Society, Politics, and Economy in Transition*. Ed. and trans. T. Huters. Cambridge, MA: Harvard University Press, 2003.

Hui, Wang, and En Liang Khong. "After the Party: An Interview with Wang Hui." *Open Democracy*, January 13, 2014. https://www.opendemocracy.net/wang-hui-en-liang-khong/after-party-interview-with-wang-hui.

Hung, Ho-fung. "America's Head-servant? The PRC's Dilemma in the Global Crisis." *New Left Review* 60 (November-December 2009): 5–25. http://newleftreview.org/II/60/ho-fung-hung-america-s-head-servant.

Jao, Y. C. "The Rise of Hong Kong as a Financial Center." *Asian Survey* 19 (7) (July 1979): 674–694.

Jia'en, Pan, and Du Jie. "Alternative Responses to 'the Modern Dream': The Sources and Contradictions of Rural Reconstruction in China." *Inter-Asia Cultural Studies* 12 (3) (2011): 454–464. http://www.tandfonline.com/doi/abs/10.1080/14649373.2011.578809.

Khan, Azizur Rahman, and Carl Riskin. *Inequality and Poverty in China in the Age of Globalization*. New York: Oxford University Press, 2001.

Kirdahy, Matthew. "Walmart Strikes Pay Deal with Chinese Union." *Forbes*, July 25, 2008. http://www.forbes.com/2008/07/25/walmart-china-unions-lead-cx_mk_0725acftu.html.

Klein, Naomi. *This Changes Everything: Capitalism vs. the Climate*. New York: Simon & Schuster, 2014.

Knox, Paul L. "Globalization and the World City Hypothesis." *Scottish Geographical Magazine* 112 (2) (1996): 124–126.

Krugman, Paul. *Geography and Trade*. Cambridge, MA: MIT Press, 1991.

Kwong, Jo Ann. *Market Environmentalism: Lessons for Hong Kong*. Hong Kong: Chinese University Press, 1989.

Lee, Eliza W. Y. "The Renegotiation of the Social Pact in Hong Kong: Economic Globalization, Socio-economic Change, and Local Politics." *Journal of Social Policy* 34 (2) (April 2005): 293–310.

Li, Shan, Samantha Masunaga, and Andrew Khouri. "California Braces for Falling Yuan's Ripples." *Los Angeles Times*, August 12, 2015. http://www.latimes.com/business/la-fi-china-yuan-california-20150813-story.html#page=1.

Loh, Christine. *Underground Front: The Chinese Communist Party in Hong Kong*. Hong Kong: Hong Kong University Press, 2010.

Loh, Christine, and Civic Exchange, eds. *At the Epicentre: Hong Kong and the SARS Outbreak*. Hong Kong: Hong Kong University Press, 2004.

Lora-Wainwright, Anna. *Fighting for Breath: Living Morally and Dying of Cancer in a Chinese Village*. Honolulu: University of Hawai'i Press, 2013.

Ma, Laurence J. C., and Allen G. Noble. *The Environment: Chinese and American Views*. New York: Methuen, 1981.

Maohong, Bao. "Environmental History in China." *Environment and History* 10 (2004): 475–499. http://www.environmentandsociety.org/sites/default/files/key_docs/maohong-10-4.pdf.

Marcuse, Peter. "Space in the Globalizing City." In *The Global Cities Reader*, ed. Neil Brenner and Roger Keil, 361–369. New York: Routledge, 2006.

McPhee, John. "Los Angeles against the Mountains." *New Yorker*, September 26, 1988, 26. http://www.newyorker.com/magazine/1988/09/26/los-angeles-against-the-mountains-i.

McWilliams, Carey. *Southern California: An Island on the Land*. Santa Barbara, CA: Peregrine Smith, 1973.

Meyer, David R. *Hong Kong as a Global Metropolis*. Cambridge: Cambridge University Press, 2000.

Meyer, David R. "Hong Kong's Transformation as a Financial Center." http://www.hkimr.org/uploads/conference_detail/585/con_paper_0_413_david-meyer-paper.pdf.

"Migrant Workers and Their Children." *China Labour Bulletin*, June 27, 2013. http://www.clb.org.hk/content/migrant-workers-and-their-children.

Mol, Arthur. "Environment and Modernity in Transitional China: Frontiers of Ecological Modernization." *Development and Change* 37 (1) (2006): 29–56.

Ngo, Tak-Wing. *Hong Kong's History: State and Society under Colonial Rule*. London: Routledge, 1999.

Olson, Scott. "Seven Questions: China's Total Toy Recall." Interview with M. Eric Johnson. *Foreign Policy*, August 22, 2007. http://www.foreignpolicy.com/articles/2007/08/21/seven_questions_chinas_total_toy_recall.

Sassen, Saskia. *The Global City: New York, London, Tokyo*. Princeton: Princeton University Press, 2001.

Sassen, Saskia. "Urban Economies and Fading Distances." In *Megacities: Exploring a Sustainable Future*, ed. Steef Buijs, Wendy Tan, and Devisari Tunas, 47–55. Rotterdam: 010 Publishers, 2010.

Schell, Orville, and John Delury. *Wealth and Power: China's Long March to the Twenty-first Century*. New York: Random House, 2013.

"Shades of Grey: Ten Years of China in the WTO." *Economist*, December 10, 2011. http://www.economist.com/node/21541408.

Shapiro, Judith. *China's Environmental Challenge*. Malden, MA: Polity Books, 2012.

Shapiro, Judith. *Mao's War against Nature: Politics and the Environment in Revolutionary China*. Cambridge: Cambridge University Press, 2001.

Shen, Jianfa. "Hong Kong under Chinese Sovereignty: Economic Relations with Mainland China, 1978–2007." *Eurasian Geography and Economics* 49 (3) (2008): 326–340. http://www.grm.cuhk.edu.hk/eng/research/RAE2011/ShenJianfa/3%20P111Y2008HKmainland1978to2007EGEV4Journal.pdf.

Shi, L., H. Sato, and T. Sicular, eds. *Rising Inequality in China: Challenges to a Harmonious Society*. New York: Cambridge University Press, 2013.

Short, John Rennie. *Global Metropolitan: Globalizing Cities in a Capitalist World*. London: Routledge, 2004.

Sito, Peggy. "Tycoon Li Ka-shing Downbeat over Hong Kong's Future." *South China Morning Post*, February 28, 2014. http://www.scmp.com/news/hong-kong/article/1437471/li-ka-shing-backs-press-freedom-after-attack-kevin-lau?page=all.

Smart, Alan, and James Lee. "Financialization and the Role of Real Estate in Hong Kong's Regime of Accumulation." *Economic Geography* 79 (2) (2003): 153–171. http://www.readcube.com/articles/10.1111%2Fj.1944-8287.2003.tb00206.x.

Smil, Vaclav. *The Bad Earth: Environmental Degradation in China.* New York: M. E. Sharpe, 1984.

Smith, Richard. "Creative Destruction: Capitalist Development and China's Environment." *New Left Review* I-222 (March-April 1997). https://newleftreview .org/I/222/richard-smith-creative-destruction-capitalist-development-and-china-s -environment.

Solinger, Dorothy J. *Contesting Citizenship in Urban China: Peasant Migrants, the State, and the Logic of the Market.* Berkeley: University of California Press, 1999.

Sung, Yun-Wing, and Kar-yiu Wong. "Growth of Hong Kong Before and After Its Reversion to China: The China Factor." Paper presented at the Joint Session of the American Economics Association and Chinese Economics Association in North America, January 3, 1998. http://faculty.washington.edu/karyiu/papers/HK -ChinaFac.pdf.

Tiejun, Wen. "China's Century-long Quest for Industrialization." *China Study Group,* November 18, 2003. (Translation from *Dushu,* no. 3, March 2001.)

Tiejun, Wen. "Deconstructing Modernization." *Chinese Sociology and Anthropology* 39 (4) (Summer 2007): 10–25. http://chinastudygroup.net/wp-content/uploads/2010/ 12/Wen_2007_Deconstructing_Modernization.pdf.

Tiejun, Wen, et al. "Ecological Civilization, Indigenous Culture, and Rural Reconstruction in China." *Monthly Review* (New York) 63 (9) (February 2012). http:// monthlyreview.org/2012/02/01/ecological-civilization-indigenous-culture-and-rural -reconstruction-in-china.

Tirschwell, Peter. "Atlantic Expectations." *Journal of Commerce,* July 16, 2010.

Tsai, Jung-fang. "From Antiforeignism to Popular Nationalism: Hong Kong between China and Britain, 1839–1911. In *Precarious Balance: Hong Kong between China and Britain, 1842–1992,* ed. M. K. Chan. 9–26. Armonk, NY: M. E. Sharpe, 1994.

Tse, Crystal. "Hong Kong Is Slowly Dimming Its Neon Glow." *New York Times,* October 13, 2015. http://www.nytimes.com/2015/10/14/world/asia/hong-kong-neon -sign-maker.html?_r=1.

Vogel, Ezra F. *Canton under Communism: Programs and Politics in a Provincial Capital, 1949–1968.* Cambridge, MA: Harvard University Press, 1969.

Vogel, Ezra F. *One Step Ahead in China: Guangdong under Reform.* Cambridge, MA: Harvard University Press, 1989.

Wang, Zhihe. "Ecological Marxism in China." *Monthly Review* (New York), February 1, 2012. http://monthlyreview.org/2012/02/01/ecological-marxism-in-china.

Wang, Zhihe, et al. "The Ecological Civilization Debate in China: The Role of Ecological Marxism and Constructive Postmodernism—beyond the Predicament of

Legislation." *Monthly Review* (New York) 66 (6) (November 2014). http://monthly
review.org/2014/11/01/the-ecological-civilization-debate-in-china/.

Wang, Zhihe, et al. "What Does Ecological Marxism Mean for China?" *Monthly Review* (New York) 64 (9) (February 2013).

Wasserstrom, Jeffrey N. *Global Shanghai, 1850–2010.* London: Routledge, 2009.

Wasserstrom, Jeffrey N. "In 'Global Cities,' You Get a View of the Future." *Los Angeles Times*, February 11, 2007. http://articles.latimes.com/2007/feb/11/opinion/op
-wasserstrom11.

Watts, Jonathan. *When a Billion Chinese Jump: How China Will Save Mankind—Or Destroy It.* New York: Scribner, 2010.

Weller, Robert. *Discovering Nature: Globalization and Environmental Culture in China and Taiwan.* Cambridge: Cambridge University Press, 2006.

Wong, Stan Hok-Wui. "Real Estate Elite, Economic Development, and Political Conflicts in Post-colonial Hong Kong." *China Review* 15 (1) (Spring 2015): 1–38.

Yu, Au Loong, and Kevin Li. "Preliminary Report on China's Going Global Strategy: A Labour, Environment, and Hong Kong Perspective." Asia Monitor Resource Center and Globalization Monitor. Capital Mobility Research Paper Series No. 3, 2011.

Yu, Henry. "Los Angeles and American Studies in a Pacific World of Migrations." *American Quarterly* 56 (3) (2004): 531–543.

Zhang, Lijia. "Factory Life Far from Home Leaves China's Migrant Workers Vulnerable." *CNN*, January 2, 2014. http://www.cnn.com/2014/01/02/world/asia/china
-migrants-mental-health/index.html.

Chapter 2

Bailey, Diane, Zach Goldman, and Maria Minjares. "Driving on Fumes: Truck Drivers Face Elevated Health Risks from Diesel Pollution." Natural Resources Defense Council (NRDC) Issue Paper, December 2007. https://www.nrdc.org/health/effects/
driving/driving.pdf.

Barboza, Tony. "The Port of L.A. Rolled Back Measures to Cut Pollution—during Its 'Green' Expansion." *Los Angeles Times*, December 15, 2015. http://www.latimes
.com/local/california/la-me-port-pollution-20151215-story.html.

Barboza, Tony. "Southland Air Regulators Seek to Hold Ports to Pollution Targets." *Los Angeles Times*, April 14, 2014. http://www.latimes.com/local/la-me-ports-air
-20140415-story.html.

Bensman, David. "Port Trucking Down the Low Road: A Sad Story of Deregulation." *Demos* (Mexico City) 3 (2009). http://www.demos.org/sites/default/files/
publications/Port%20Trucking%20Down%20the%20Low%20Road.pdf.

Bonacich, Edna, and Jake B. Wilson. *Getting the Goods: Ports, Labor, and the Logistics Revolution.* Ithaca: Cornell University Press, 2008.

Booth, Veronica, and Christine Loh. "Reducing Vessel Emissions: Science, Policy and Engagement in the Hong Kong–Pearl River Delta Region." Civic Exchange, Hong Kong, April 2012.

Burnson, Patrick. "Panama Canal Expansion Update: When the 'Tipping Point' Becomes Reality." *Logistics Management,* July 1, 2013. http://www.logisticsmgmt.com/article/panama_canal_expansion_update_when_the_tipping_point_becomes_reality.

Cain, Cindy Wojdyla. "Grain Exports Are Booming." *Southtown Star,* November 18, 2011. Accessed June 13, 2014. http://southtownstar.suntimes.com/news/8931084-418/grain-exports-are-booming.html#.U00-PH7n_IU.

California Air Resources Board. "ARB Health Risk Assessment Guidance for Rail Yard and Intermodal Facilities." 2006. http://www.arb.ca.gov/railyard/hra/1107hra_guideline.pdf.

California Air Resources Board. "Quantification of the Health Impacts and Economic Impacts of Air Pollution from Ports and Goods Movement in California." Appendix A to "Emission Reduction Plan for Ports and Goods Movement." August 31, 2011. http://www.arb.ca.gov/planning/gmerp/plan/appendix_a.pdf.

Center for Community Action and Environmental Justice. "Inland Ports of Southern California—Warehouses, Distribution Centers, Intermodal Facilities— Impacts, Costs, and Trends." 2012. www.ccaej.org.

Chen, Shu-Ching Jean. "Choi's Toys." *Forbes,* January 18, 2008. http://www.forbes.com/global/2008/0128/026.html.

Chiu, Joanne. "Hong Kong's Port Is Caught in a Storm." *Wall Street Journal,* February 16, 2016. http://www.wsj.com/articles/hong-kong-ports-steady-drop-persists-1455654601.

Cristea, Anca D., et al. "Trade and the Greenhouse Gas Emissions from International Freight Transport." National Bureau of Economic Research, Working Paper 17117, June 2011. http://www.nber.org/papers/w17117.

Cummings, Scott L. *Preemptive Strike: Law in the Campaign for Clean Trucks.* Los Angeles: UCLA School of Law, 2014.

DePillis, Lydia. "Ports Are the New Power Plants—At Least in Terms of Pollution." *Washington Post,* November 24, 2015. https://www.washingtonpost.com/news/wonk/wp/2015/11/24/ports-are-the-new-power-plants-at-least-in-terms-of-pollution/.

Dolan, Jack, and Tony Barboza. "Port of LA Helped Pay for Cleaner China Shipping Vessels—Which Later Stopped Docking in LA." *Los Angeles Times,* March 24, 2016.

http://www.latimes.com/local/california/la-me-china-shipping-20160324-story
.html.

Environmental News Service. "Port of Los Angeles Hosts First Plugged-In Container Ship." *Environmental News Service*, June 21, 2004. http://www.ens-newswire.com/ens/jun2004/2004-06-21-04.asp.

Erie, Steven P. *Globalizing L.A.: Trade, Infrastructure, and Regional Development*. Stanford: Stanford University Press, 2004.

Garcia, Analilia, et al. "THE (Trade, Health Environment) Impact Project: A Community- Based Participatory Research Environmental Justice Case Study." *Environmental Justice* 6 (1) (2013): 17–26. http://lbaca.org/wp-content/uploads/2013/07/LBACA-Trade-Health-Environment-Impact-Project.pdf.

George, Rose. *Ninety Percent of Everything: Inside Shipping, the Invisible Industry that Puts Clothes on Your Back, Gas in Your Car, and Food on Your Plate*. New York: Metropolitan Books, 2013.

Gough, Neil. "Hong Kong Strike Clogs Shipping Traffic." *New York Times*, April 3, 2013. http://www.nytimes.com/glogin?URI=http%3A%2F%2Fwww.nytimes.com%2F2013%2F04%2F04%2Fbusiness%2Fglobal%2Fhong-kong-strike-clogs-shipping-traffic.html%3F_r%3D1.

Grassman, Curtis. "The Los Angeles Free Harbor Fight Controversy and the Creation of a Progressive Coalition." *Southern California Quarterly* 55 (1973): 445–467.

Grescoe, Taras. "The Dirty Truth about 'Clean Diesel.'" *New York Times*, January 3, 2016.

Guerra, Ferdinando. "Growing Together: China and Los Angeles County." Los Angeles Economic Development Corporation, Los Angeles, June 2014. http://laedc.org/wp-content/uploads/2014/05/2014-Growing-Together-China-and-LA-County.pdf.

Hedley, Anthony J., et al. "Air Pollution: Costs and Paths to a Solution in Hong Kong—Understanding the Connections among Visibility, Air Pollution, and Health Costs in Pursuit of Accountability, Environmental Justice, and Health Prevention." *Journal of Toxicology and Environmental Health, Part A* 71 (2008): 544–554.

"Hong Kong Wants Ship Emission Standards Now." *Maritime Executive*, March 19, 2015. http://www.maritime-executive.com/article/hong-kong-wants-ship-emission-standards-now.

Houston, Douglas, Wei Li, and Jun Wu. "Disparities in Exposure to Automobile and Truck Traffic and Vehicle Emissions near the Los Angeles–Long Beach Port Complex." *American Journal of Public Health* 104 (1) (January 2014): 156–164. http://www.ncbi.nlm.nih.gov/pmc/articles/PMC3910024/.

Hricko, Andrea. "Progress and Pollution: Port Cities Prepare for the Panama Canal Expansion." *Environmental Health Perspectives* 120 (December 2012): 470–473. http:// ehp.niehs.nih.gov/120-a470/.

Hricko, Andrea, et al. "Global Trade, Local Impacts: Lessons from California on Health Impacts and Environmental Justice Concerns for Residents Living Near Freight Yards." *International Journal of Environmental Research and Public Health* 11 (2014): 1914–1941.

"The Huntington Gall." *Los Angeles Times*, April 6, 1896.

Intermodal Container Transfer Facility. "Joint Powers Authority." Accessed August 10, 2016. http://www.ictf-jpa.org/.

Keil, Roger. *Los Angeles: Globalization, Urbanization and Social Struggles*. New York: John Wiley & Sons, 1998.

Keqiang, Li. "China Deepens Strategy of Domestic Demand Expansion in the Course of Reform and Opening Up." *China.org.cn*, March 4, 2012. http://www.china.org.cn/ china/2012-03/04/content_24801231.htm.

Kirkham, Chris. "Warehouse Empire." *Los Angeles Times*, April 19, 2015. http:// www.latimes.com/business/la-fi-inland-empire-warehouses-20150419-story.html #page=1.

Kirkham, Chris, and Andrew Khouri. "Cargo Crush." *Los Angeles Times*, June 2, 2015. http://www.pressreader.com/usa/los-angeles-times/20150602/281921656653544/ TextView.

Kitroeff, Natalie. "L.A. Ports Face a Sea of Rivals." *Los Angeles Times*, April 27, 2016. http://www.latimes.com/business/la-fi-la-ports-competition-20160427-story.html.

Kravtsov, Grigory. "Ship Emissions Blamed for Worsening Pollution in Hong Kong." *CNN*, December 20, 2013. http://www.cnn.com/2013/12/19/world/asia/hong-kongs -worsening-ship-pollution/.

Kutalik, Chris. "As Immigrants Strike, Truckers Shut Down Nation's Largest Port." *Labor Notes*, September 28, 2006. http://labornotes.org/2006/09/immigrants-strike -truckers-shut-down-nation%E2%80%99s-largest-port.

Lai, Hak-Kan, et al. "Health Impact Assessment of Measures to Reduce Marine Ship- ping Emissions." Final report. University of Hong Kong, Department of Community Medicine, School of Public Health, July 2012.

Laird, Gordon. *The Quest for Cheap and the Death of Globalization*. New York: Palgrave Macmillan, 2009.

LaSalle, Jones Lang. "Growing U.S. Exports Will Drive Inland Ports." *Perspectives on Global Supply Chain Dynamics*. Summer 2012. http://www.us.am.joneslanglasalle

.com/ResearchLevel1/Growing%20US%20exports%20will%20drive%20Inland%20 Ports.pdf.

Lavigne, Grace M. "Port Congestion Worse at LB than LA, CargoSmart Says." *JOC. com*. November 4, 2014. http://www.joc.com/port-news/us-ports/port-los-angeles/ port-congestion-worse-lb-la-cargosmart-says_20141104.html.

Levinson, Marc. *The Box: How the Shipping Container Made the World Smaller and the World Economy Bigger*. Princeton: Princeton University Press, 2006.

Li, Keqiang. "China Deepens Strategy of Domestic Demand Expansion in the Course of Reform and Opening Up." *China.org.cn*, March 4, 2012. (Excerpt of an article by Li Keqiang, shortly before he became prime minister, published in Chinese in *Qiushi* 4 [2012], translated into English by He Shan and Chen Xia, http://www.china.org.cn/ china/2012-03/04/content_24801231.htm.)

Lopez, Ricardo. "Truck Drivers Set for Two-day Strike at Ports." *Los Angeles Times*, April 27, 2014. http://www.latimes.com/business/la-me-port-truck-driver-strike -20140427,0,3804654.story#axzz3071D1oKC.

Los Angeles County Economic Development Corporation. "International Trade Outlook: The Southern California Region, 2013-2014." Kyser Center for Economic Research. http://cdn.laedc.org/wp-content/uploads/2013/05/ITO-2013-REPORT -FINAL.pdf.

Maersk Lines. "Skies Clearing in Hong Kong." January 31, 2013. http://www .maerskline.com/en-us/countries/int/news/news-articles/2013/01/skies-clearing.

Magretta, Joan. "Fast, Global, and Entrepreneurial: Supply Chain Management, Hong Kong Style: An Interview with Victor Fung." *Harvard Business Review* 76 (5) (September-October 1998): 103–114. http://hbr.org/1998/09/fast-global-and -entrepreneurial-supply-chain-management-hong-kong-style/ar/1.

Matson, Clarence. "The Port of Los Angeles: Historical." In *The Story of Los Angeles Harbor: Its History, Development, and Growth of Its Commerce*. Los Angeles: Department of Foreign Commerce and Shipping, Los Angeles Chamber of Commerce, 1935.

Matsuoka, Martha, et al. "Global Trade Impacts: Addressing the Health, Social, and Environmental Consequences of Moving Freight through our Communities." Los Angeles: Occidental College and the University of Southern California, March 2011. http://scholar.oxy.edu/cgi/viewcontent.cgi?article=1410&context=uep_faculty.

Matsuoka, Martha M., and Robert Gottlieb. "Environmental and Social Justice Movements and Policy Change in Los Angeles: Is an Inside-outside Game Possible?" In *New York and Los Angeles: The Uncertain Future*, ed. David Hale and Andrew A. Beveridge, 445–466. New York: Oxford University Press, 2013.

McKinnon, Alexander. "Hong Kong and Shenzhen Ports: Challenges, Opportunities, and Global Competitiveness." Hong Kong Centre for Maritime and Transportation Law, City University of Hong Kong, Working Paper Series, 2011. http://www.cityu .edu.hk/slw/HKCMT/Doc/Working_Paper_3_-_Shenzhen_-_Final_(v6).pdf.

Molinski, Dan. "Ports, Shipping Companies Retool before Panama Canal Expansion." *Wall Street Journal*, February 17, 2014. http://online.wsj.com/news/articles/ SB10001424052702304675504579387851171500882?mg=reno64-wsj&url=http %3A%2F%2Fonline.wsj.com%2Farticle%2FSB10001424052702304675504579387855 1171500882.html.

Monaco, Kristen. "Incentivizing Truck Retrofitting in Port Drayage: A Study of Drivers at the Ports of Los Angeles and Long Beach." Final Report, Metrans Project 06-02, February 2008. https://www.metrans.org/sites/default/files/research-project/06-02 %20Final%20Report_0_0.pdf.

Mongelluzzo, Bill. "Mega-ships Dealing Worst Congestion Hand to LA-LB, NY-NJ." *Journal of Commerce (JOC.com)*, July 1, 2015. http://www.joc.com/port-news/us -ports/port-new-york-and-new-jersey/mega-ships-dealing-worst-congestion-hand-la -lb-ny-nj_20150701.html

Mongelluzzo, Bill. "The Port Moves Inland." *Journal of Commerce*, September 13, 2010.

Ng, Simon K. W., et al. "Marine Vessel Smoke Emissions in Hong Kong and the Pearl River Delta." Final Report, Hong Kong University of Science and Technology, Atmospheric Research Center, March 2012.

Ng, Simon K. W., et al. "Policy Change Driven by an AIS-assisted Marine Emission Inventory in Hong Kong and the Pearl River Delta." *Atmospheric Environment* 76 (September 2013): 102–112.

O'Connor, Rose Ellen. "Widened Panama Canal Threatens the Environment." *Natural Resources News Service*, December 4, 2013. http://www.dcbureau.org/ 201312049376/natural-resources-news-service/the-new-panama-canal-economic -promises-vs-environmental-degradation.html.

Parsons, Christi, and Don Lee. "Ports Eye Panama Canal Project." *Los Angeles Times*, November 29, 2013.

People's Republic of China, State Council. "China's Foreign Trade." Information Office of the State Council, Beijing, December 2011. http://english.gov.cn/archive/ white_paper/2014/08/23/content_281474983043184.htm.

Phillips, Erica E. "Developers Dig into Distribution: Southern California's Inland Empire Illustrates Speculative Building Surge of Warehouse Hubs." *Wall Street Journal*, August 13, 2013. http://online.wsj.com/news/articles/SB1000142412788732344 6404579010853015245422.

Poon, Alice. *Land and the Ruling Class in Hong Kong*. Hong Kong: Enrich Professional Publishing, 2010.

Poon, Daniel. "Toy Industry in Hong Kong," HKTDC Research, Hong Kong Trade Development Council, February 10, 2014. http://hong-kong-economy-research .hktdc.com/business-news/article/Hong-Kong-Industry-Profiles/Toy-Industry-in -Hong-Kong/hkip/en/1/1X000000/1X001DGH.htm.

Port of Long Beach. "Facts at a Glance." http://www.polb.com/about/facts.asp.

Port of Long Beach. "Gearing Up to Shut Down: Shore Power at the Port of Long Beach." Thomas Jelenic, presentation, Shanghai, December 13, 2012. http://www .theicct.org/sites/default/files/Thomas%20Jelenic_Shore%20Power_En.pdf.

Port of Long Beach. "Port of Long Beach Cargo Statistics." http://www.polb.com/ civica/filebank/blobdload.asp?BlobID=3945.

Port of Long Beach. "2012 Air Emissions Inventory." 2012. http://www.polb.com/ civica/filebank/blobdload.asp?BlobID=11373.

Port of Los Angeles. "About the Port." Accessed August 10, 2016. https://www .portoflosangeles.org/idx_about.asp.

Port of Los Angeles. "Alternative Marine Power." https://www.portoflosangeles.org/ environment/amp.asp.

Port of Los Angeles. "Final Environmental Impact Report, BNSF Southern California International Gateway." http://www.portoflosangeles.org/EIR/SCIG/FEIR/feir_scig .asp.

Port of Los Angeles. "Inventory of Air Emissions 2012." 2012. http://www.portoflos angeles.org/pdf/2012_Air_Emissions_Inventory.pdf.

Port of Los Angeles. "2013 Los Angeles Trade Numbers." 2013. http://www.portoflo sangeles.org/pdf/Los-Angeles-Trade-Numbers-2013.pdf.

Quennan, Charles F. *Long Beach and Los Angeles: A Tale of Two Ports*. Northridge, CA: Windsor Publications, 1986.

Queenan, Charles F. *The Port of Los Angeles: From Wilderness to World Port*. Los Angeles: Los Angeles Harbor Department, 1983.

Rivoli, Pietra. *The Travels of a T-shirt in the Global Economy*. Hoboken, NJ: John Wiley & Sons, 2005.

Roosevelt, Margot. "Clean Fuel-Ship Zone Is Widened." *Los Angeles Times*, June 24, 2011. http://articles.latimes.com/print/2011/jun/24/local/la-me-ships-20110624.

Roosevelt, Margot. "A New Crop of Eco-Warriors Take to Their Own Streets." *Los Angeles Times*, September 24, 2009. http://www.latimes.com/local/la-me-air -pollution24-2009sep24,0,7165289.story#axzz2z3tj5vG0.

Rosenberg, Paul. "Activists Defeat SCIG." *Random Lengths*, April 14, 2016. http://www.randomlengthsnews.com/activists-defeat-scig/.

Shannon, Brad. "State EIS on Cherry Point Coal-Exports Facility to Consider 'End Use' Coal Burning in China as Well as Regional Rail Impacts." *Olympian*, July 31, 2013. Accessed November 13, 2015. http://www.theolympian.com/2013/07/31/2650018/state-eis-on-cherry-point-coal.html.

Snyder, Francis. "Global Economic Networks and Global Legal Pluralism." EUI Working Paper LAW No. 99/6, European University Institute, Department of Law, San Domenico, Italy, August 1999. http://cadmus.eui.eu/bitstream/handle/1814/151/law99_6.pdf.

Song, Dong-Wook. "Regional Port Competition and Cooperation: The Case of Hong Kong and South China." *Journal of Transport Geography* 10 (2002): 99–110.

Southern California Association of Governments. "Profile of the City of Los Angeles: Local Profiles Report 2015." Los Angeles, 2015. http://www.scag.ca.gov/Documents/LosAngeles.pdf

Steinberg, Jim. "2015 a Big Year for Warehouse Development in the Inland Empire." *San Bernardino Sun*, June 6, 2015. http://www.sbsun.com/business/20150606/2015-a-big-year-for-warehouse-development-in-the-inland-empire

Tempest, Rone. "Barbie and the World Economy." *Los Angeles Times*, September 22, 1996. http://articles.latimes.com/1996-09-22/news/mn-46610_1_hong-kong.

United States Department of Agriculture Agricultural Marketing Service. "Impact of Panama Canal Expansion on the U.S. Intermodal System." January 2010. https://www.ams.usda.gov/sites/default/files/media/Impact%20of%20Panama%20Canal%20Expansion%20on%20the%20U.S.%20Intermodal%20System.pdf.

Wan, Zheng, et al. "Pollution: Three Steps to a Green Shipping Industry." *Nature* 530 (7590)(February 17, 2016): 275–277. http://www.nature.com/news/pollution-three-steps-to-a-green-shipping-industry-1.19369.

Wang, Haifeng. "Hong Kong Takes an Important First Step in Regulating Shipping Emissions." International Council on Clean Transportation Blog, April 19, 2013. http://theicct.org/blogs/staff/hong-kong-takes-important-first-step-regulating-shipping-emissions.

Wang, Jasmine, and Simon Lee. "Li Ka-Shing's Dockers Accept Pay Offer to End Longest Strike." *Bloomberg News*, May 7, 2013. http://www.bloomberg.com/news/2013-05-06/li-ka-shing-s-dockers-accept-pay-offer-to-end-longest-strike-1-.html.

Wang, Yuohong, et al. "The Role of Feeder Shipping in Chinese Container Port Development." *Transportation Journal* 53 (2) (Spring 2014): 253–267.

Weikel, Dan. "Campaign Questions Status of Port Truck Drivers." *Los Angeles Times*, February 20, 2014. http://www.latimes.com/local/lanow/la-me-ln-port-trucker -conditons-20140219,0,2221465.story#axzz2tsMq5qOi.

Weikel, Dan. "Freighters Enter the Age of the Mega-ship." *Los Angeles Times*, June 15, 1999. http://articles.latimes.com/1999/jun/15/news/mn-46667

Weikel, Dan. "Judge Tosses Environmental Report on L.A. Port Rail Yard." *Los Angeles Times*, March 31, 2016. http://www.pressreader.com/usa/los-angeles-times/ 20160331/281745563526190

Weikel, Dan. "L.A., Long Beach Ports Will See More Cargo Volume." *Los Angeles Times*, December 18, 1998. http://articles.latimes.com/1998/dec/18/business/ fi-55161

Welitzkin, Paul. "China's Toy Story." *China Daily Asia*, March 11–13, 2016.

White, Ronald D. "Immigration Rallies Fuel Resolve of Port Truckers." *Los Angeles Times*, May 4, 2006. http://articles.latimes.com/2006/may/04/business/fi-truckers4.

"The World's Richest Billionaires: #17 Li Ka-shing." *Forbes*. Accessed March 15, 2015. http://www.forbes.com/profile/li-ka-shing/.

Yam, Richard C. M., and Esther P. Y. Tang. "Transportation Systems in Hong Kong and Southern China: A Manufacturing Industries Perspective." *International Journal of Physical Distribution and Logistics Management* 26 (10) (1996): 46–59. http://202.116.197.15/cadalcanton/Fulltext/20977_2014318_11419_6.pdf.

Yang, Jing. "Port Assets Losing Allure for Li Ka-shing's Hutchison Whampoa." *South China Morning Post*, February 6, 2015. http://www.scmp.com/business/companies/ article/1703279/port-assets-losing-allure-li-ka-shings-hutchison-whampoa.

Yang, Jing. "Shenzhen in 200m Yuan Push for Green Shipping." *South China Morning Post*, September 25, 2014. http://www.scmp.com/business/economy/ article/1599745/shenzhen-200m-yuan-push-green-shipping.

Yin, David. "Li and Fung Falls out of Favor." *Forbes Asia*, September 2, 2013. http:// www.forbes.com/sites/forbesasia/2013/08/28/li-fung-falls-out-of-favor/.

Yin, Kwong Sum. "For Cleaner Air, Marine Emissions Should Be Hong Kong's Next Target." *South China Morning Post*, January 8, 2014. http://www.scmp.com/ comment/insight-opinion/article/1399889/cleaner-air-marine-emissions-should-be -hong-kongs-next.

Yong, Wang. "WTO Accession, Globalization, and a Changing China." *China Business Review*, October 1, 2011. http://www.chinabusinessreview.com/wto-accession -globalization-and-a-changing-china/.

Yu, Henry. "Los Angeles and American Studies in a Pacific World of Migrations." *American Quarterly* 56 (3) (September 2004): 531–543. https://muse.jhu.edu/ article/172850

Zhu, Julie. "China's Port Operators Harbour Global Ambitions." *Finance Asia*, October 30, 2015. http://www.financeasia.com/News/403368,chinas-port-operators -harbour-global-ambitions.aspx

Chapter 3

Barboza, Tony. "China's Industry Exporting Air Pollution to US, Study Says." *Los Angeles Times*, January 20, 2014. http://www.latimes.com/science/sciencenow/la-sci -sn-china-exports-air-pollution-united-states-20140120,0,1142263.story#axzz2rpv kaxz2.

Barboza, Tony. "An Ill Wind: Scientist in Western States Measure Pollution That Travels from China and Other Fast-Growing Asian Countries across the Pacific." *Los Angeles Times*, February 1, 2015.

Barboza, Tony. "New Map Could Refocus State's Pollution Battles." *Los Angeles Times*, April 23, 2014. http://www.latimes.com/science/la-me-0423-pollution-neigh borhoods-20140423,0,1487851.story#axzz2zofhFPw5.

Barboza, Tony. "State Scolds Local Smog Regulator." *Los Angeles Times*, January 9, 2016. http://www.latimes.com/science/la-me-0109-state-aqmd-20160109-story .html.

Bienkowski, Brian. "China's Babies at Risk From Soot, Smog." *Environmental Health News*, April 17, 2014. http://www.environmentalhealthnews.org/ehs/news/2014/ apr/chinas-babies.

Breines, Marvin. "The Fight against Smog in Los Angeles: 1943–1957." PhD diss., University of California at Davis, 1975.

Bushak, Lecia. "US Air Pollution Tied to Chinese Exports: Strong Winds Blow Nitro- gen Oxides, Carbon Monoxide across Ocean." *Medical Daily*, January 20, 2014. http://www.medicaldaily.com/us-air-pollution-tied-chinese-exports-strong-winds -blow-nitrogen-oxides-carbon-monoxide-across-ocean.

BusinessGreen. "Panasonic Staff Earn Hazard Pay in Polluted Chinese Cities." *Green- Biz.com*, March 14, 2014. https://www.greenbiz.com/blog/2014/03/14/panasonic -hazard-pay-polluted-chinese-cities.

Cadarso, Maria-Angeles, et al. "CO_2 Emissions of International Freight Transport and Offshoring: Measurement and Allocation." *Ecological Economics* 69 (2010), 1682–1694. http://www.sciencedirect.com/science/article/pii/S0921800910001151.

Chee-hwa, Tung. "Quality People, Quality Home: Positioning Hong Kong for the Twenty-first Century." The 1999 Policy Address. http://www.policyaddress.gov.hk/ pa99/english/espeech.pdf.

Chen, Stephen. "Agriculture Feels the Choke as China Smog Starts to Foster Disastrous Conditions." *South China Morning Post*, February 25, 2014. http://www.scmp.com/news/china/article/1434700/china-smog-threatens-agriculture-nuclear-fallout-conditions-warn.

Chen, Zhu, et al. "China Tackles the Health Effects of Air Pollution." *Lancet* 382 (December 14, 2013): 1959–1960.

Clean Air Asia. "Do Not Adopt to Air Pollution—Clean Air Asia Launches Hairy Nose Campaign." December 5, 2012. http://cleanairasia.org/node11316/

Cone, Marla. "Smog Agency Bows to L.A., Defers Some Stricter Limits." *Los Angeles Times*, September 10, 1994. http://articles.latimes.com/1994-09-10/news/mn-36870_1_clean-air-plan.

"Could 'Fresh Canned Air' Business Work in China?" *Want China Times*, May 7, 2014.

Dallmann, Timothy R., and Robert A. Harley. "Evaluation of Mobile Source Emission Trends in the United States." *Journal of Geophysical Research* 115 (2010). http://onlinelibrary.wiley.com/doi/10.1029/2010JD013862/pdf.

Demick, Barbara. "Lung Cancer Is a Cloud over China." *Los Angeles Times*, December 24, 2013. http://www.latimes.com/world/la-fg-china-lung-cancer-20131224,0,3058283.story#axzz2ohBNYMbj.

Dewey, Scott Hamilton. *Don't Breathe the Air: Air Pollution and U.S. Environmental Politics, 1945–1970*. College Station: Texas A&M University Press, 2000.

Dockery, Douglas W., et al. "An Association between Air Pollution and Mortality in Six U.S. Cities." *New England Journal of Medicine* 329 (24)(December 9, 1993): 1753–1759. http://www.nejm.org/doi/full/10.1056/NEJM199312093292401.

Doty, Robert A., and Leonard Levine. *Profile of an Air Pollution Controversy*. Riverside, CA: Clean Air Now, 1974.

Driver, Carol. "China Sells BOTTLED AIR to Tourists as Smog Described as 'Environmental Crisis.'" *Daily Mail*, March 25, 2014. http://www.dailymail.co.uk/travel/article-2588788/China-sells-BOTTLED-AIR-tourists.html.

Dunsby, Joshua. "Localizing Smog: Transgressions in the Therapeutic Landscape." In *Smoke and Mirrors: The Politics and Culture of Air Pollution*, ed. E. Melanie DuPuis, 170–200. New York: New York University Press, 2004.

Elkind, Sarah S. *How Local Politics Shape Federal Policy: Business, Power, and the Environment in Twentieth-Century Los Angeles*. Chapel Hill: University of North Carolina Press, 2011.

FlorCruz. Michelle, "Chinese Province of Guizhou Selling Canned 'Fresh Air' in New Tourism Gimmick." *International Business Times*, March 21, 2014. http://www

.ibtimes.com/chinese-province-guizhou-selling-canned-fresh-air-new-tourism -gimmick-1562954.

Gardner, Beth. "Air of Revolution: How Activists and Social Media Scrutinize City Pollution." *Guardian*, January 31, 2014. http://www.theguardian.com/cities/2014/ jan/31/air-activists-social-media-pollution-city.

Gauderman, James W., et al. "Association of Improved Air Quality with Lung Development in Children." *New England Journal of Medicine* 372 (10) (March 5, 2015): 905–913. http://www.nejm.org/doi/full/10.1056/NEJMoa1414123#t=article.

Gauderman, James W., et al. "Effect of Exposure to Traffic on Lung Development from Ten to Eighteen Years of Age: A Cohort Study." *Lancet* 369 (9561) (2007): 571–577.

Geall, Sam. "Interpreting Ecological Civilization (Part One)." *China Dialogue*, July 6, 2015. https://www.chinadialogue.net/article/show/single/en/8018-Interpreting -ecological-civilisation-part-one.

Greer, Linda. "China Fights Back against Airpocalypse: A New Air Pollution Initiative that Just Might Work!" January 12, 2014, Linda Greer's Blog, Natural Resources Defense Council. http://switchboard.nrdc.org/blogs/lgreer/china_fights _back_against_airpocalypse_embarking_on_a_new_air_pollution_initiative_that _just_might_work.html.

Guan, Dabo, et al. "The Socioeconomic Drivers of China's Primary PM2.5 Emissions." *Environmental Research Letters* 9 (2) (February 2014). http://iopscience.iop .org/1748-9326/9/2/024010/article.

Haagen-Smit, A. J. "Chemistry and Physiology of Los Angeles Smog." *Industrial and Chemical Engineering* 44 (6) (1952): 1342–1346. http://cires1.colorado.edu/jimenez/ AtmChem/2013/Haagen-Smit_1952_IndEngChem_LA_smog_paper.pdf.

Hahn, Kenneth. "Health Warning: Smog: Record of Correspondence between Kenneth Hahn, Los Angeles County Supervisor, and the Presidents of General Motors, Ford, and Chrysler on Controlling Air Pollution." Los Angeles County Board of Supervisors, May 1972.

Hedley, Anthony J. "Air Pollution and Public Health." University of Hong Kong, School of Public Health, February 2009, http://www.legco.gov.hk/yr08-09/english/ panels/ea/ea_iaq/papers/ea_iaq0212cb1-733-2-e.pdf.

Hedley, Anthony J., et al. "Air Pollution: Costs and Paths to a Solution in Hong Kong—Understanding the Connections among Visibility, Air Pollution, and Health Costs in Pursuit of Accountability, Environmental Justice, and Health Protection." *Journal of Toxicology and Environmental Health. Part A* 71 (2008): 544–554.

Hedley, Anthony J., et al. "Cardiorespiratory and All-Cause Mortality after Restrictions on Sulphur Content of Fuel in Hong Kong: An Intervention Study." *Lancet* 360 (2002): 1646–1652.

Hong Kong Special Administrative Region, Environmental Protection Department. "HK and Guangdong Strengthen Co-Operation on Cleaner Production to Improve Regional Environmental Quality." October 29, 2015. http://www.info.gov.hk/gia/general/201510/29/P201510290629.htm.

Hsin, Huang. "Wealthy Fleeing from China's Smoggy Cities." *Want China Times*, May 5, 2014.

Hu, Shishan, et al. "Observation of Elevated Air Pollutant Concentrations in a Residential Neighborhood of Los Angeles, California, Using a Mobile Platform." *Atmospheric Environment* 51 (May 1, 2012): 311–319.

International Agency for Research on Cancer. "Diesel Engine Exhaust Carcinogenic." June 12, 2012. http://www.iarc.fr/en/media-centre/pr/2012/pdfs/pr213_E.pdf.

Jacobs, Chip, and William J. Kelly. *Smogtown: The Lung-Burning History of Pollution in Los Angeles*. Woodstock, NY: Overlook Press, 2008.

Jahn, Heiko J., et al. "Ambient and Personal $PM_{2.5}$ Exposure Assessment in the Chinese Megacity of Guangzhou." *Atmospheric Environment* 74 (2013): 402–411.

Jinqiang, Liu. "China's New Environmental Law Looks Good on Paper." *China Dialogue*, April 24, 2014. https://www.chinadialogue.net/blog/6937-China-s-new-environmental-law-looks-good-on-paper/en.

Jun, Ma, et al. "Hong Kong's Role in Mending the Disclosure Gap." Institute of Public and Environmental Affairs, Beijing, and Civic Exchange, Hong Kong, March 2010.

Kaiman, Jonathan. "China's Toxic Air Pollution Resembles Nuclear Winter, Say Scientists." *Guardian*, February 25, 2014. http://www.theguardian.com/world/2014/feb/25/china-toxic-air-pollution-nuclear-winter-scientists.

Kilburn, Mike. "Air Quality Report Card of the Donald Tsang Administration (2005–2012)." Civic Exchange, Hong Kong, January 2012.

Kilburn, Mike, and Christine Loh. "Principles and Measures to Improve Air Quality: Policy Recommendations for a New Administration." Civic Exchange, Hong Kong, January 2012. http://www.civic-exchange.org/en/publications/164986972.

Koster, Betty. "A History of Air Pollution Control Efforts in Los Angeles County." Public Education and Information Division, Los Angeles County Air Pollution District, August 31, 1956.

Krier, James E., and Edmund Ursin. *Pollution and Policy: A Case Essay on California and Federal Experience with Motor Vehicle Air Pollution 1940–1973*. Berkeley: University of California Press, 1977.

Lambauch, Robert J., H. M. Kipen, K. Kelly-McNeil, L. Zhang, P. J. Lioy, P. Ohman-Strickland, J. Gong, A. Kusnecov, and N. Fiedler. "Sickness Response Symptoms among Healthy Volunteers after Controlled Exposures to Diesel Exhaust and Psychological Stress." *Environmental Health Perspectives* (2011). PMID: 21330231.

Lau, Alexis, et al. "Relative Significance of Local vs. Regional Sources: Hong Kong's Air Pollution." Hong Kong University of Science and Technology Institute of the Environment and Civic Exchange, 2007. http://www.hongkongcan.org/doclib/200703_RelativeSignificanceofLocalvsRegionalSources_Hong%20KongsAir Pollution.pdf (English and Chinese).

Leung, Gabriel M., Anthony J. Hedley, Edith M. C. Lau, and Tai-Hing Lam. "The Public Health Viewpoint." In *At the Epicentre: Hong Kong and the SARS Outbreak*, ed. Christine Loh and Civic Exchange, 55–80. Hong Kong: Hong Kong University Press, 2004.

Levin, Dan. "In Beijing, Complaints about Smog Grow Louder, and Retaliation Grows Swifter." *New York Times (Sinosphere blogs)*, February 25, 2014. http://sinosphere.blogs.nytimes.com/2014/02/25/in-beijing-complaints-about-smog-grow-louder-and-retaliation-grows-swifter/?_php=true&_type=blogs&_r=0.

Li, Meng, and Jiahua Pan. "Achievements and Challenges: $PM_{2.5}$ Control in China." *Chinese Journal of Urban and Environmental Studies* 2 (1) (2014): 1–15. http://www.worldscientific.com/doi/abs/10.1142/S2345748114500080?journalCode=cjues.

Lin, Jintai, et al. "China's International Trade and Air Pollution in the United States." *Proceedings of the National Academy of Sciences of the United States of America* 111 (5) (February 4, 2014): 1736–1741. http://www.pnas.org/content/111/5/1736.full.pdf.

Lin, Zhen, and Yin Guan. "Environmental NGOS and Participative Governance: The Case of the PM2.5 Incident." In *Chinese Environmental Governance: Dynamics, Challenges, and Prospects in a Changing Society*, ed. Bingqiang Ren and Huisheng Shou, 175–187. New York: Palgrave Macmillan, 2013.

Lurmann, Fred, Ed Avol, and Frank Gilliland. "Emissions Reduction Policies and Recent Trends in Southern California's Ambient Air Quality." *Journal of the Air and Waste Management Association* 65 (3) (2015): 324–335. http://www.tandfonline.com/doi/abs/10.1080/10962247.2014.991856?journalCode=uawm20#preview.

Makinen, Julie. "Artists Finding Inspiration in China's Bad Air." *Los Angeles Times*, May 7, 2014. http://www.latimes.com/world/la-fg-c1-china-art-pollution-20140507-story.html.

Makinen, Julie. "Beijingers Shrug Off Smog." *Los Angeles Times*, February 26, 2014.

Malm, Andreas. "China as Chimney of the World: The Fossil Capital Hypothesis" [abstract]. *Organization and Environment* 25 (2) (2012): 146–177. http://oae.sagepub.com/content/25/2/146.

Masters, Nathan. "L.A.'s Smoggy Past, in Photos." *KCET SoCal Focus*, March 17, 2011. http://www.kcet.org/updaily/socal_focus/history/los-angeles-smoggy-past-photos-31321.html

McIntyre, Douglas. "The 10 Cities with the World's Worst Air." *Daily Finance*, November 29, 2010. http://www.dailyfinance.com/2010/11/29/10-cities-with-worlds-worst-air/

Metcalfe, John. "How the Western World Enables China's Air Pollution." *Atlantic Cities*, January 20, 2014.

Murray, Lisa, and Angus Grigg. "Coca-Cola Offers Expats China Pollution Hazard Pay." *Australian Financial Review*, July 8, 2014. http://www.afr.com/news/world/asia/cocacola-offers-expats-china-pollution-hazard-pay-20140708-j04mk

Neuman, Scott. "What's a Breath of Fresh Air Worth? In China, about $860." *NPR*, April 10, 2014. http://www.npr.org/blogs/thetwo-way/2014/04/10/301504334/whats-it-worth-for-a-breath-of-fresh-air-in-china-about-860

Peters, Glen P., et al. "Growth in Emission Transfers via International Trade from 1990 to 2008." *Proceedings of the National Academy of Sciences of the United States of America* 108 (21) (2011): 8903–8908. http://www.pnas.org/content/108/21/8903.full.

Roberts, David. "How the US Embassy Tweeted to Clear Beijing's Air." *Wired*, March 6, 2015. http://www.wired.com/2015/03/opinion-us-embassy-beijing-tweeted-clear-air/.

Rowangould, Gregory. "A Census of the US Near-Roadway Population: Public Health and Environmental Justice Considerations." *Transportation Research Part D, Transport and Environment* 25 (December 2013): 59–67.

Rusco, F. W., and W. D. Walls. *Clearing the Air: Vehicular Emissions Policy for Hong Kong.* Hong Kong: Chinese University Press of Hong Kong, 1995.

Schiller, Ronald. "The Los Angeles Smog." *National Municipal Review* (December 1955): 558–564.

Shen, Feifei. "Chinese Man Sues His Polluted Hometown for Cost of Fighting Smog." *Bloomberg News*, February 24, 2014. http://www.bloomberg.com/news/2014-02-25/chinese-man-sues-his-polluted-hometown-for-cost-of-fighting-smog.html.

South Coast Air Quality Management District. "Smog and Health: Historical Information." 1996. http://www.aqmd.gov/home/library/public-information/publications/smog-and-health-historical-info#Historic%20Air%20Pollution%20Disasters.

South Coast Air Quality Management District. "The Southland's War on Smog: Fifty Years of Progress toward Clean Air (through May 1997)." http://www.aqmd.gov/

home/library/public-information/publications/50-years-of-progress#Formation%20 of%20the%20South%20Coast%20Air%20Quality%20Management%20District

Stanford Research Institute. "The Smog Problem in Los Angeles County: A Report by the Stanford Research Institute on Studies to Determine the Nature and Causes of Smog." Western Oil and Gas Association, Los Angeles, January 1954.

Stewart, Frank. "Smog Circus in L.A." *Frontier*, December 1954, pp. 12–13.

Stradling, David. *Smokestacks and Progressives: Environmentalists, Engineers, and Air Quality in America, 1881–1951*. Baltimore: Johns Hopkins University Press, 1999.

Subbaraman, Nidhi. "Beijing's Smogpocalypse: China's Air Crisis by the Numbers." *NBC News Blog,* February 25, 2014. http://www.nbcnews.com/science/environment/ beijings-smogpocalypse-chinas-air-crisis-numbers-n38251

Tatlow, Didi Kirsten. "In China, Dreaming of a Less Smoggy Christmas." *New York Times*, December 24, 2013. http://sinosphere.blogs.nytimes.com/2013/12/24/ in-china-dreaming-of-a-less-smoggy-christmas/?_r=0

Taylor, Alan. "Beijing's Toxic Sky." *Atlantic* 4 (March) (2015). http://www .theatlantic.com/photo/2015/03/beijings-toxic-sky/386824/.

Unger, Nancy C. *Beyond Nature's Housekeepers: American Women in Environmental History*. New York: Oxford University Press, 2012.

University of California, Los Angeles, Environmental Science and Engineering. "Research Report: Air Pollution and City Planning, Case Study of a Los Angeles District Plan." 1972.

Wang, Xuemei, et al. "Progress in Understanding the Formation of Fine Particulate Matter and Ground-Level Ozone in Pearl River Delta Region, China." *Atmospheric Environment* 122 (2015): 88.

Wen, Zhao. "Parents Push Schools to Install Air Purifiers." *Shanghai Daily*, December 10, 2013. http://www.shanghaidaily.com/Metro/education/Parents-push-schools-to -install-air-purifiers/shdaily.shtml.

Wong, Edward. "China Exports Pollution to US, Study Finds." *New York Times*, January 20, 2014. http://www.nytimes.com/2014/01/21/world/asia/china-also-exports -pollution-to-western-us-study-finds.html?_r=0).

Wong, Edward. "China's Plan to Curb Air Pollution Sets Limits on Coal Use and Vehicles." *New York Times*, September 12, 2013.

Wong, Edward. "Urbanites Flee China's Smog for Blue Skies." *New York Times*, November 22, 2013. http://www.nytimes.com/2013/11/23/world/asia/urbanites -flee-chinas-smog-for-blue-skies.html?pagewanted=all&_r=0.

Wong, Kandy. "Beijing's Record Smog Poses Health Nightmare." *Scientific American*, February 5, 2013, 9. http://www.scientificamerican.com/article/beijings-record -smog-poses-health-nightmare/.

Yue, Zhang. "School Beats Heavy Smog with Online Classes." *China Daily*, January 2, 2014. http://www.chinadaily.com.cn/china/2014-01/02/content_17209857.htm.

Yunfeng, Yan, and Laike Yang. "China's Foreign Trade and Climate Change: A Case Study of CO2 Emissions." *Energy Policy* 38 (1) (January 2010): 350–356.

Chapter 4

"All Dried Up." *Economist*, October 12, 2013. http://www.economist.com/news/ china/21587813-northern-china-running-out-water-governments-remedies-are -potentially-disastrous-all.

Beckman, David. "The Threats to Our Drinking Water." *New York Times*, August 7, 2014. http://www.nytimes.com/2014/08/07/opinion/the-threats-to-our-drinking -water.html?_r=0.

Bigger, Richard, and James D. Kitchen. *How the Cities Grew: A Century of Municipal Independence and Expansionism in Metropolitan Los Angeles*. Los Angeles: Bureau of Governmental Research, University of California at Los Angeles, 1952.

Bissell, C. A., ed. *Metropolitan Water District of Southern California: History and First Annual Report, for the Period Ending June 30, 1938*. Los Angeles: Metropolitan Water District of Southern California, 1939.

Boxall, Bettina. "Seawater Desalination Plant Might Be Just a Drop in the Bucket." *Los Angeles Times*, February 17, 2013. http://articles.latimes.com/2013/feb/17/local/ la-me-carlsbad-desalination-20130218.

Chan, Paul. "LCQ17: Total Water Management Strategy and Related Measures." Written reply recorded in the Legislative Council by Paul Chan, Secretary for Development, Hong Kong Special Administrative Region, April 17, 2013. http://www .info.gov.hk/gia/general/201304/17/P201304170456.htm.

Chan, Wai-Shin. "Water Pollution Could Lead to More Trade." *China Water Risk*, April 9, 2014. http://chinawaterrisk.org/opinions/water-pollution-could-lead-to -more-trade/.

"China's Underground Water Quality Worsens: Report." *Xinhua*, April 22, 2014. http://news.xinhuanet.com/english/china/2014-04/22/c_126421022.htm.

"China's Water Crisis: Part 1—Introduction." *China Water Risk*, March 2010. http:// chinawaterrisk.org/wp-content/uploads/2011/06/Chinas-Water-Crisis-Part-1.pdf.

Ching, Lam. "The Privatization of Water Supply in China." *Globalization Monitor*, August 11, 2012. http://www.globalmon.org.hk/content/privatization-water-supply -china

City of Los Angeles Department of Public Works, Bureau of Sanitation, and Department of Water and Power. "Water Integrated Resources Plan: Five Year Review." June 2012. http://www.lacity-irp.org/documents/FINAL_IRP_5_Year_Review_Docu ment.pdf.

Cohen, Andrew. "Water Supply and Land Use Planning: Making the Connection." *Land Use Forum*, Fall 1992, 341–344.

Cooley, Heather, Peter Gleick, and Kristina Donnelly. "Insights into Proposition 1: the 2014 California Water Bond." Pacific Institute, San Francisco, October 2014. http://pacinst.org/wp-content/uploads/sites/21/2014/10/Insights-into-Prop-1-full -report.pdf.

Council of Environmental Quality. *Second Annual Report*. Washington, DC: US Government Printing Office, 1971.

Davis, Mike. "The Case for Letting Malibu Burn." *Environmental History Review* 19 (2) (Summer 1995): 1–36.

Deng, Hongmei, et al. 2006. "Distribution and Loadings of Polycyclic Aromatic Hydrocarbons in the Xijiang River in Guangdong, South China." *Chemosphere* 64 (2006): 1401–1411.

Dodd, Chris. "France's Suez Hoping to Clean Up in China." *Finance Asia*, February 25, 2014. http://www.financeasia.com/News/373254,frances-suez-hoping-to-clean- up-in-china.aspx.

Dow Chemical. Water and Process Solutions. "China's Thirst for Water." April 2011. Accessed June 14, 2014. http://www.futurewecreate.com/water/includes/DOW072 _China%20White_Opt1_Rev1.pdf.

"Down to the Very Last Drop." *South China Morning Post*, April 18, 2007. http:// www.scmp.com/article/589365/down-very-last-drop.

Du, Qing-ping, et al. "Chlorobenzenes in Waterweeds from the Xijiang River (Guangdong Section) of the Pearl River." *Journal of Environmental Sciences (China)* 19 (2007): 1171–1177. http://www.jesc.ac.cn/jesc_en/ch/reader/view_abstract.aspx?file _no=2007191004.

Ediger, Laura, and Linda Hwang. "Water Quality and Environmental Health in Southern China." July 2009. BSR Forum, Guangzhou, China, May 15, 2009. http:// www.bsr.org/reports/BSR_Southern_China_Water_Quality_Environmental_Health _Forum.pdf.

"Eight Things You Should Know about Hong Kong Water." *China Water Risk*, January 12, 2012. http://chinawaterrisk.org/resources/analysis-reviews/8-things-you-should-know-about-hong-kong-water/.

Erie, Steven P. *Beyond Chinatown: The Metropolitan Water District and the Environment in Southern California*. Stanford: Stanford University Press, 2006.

Executive Department, State of California. "Executive Order B-29-15." April 1, 2015. http://gov.ca.gov/docs/4.1.15_Executive_Order.pdf.

Fu, Jiamo, et al. "Persistent Organic Pollutants in Environment of the Pearl River Delta, China: An Overview." *Chemosphere* 52 (2003): 1411–1422.

Fung Business Intelligence Centre. "China Retail: Hypermarkets and Supermarkets." Kowloon, Hong Kong, February 2015. http://www.funggroup.com/eng/knowledge/research/china_dis_issue125.pdf.

Globalization Monitor. "Why Bottled Water Is NOT the Solution for China's Water Crisis." June 24, 2014. http://globalmon.org.hk/content/why-bottled-water-not-solution-china's-drinking-water-crisis.

Gold, Mark. "Keeping L.A.'s Taps Flowing." *Los Angeles Times*, July 15, 2011. http://articles.latimes.com/2011/jul/15/opinion/la-oe-gold-dwp-rate-hike-20110715.

Gottlieb, Robert. "For the MWD, It Isn't Easy Being Green." *Los Angeles Times*, December 3, 1992. http://articles.latimes.com/1992-12-03/local/me-1607_1_water-industry.

Gottlieb, Robert. *A Life of Its Own: The Politics and Power of Water*. San Diego: Harcourt Brace Jovanovich, 1988.

Gottlieb, Robert, and Andrew Cohen. "Water and Growth: Restructuring the Relationship." P.O.W.E.R. position paper. Los Angeles: Public Officials for Water and Environmental Reform, 1991.

Gottlieb, Robert, and Margaret FitzSimmons. *Thirst for Growth: Water Agencies as Hidden Government in California*. Tucson: University of Arizona Press, 1989.

Gottlieb, Robert, and Don Villarejo. "Urban, Rural Unite for a New Water Ethic." *Los Angeles Times*, April 17, 1989. http://articles.latimes.com/1989-04-17/local/me-1877_1_water-users-water-industry-california-water-agencies.

Government of Hong Kong. "Drinking Water Quality in Hong Kong." Accessed February 15, 2015. http://www.gov.hk/en/residents/environment/water/drinkingwater.htm.

Government Waterworks Professionals Association. "Hong Kong: A Role Model of Public-Operated Water Supply Services." In "Reclaiming Public Water: Achievements, Struggles and Visions Around the World." Transnational Institute, Chinese edition, September 2006. https://www.tni.org/files/waterchina.pdf.

Greenpeace. "Poisoning the Pearl: An Investigation into Industrial Water Pollution in the Pearl River Delta." 2nd ed. January 2010. http://www.greenpeace.org/eastasia/ Global/eastasia/publications/reports/toxics/2010/poisoning-the-pearl.pdf.

Guangwei, He. "China's Dirty Pollution Secret: The Boom Poisoned Its Soil and Crops." *Yale Environment 360*, Yale School of Forestry and Environmental Studies, June 30, 2014. http://e360.yale.edu/feature/chinas_dirty_pollution_secret_the _boom_poisoned_its_soil_and_crops/2782/.

Hackman, Rose. "California Drought Shaming Takes on a Class Conscious Edge." *Guardian*, May 16, 2015. http://www.theguardian.com/us-news/2015/may/16/ california-drought-shaming-takes-on-a-class-conscious-edge.

Haibo, Zhang, et al. "PCB Contamination in Soils of the Pearl River Delta, South China: Levels, Sources, and Potential Risks." *Environmental Science and Pollution Research International* 20 (8) (August 2013): 5150–5159.

Hallwachs, Rob. "Water Scents." *People.Interactive: Metropolitan's Employee Magazine*, May 2008. http://www1.mwdh2o.com/Peopleinteractive/archive_08/May_08/ article01.html.

Hildebrand, Carver E. "The Relationship between Urban Water Demand and the Price of Water." Metropolitan Water District of Southern California, Los Angeles, February 1984.

Ho, K. C., et al. "Chemical and Microbiological Qualities of the East River (Dong-jiang) Water, with Particular Reference to Drinking Water Supply in Hong Kong." *Chemosphere* 52 (2003): 1441–1450.

Ho Pui Yin. *Water for a Barren Rock: 150 Years of Water Supply in Hong Kong*. Hong Kong: Commercial Press, 2001.

Hong Kong Special Administrative Region, Environmental Protection Department. "River Water Quality in Hong Kong in 2012." 2013. http://wqrc.epd.gov.hk/pdf/ water-quality/annual-report/Report_2012_Eng_Combined.pdf.

Hong Kong Special Administrative Region, Environmental Protection Department. "Twenty Years of River Water Quality Monitoring in Hong Kong, 1986–2005." 2006. http://www.epd.gov.hk/epd/misc/river_quality/1986-2005/eng/1_background _menu.htm.

Hong Kong Special Administrative Region, Water Supplies Department. "Dongjiang Raw Water." August 2014. http://www.wsd.gov.hk/en/water_resources/raw_water _sources/dongjiang_raw_water/index.html.

Hong Kong Special Administrative Region, Water Supplies Department. "Plover Cove Reservoir." 2009. http://www.wsd.gov.hk/en/education/fun_of_fishing_in _hong_kong/brief_introduction_of_reservoirs/plover_cove_reservoir/.

Hong Kong Special Administrative Region, Water Supplies Department. "Total Water Management in Hong Kong: Toward Sustainable Use of Water Resources." 2008. http://www.wsd.gov.hk/filemanager/en/share/pdf/TWM.pdf.

Hong Kong Special Administrative Region, Water Supplies Department. "Water from Dongjiang at Guangdong: The Major Source of Supply." June 2014. http://www.wsd.gov.hk/en/water_resources/raw_water_sources/water_sources_in_hong_kong/water_from_dongjiang_at_guangdong/index.html.

Hongqiao, Liu. "The Polluted Legacy of China's Largest Rice-growing Province." *China Dialogue*, May 30, 2014. https://www.chinadialogue.net/article/show/single/en/7008-The-polluted-legacy-of-China-s-largest-rice-growing-province.

Hu, Feng, Debra Tan, and Inna Lazareva. "Eight Facts on China's Wastewater." *China Water Risk*, March 12, 2014. http://chinawaterrisk.org/resources/analysis-reviews/8-facts-on-china-wastewater/.

Jingxi, Xu. "Guangdong Project to Ensure Fresh Water for Hong Kong." *China Daily Asia*, February 20, 2014. http://www.chinadailyasia.com/news/2014-02/20/content_15119638.html.

Jun, Ma. *China's Water Crisis (Zhongguo shui weiji)*. Norwalk, CT: Eastbridge, 2004.

Kaiman, Jonathan. "China Says That More Than Half of Its Groundwater Is Polluted." *Guardian*, April 23, 2014. http://www.theguardian.com/environment/2014/apr/23/china-half-groundwater-polluted.

Kallis, Giorgis, et al. "Public versus Private: Does It Matter for Water Conservation? Insights from California." *Environmental Management* 45 (1) (January 2010): 177–191. http://www.ncbi.nlm.nih.gov/pmc/articles/PMC2815296/.

Keqiang, Li. "Report on the Work of the Government." Delivered at the second session of the Twelfth National People's Congress, Beijing, March 5, 2014. http://online.wsj.com/public/resources/documents/2014GovtWorkReport_Eng.pdf.

Keyi, Sheng. "China's Poisonous Waterways." *New York Times*, May 6, 2014. http://www.nytimes.com/2014/04/05/opinio1n/chinas-poisonous-waterways.html?_r=0.

Lao Man-lei, Mandy, and Carine Lai. "Reducing Plastic Waste in Hong Kong: Public Opinion Survey of Bottled Water Consumption and Attitudes towards Plastic Waste." Civic Exchange, Hong Kong, April 2015.

Lee, Jane G. "Driven by Climate Change, Algae Blooms behind Ohio Water Scare Are New Normal." *National Geographic*, August 6, 2014. http://news.nationalgeographic.com/news/2014/08/140804-harmful-algal-bloom-lake-erie-climate-change-science/.

Lee, Nelson K. "The Changing Nature of Border, Scale, and the Production of Hong Kong's Water Supply System since 1959." *International Journal of Urban and Regional Research* 38 (3) (May 2014): 903–921.

Lee, S. C., et al. "Multipathway Risk Assessment on Disinfection Byproducts of Drinking Water in Hong Kong." *Environmental Research* 94 (2004): 47–56.

Leslie, Jacques. "Los Angeles, City of Water." *New York Times*, December 6, 2014. http://www.nytimes.com/2014/12/07/opinion/sunday/los-angeles-city-of-water .html?_r=0.

Lewis, Joanna I. *Green Innovation in China: China's Wind Power Industry and the Global Transition to a Low-Carbon Economy.* New York: Columbia University Press, 2013.

Linton, Joe. *Down by the Los Angeles River: Friends of the Los Angeles River's Official Guide.* Los Angeles: Wilderness Press, 2005.

Linton, Joe. "Kayaking the L.A. River: Day 1" and "Kayaking the L.A. River: Day 2." *L.A. Creek Freak Blog,* July 26–27, 2008. http://lacreekfreak.wordpress.com/2008/ 07/26/kayaking-the-los-angeles-river-day-1/ and http://lacreekfreak.wordpress.com/ 2008/07/27/kayaking-the-los-angeles-river-day-2/.

Liu, Su. "Liquid Assets IV: Hong Kong's Water Resources Management under 'One Country, Two Systems.'" Civic Exchange, Hong Kong, November 13, 2013. http:// www.civic-exchange.org/en/publications/4292629.

Liu, Su, and Berton Bian. "Liquid Assets II: Industrial Relocation in Guangdong Province: Avoid Repeating Mistakes." Civic Exchange, Hong Kong, January 10, 2012. http://www.civic-exchange.org/en/publications/4292630.

Liu, Su, and Jessica Williams. "Liquid Assets V: The Water Tales of Hong Kong and Singapore: Divergent Approaches to Water Dependency." Civic Exchange, Hong Kong, January 2014. http://www.civic-exchange.org/en/publications/164987052.

Liu, Su, et al. "Liquid Assets IIIB: Dongjiang Overloaded: A Photographic Report of the 2011 Dongjiang Expedition." Civic Exchange, Hong Kong, May 30, 2012. http:// www.civic-exchange.org/en/publications/4292633.

Los Angeles Department of Water and Power. "Urban Water Management Plan 2010." http://www.water.ca.gov/urbanwatermanagement/2010uwmps/Los%20 Angeles%20Department%20of%20Water%20and%20Power/LADWP%20 UWMP_2010_LowRes.pdf.

Metropolitan Water District of Southern California. "Groundwater Assessment Study." Report No. 1308. September 2007.

Metropolitan Water District of Southern California. "Integrated Water Resources Plan." June 2016. http://mwdh2o.com/Reports/2.4.1_Integrated_Resources_Plan .pdf.

Metropolitan Water District of Southern California. "Metropolitan Board Approves Nation's Largest Conservation Program to Meet Unprecedented Consumer Demand in Drought's Fourth Year." News release, May 26, 2015.

Metropolitan Water District of Southern California. "2015 Urban Water Management Plan." June 2016. http://www.mwdh2o.com/PDF_About_Your_Water/2.4.2_Regional_Urban_Water_Management_Plan.pdf.

Metropolitan Water District of Southern California. "Water Tomorrow: Integrated Water Resources Plan 2015 Update." Report No. 1518, January 2016. http://www.mwdh2o.com/PDF_About_Your_Water/2015%20IRP%20Update%20Report%20(web).pdf.

Minter, Adam. "If You Thought China's Air Was Bad, Try the Water." *Bloomberg News*, April 14, 2014. https://www.bloomberg.com/view/articles/2014-04-14/if-you-thought-china-s-air-was-bad-try-the-water.

Muir, Frederick. "MWD Looking at Rate, Tax and Fee Hikes." *Los Angeles Times*, September 27, 1991.

O'Connor, Dennis E. "The Governance of the Metropolitan Water District of Southern California: An Overview of the Issues." California Research Bureau, CRB-98-013, August 1998. https://www.library.ca.gov/crb/98/13/98013.pdf.

Office of the South-to-North Water Diversion Commission of the State Council. "Biggest Migration in History." http://www.nsbd.gov.cn/zx/gczz/201106/t20110630_188237.html (Chinese only).

Pacific Institute and the Natural Resources Defense Council. June 2014. "The Untapped Potential of California's Water Supply: Efficiency, Reuse, and Stormwater." Issue Brief, Pacific Institute and NRDC. http://pacinst.org/wp-content/uploads/sites/21/2014/06/ca-water-capstone.pdf.

People's Government of Guangdong Province. "Guangdong Province Urban-Rural Sewage Treatment and Water Reuse Facilities Infrastructure under Twelfth Five-Year Guideline of People's Republic of China." People's Republic of China, 2013 (Chinese).

Rissien, Adam. "Toledo's Water Crisis Points to Continued Dangers to Drinking Supply." *Cleveland.com*, August 18, 2014. http://www.cleveland.com/metro/index.ssf/2014/08/toledos_water_crisis_points_to.html.

Rodrigo, Dan. "Lessons Learned from Southern California's Integrated Resources Plan." Metropolitan Water District of Southern California, 2011. http://opensiuc.lib.siu.edu/cgi/viewcontent.cgi?article=1316&context=jcwre.

Roraback, Dick. "Up a Lazy River, Seeking the Source: Your Explorer Follows in Footsteps of Gaspar de Portola." *Los Angeles Times*, October 20, 1985.

Rosenthal, Elisabeth. "Pollution Victims Start to Fight Back in China." *New York Times*, May 16, 2000. http://www.nytimes.com/2000/05/16/world/pollution-victims-start-to-fight-back-in-china.html?pagewanted=all&src=pm.

San Diego County Water Authority. "Water Authority–Imperial Irrigation District Water Transfer." http://sdcwa.org/water-transfer.

Shen, Ying, and Debra Tan. "Groundwater Crackdown—Hope Springs." *China Water Risk*, July 11, 2013. http://chinawaterrisk.org/resources/analysis-reviews/groundwater-crackdown-hope-springs/.

State Council of the People's Republic of China. "Action Plan for Water Pollution Prevention." April 2, 2015.

Sze-Man, Natalie Ng. "Why Bottled Water Is NOT the Solution for China's Drinking Water Crisis?" *Globalization Monitor*, October 2013. http://globalmon.org.hk/sites/default/files/attachment/Why%20Bottled%20Water%20is%20not%20the%20solution%20for%20Water%20Crisis_SI.pdf.

Tan, Debra. "Just What Is Bottled Water?" *China Water Risk*, March 11, 2011. http://chinawaterrisk.org/opinions/just-what-is-bottled-water/.

Tan, Debra. "The War on Water Pollution." *China Water Risk*, March 12, 2014. http://chinawaterrisk.org/resources/analysis-reviews/the-war-on-water-pollution/.

Torres-Rouff, David S. "Water Use, Ethnic Conflict, and Infrastructure in Nineteenth-Century Los Angeles." *Pacific Historical Review* 75 (1) (2006): 119–140. http://web.mit.edu/people/spirn/Public/Granite%20Garden%20Research/Urban%20Environmental%20History/Torres-Rouff%202006%20Water%20Conflict%20Los%20Angeles.pdf.

United States Environmental Protection Agency, Pacific Southwest Media Center (Region 9). "EPA Takes Action to Strengthen Environmental and Public Health Protection for the L.A. River Basin." Updated November 2, 2012. https://archive.epa.gov/region9/mediacenter/web/html/index-24.html.

United States Environmental Protection Agency, Pacific Southwest Media Center (Region 9). "San Gabriel Valley (Area 4), City of Industry, Puente Valley." Accessed July 30, 2016. https://yosemite.epa.gov/r9/sfund/r9sfdocw.nsf/BySite/San%20Gabriel%20Valley%20(area%204)%20City%20Of%20Industry,%20Puente%20Valley?OpenDocument#.

Wallach, Mark. "Water: Who's Getting Soaked." *Los Angeles's Free Weekly* 14 (12) (January 3, 1992).

Wang, Yunpeng, et al. "Water Quality Change in Reservoirs of Shenzhen, China: Detection Using LANDST/TM data." *Science of the Total Environment* 328 (2004): 195–206.

Water Research Foundation. "Advancing the Science of Water: WRF and Research on Taste and Odor in Drinking Water." 2014. http://www.waterrf.org/resources/StateOfTheScienceReports/TasteandOdorResearch.pdf.

Water Resources Group. 2010. "Charting Our Water Future." http://www.2030waterresourcesgroup.com/water_full/Charting_Our_Water_Future_Final.pdf.

Wee, Sui-Lee. "Chinese Court Dismisses Water Pollution Lawsuit." *Reuters*, April 15, 2014. http://www.reuters.com/article/2014/04/15/us-china-water-veolia-idUSBREA3E05P20140415.

Wong, Olga. "Seawater Flushing to Be Extended to More Hong Kong Toilets Next Year." *South China Morning Post*, April 3, 2014. http://www.scmp.com/news/hong-kong/article/1463333/seawater-flushing-be-extended-more-hong-kong-toilets-next-year.

World Bank. "Addressing China's Water Scarcity: Recommendations for Selected Water Resource Management Issues." 2009. http://www-wds.worldbank.org/external/default/WDSContentServer/WDSP/IB/2009/01/14/000333037_20090114011126/Rendered/PDF/471110PUB0CHA0101OFFICIAL0USE0ONLY1.pdf.

World Bank and State Environmental Protection Administration. "Cost of Pollution in China: Economic Estimates of Physical Damage." February 2007. http://siteresources.worldbank.org/INTEAPREGTOPENVIRONMENT/Resources/China_Cost_of_Pollution.pdf.

Wright, Jim. *The Coming Water Famine*. New York: Coward-McCann, 1966.

Yang, Liang, Chunxiao Zhang, and Grace W. Ngaruiya. "Water Supply Risks and Urban Responses under a Changing Climate: A Case Study of Hong Kong." *Pacific Geographies* 39 (January/February 2013): 9–15. http://qa-pubman.mpdl.mpg.de/pubman/faces/viewItemFullPage.jsp?itemId=escidoc%3A2034671%3A2.

Yang, Liang, et al. "Water Supply Risks and Urban Responses under a Changing Climate: A Case Study of Hong Kong." *Pacific Geographies* 39 (January/February 2013): 9–15. http://www.pacific-news.de/pg39/pg39_Liang%20Yang_et_al.pdf.

Yardley, Jim. "Rules Ignored, Toxic Sludge Sinks Chinese Village." *New York Times*, September 4, 2006. http://www.nytimes.com/2006/09/04/world/asia/04pollution.html?pagewanted=print&_r=0.

Yu, Au Loong, and Yan-Zhu Wei. "The Privatization of Water Supply in China." *Globalization Monitor*, December 8, 2006. http://www.globalmon.org.hk/content/privatization-water-supply-china-0.

Zhang, Haibo, et al. "PCB Contamination in Soils of the Pearl River Delta, South China: Levels, Sources, and Potential Risks." *Environmental Science and Pollution Research International* 20 (8) (2013): 5150–5159.

Zhong, Lijin, Arthur P. J. Mol, and Tao Fu. "Public-Private Partnerships in China's Urban Water Sector." *Environmental Management* 41 (6) (June 2008): 863–877. http://www.ncbi.nlm.nih.gov/pmc/articles/PMC2359833/.

Zuo, Mandy. "South-North Water Diversion Project Not Cause of Hubei and Henan Droughts, Say Officials." *South China Morning Post*, October 25, 2014. http://www .scmp.com/news/china/article/1580527/south-north-water-diversion-project-not -cause-hubei-and-henan-droughts.

Chapter 5

Allen, Katie. "China Overtakes US as World's Biggest Grocery Market." *Guardian*, April 3, 2012. http://www.theguardian.com/world/2012/apr/04/china-biggest -grocery-market-world.

Azrilian, Jesse, et al. "Creating Healthy Corner Stores: An Analysis of Factors Necessary for Effective Corner Store Conversions." Los Angeles Food Policy Council, May 2012. http://goodfoodla.org/wp-content/uploads/2013/06/Pages-from-Creating -Healthy-Corner-Stores-Report-prepared-for-LAFPC.11.pdf.

Balzano, Joe. "Lingering Food Safety Regulatory Issues for China in 2016." *Forbes Asia,* January 10, 2016. http://www.forbes.com/sites/johnbalzano/2016/01/10/ lingering-food-safety-regulatory-issues-for-china-in-2016/#4885fa0e5f5a.

Benjamin, Laura. "Growing a Movement: Community Gardens in Los Angeles County." Urban and Environmental Policy Department, Occidental College, May 2008. http://www.oxy.edu/sites/default/files/assets/UEP/Comps/2008/benjamin CommunityGardens.pdf.

Bittman, Mark. "On Becoming China's Farm Team." *New York Times*, November 5, 2013. http://www.nytimes.com/2013/11/06/opinion/bittman-on-becoming-chinas -farm-team.html.

Black, Jane. "Food Deserts versus Food Swamps: The USDA Weighs In." *Washington Post*, June 25, 2009. http://voices.washingtonpost.com/all-we-can-eat/food-politics/ food-deserts-vs-swamps-the-usd.html.

Blake, Kathleen Megan. "Ordinary Food Places in a Global City: Hong Kong." *Streetnotes* 21 (2013): 1–12. http://escholarship.org/uc/item/11d2j987#page-1.

Bloomberg News. "China 'Catastrophe' Hits 114 Million as Diabetes Spreads." *Bloomberg*, September 3, 2013. http://www.bloomberg.com/news/2013-09-03/china -catastrophe-hits-114-million-as-diabetes-spreads.html.

Bose, Nandita, and Adam Rose. "Walmart's China Syndrome a Symptom of International Woes." *Reuters*, February 21, 2014. http://www.reuters.com/article/2014/ 02/21/us-walmart-emerging-idUSBREA1K08220140221.

Bowman, Elizabeth. "Growing Los Angeles' Urban Agriculture Policy." Antioch Urban Sustainability Program, Antioch University, Los Angeles, March 2012.

Bray, Chad, and Neil Gough. "Tesco Plans Venture with Leading Chinese Grocery Chain." *New York Times*, October 1, 2013. http://dealbook.nytimes.com/2013/10/01/

tesco-to-pay-550-million-in-supermarket-tie-up-in-china/?_php=true&_type =blogs&_r=0.

Brown, Lester R. *Who Will Feed China? Wake Up Call for a Small Planet.* New York: W. W. Norton, 1995.

Burgdorfer, Bob. "U.S. Chicken Feet Are Being Booted Out of China." *Reuters,* July 7, 2009. http://www.reuters.com/article/2009/07/07/china-chicken-idUSN07334744 20090707.

Burkitt, Laurie. "China Retailer CRE Adopts Rivals' Western Ways." *Wall Street Journal,* October 23, 2012. http://online.wsj.com/news/articles/SB100008723963904432 94904578046103441443928.

Burkitt, Laurie, and Peter Evans. "Tesco, China Resources in Talks about Joint Supermarket Venture in China." *Wall Street Journal,* August 9, 2013. http://online.wsj .com/news/articles/SB10001424127887323977304579001752585974952.

Cai, Jianming. "Periurban Agriculture Development in China." Resource Centers on Urban Agriculture & Food Security (RUAF), 40–42. http://www.ruaf.org/sites/default/ files/Periurban%20Agriculture%20Development%20in%20China.pdf.

Cain, Cindy Wojdyla. "Grain Exports Are Booming." *Herald-News* [Joliet], November 20, 2011: n. pag.

Calinoff, Jordan. "China's New Organic Industry." *Global Post,* February 19, 2009. http://www.globalpost.com/dispatch/china-and-its-neighbors/090217/chinas -new-organic-industry.

Centre for Food Safety. "Hong Kong Population-Based Food Consumption Survey: 2005–2007." Final Report. Chinese University of Hong Kong Food and Environmental Hygiene Department, April 2010. http://www.cfs.gov.hk/english/programme/ programme_firm/files/FCS_final_report.pdf.

Centre for Health Protection. "Assessment of Dietary Patterns in Primary Schools 2008." Hong Kong Department of Health, July 2009. http://www.chp.gov.hk/files/ pdf/report_part1_english.pdf.

Chan, Kahon. "War of the Papayas." *China Daily,* September 8, 2011. http://www .chinadaily.com.cn/hkedition/2011-09/08/content_13645581.htm.

Chan, Oswald. "Mighty Li Takes Stock of Empire." *China Daily USA,* January 15–17, 2016.

Chan, Samuel. "No Plan to Lift Two-Can Milk Formula Limit at the Border Says Hong Kong Health Minister." *South China Morning Post,* September 8, 2015. http:// www.scmp.com/news/hong-kong/health-environment/article/1856056/hong-kong -health-minister-says-no-plan-ease-ban.

Chan, Sucheng. *This Bitter-sweet Soil: The Chinese in California Agriculture, 1860–1910*. Berkeley: University of California Press, 1986.

Chan, Vinicy, and Eleni Himaras. "Li Ka-Shing Watson Spinoff May Bring Biggest Asia IPO in Three Years." *Bloomberg,* October 21, 2013. http://www.bloomberg.com/news/2013-10-18/hutchison-scraps-parknshop-private-sale-after-review.html.

Chemical Inspection and Regulatory Service. "The Strictest 'Food Safety Law' in China." April 2015. http://www.cirs-reach.com/news/The_Strictest_Food_Safety _Law_in_China.html.

Chen, Aijuan, and Steffanie Scott. "Rural Development Strategies and Government Roles in the Development of Farmers' Cooperatives in China." *Journal of Agriculture, Food Systems, and Community Development* 4 (4) (2014): 35–55. http://www.agdev journal.com/volume-4-issue-4/471-farmers-cooperatives-china.html.

Chen, Michelle. "Walmart Empire Clashes with China." *Progressive*, April 2014. http://www.progressive.org/walmart-empire-clashes-with-china.

Cheng, Siu Kei. "Adopting a New Lifestyle: Formation of a Local Organic Food Community in Hong Kong." Master's thesis, Hong Kong University of Science and Technology, August 2009.

Child, Peter N. "Carrefour China: Lessons from a Global Retailer." Reprint from McKinsey & Co interview with Carrefour China head Jen-Luc Chereau. *Forbes*, October 25, 2006. http://www.forbes.com/2006/10/25/carrefour-china-chereau-qanda -biz-cx_pnc_1025mckinsey.html.

China Resources Enterprise. "China Resources Enterprise Announces Annual Results for 2013: 'Get Ready for the Next Journey.'" Press release, 2014. http://www.cre .com.hk/home_15152/infocenter/compantnews/company2014/201511/ P020151109389703678022.pdf.

China Retail Research Center. "Walmart 'Made in America' Drive Follows Suppliers Lead." September 26, 2013. Tsinghua University. http://crrc.sem.tsinghua.edu.cn/ inews/256652.jhtml.

"Chinese Grocers Go on a Shopping Spree." *China Economic Review*, August 22, 2013. http://www.chinaeconomicreview.com/Tesco-CRE-CR-Vanguard-supermarkets -grocers-chains-retail.

Chui, Timothy, and Agnes Lu. "A Fading Dream of Affordable Housing." *China Daily Asia*, September 23, 2014. http://www.chinadailyasia.com/focus/2014-09/23/ content_15169340.html.

Chun, Zhang. "China's Declining Crop Diversity Threatens Its Food Sovereignty." *China Dialogue*, October 4, 2014. https://www.chinadialogue.net/article/show/ single/en/6872-China-s-declining-crop-diversity-threatens-its-food-sovereignty.

Clark, D. J. "The Looming Food Crisis in Asia: Gardens of Beijing." *China Daily*, July 20, 2011. http://usa.chinadaily.com.cn/video/2011-07/20/content_12946094.htm.

Cockrall-King, Jennifer. *Food and the City: Urban Agriculture and the New Food Revolution*. Amherst, NY: Prometheus Books, 2012.

Cone, Cynthia Abbott, and Ann Kakaliouras. "Community Supported Agriculture: Building Moral Community or an Alternative Consumer Choice." *Culture and Agriculture* 15 (1995): 28–31.

Couling, Stephen. "Organic Farming in China: Reflections from a Study Tour." July 22, 2012. http://postgrowth.org/organic-farming-in-urban-china-reflections-from -a-study-tour/.

Credeur, Mary Jane. "Chicken-Feet Exports Help Savannah Lead U.S. Port Growth." *Bloomberg*, December 6, 2012. http://www.bloomberg.com/news/2012-12-07/ chicken-feet-exports-help-savannah-lead-u-s-port-growth.html.

Crespí, Juan. *A Description of Distant Roads: Original Journals of the First Expedition into California, 1769–1770*. Ed. and trans. A. K. Brown. San Diego: San Diego State University Press, 2001.

Day, Alexander F. *The Peasant in Postsocialist China: History, Politics, and Capitalism*. Cambridge: Cambridge University Press, 2013.

Delwiche, Alexa, and Joann Lo. "Los Angeles' Good Food Purchasing Policy: Worker, Farmer and Nutrition Advocates Meet … and Agree!" *Progress in Planning* (Fall 2013): 24–28.

Demick, Barbara. "In China, What You Eat Tells Who You Are." *Los Angeles Times*, September 16, 2011. http://articles.latimes.com/2011/sep/16/world/la-fg-china-elite -farm-20110917.

Ferdman, Roberto A., and Jiaxi Lu. "A Quarter of McDonald's Restaurants in China May Have Been Serving Expired Meat." *Washington Post*, July 21, 2014. http://www .washingtonpost.com/blogs/wonkblog/wp/2014/07/21/a-quarter-of-mcdonalds -restaurants-in-china-have-been-serving-expired-meat/.

Flander, Scott. "Au Revoir, Carrefour Carrefour to U.S.: Au Revoir! Two Area Stores to Close Soon." *Philadelphia Daily News* (*Philly.Com*), September 7, 1993. http://articles .philly.com/1993-09-07/news/25984620_1_stores-low-prices-philadelphia-area.

French, Paul, and Matthew Crabbe. *Fat China: How Expanding Waistlines Are Changing a Nation*. London: Anthem Press, 2010.

Fromartz, Samuel. *Organic Inc.: Natural Foods and How They Grew*. Boston: Houghton Mifflin Harcourt, 2007.

Fung, Fanny W. Y. "Farmers Question Plan for Hi-Tech Agricultural Park in Hong Kong." *South China Morning Post*, March 8, 2015. http://www.scmp.com/news/hong -kong/article/1732187/farmers-question-plan-hi-tech-agricultural-park-hong-kong.

Gale, H. Frederick. "Building Trust in Food." *China Dialogue*, April 4, 2011. https://www.chinadialogue.net/article/show/single/en/4207-Building-trust-in-food.

Gaoming, Jiang. "Beware the GM Giants." *China Dialogue*. http://www.chinadialogue.net/article/show/single/en/3463-Beware-the-GM-giants.

Garcia, Matt. *A World of Its Own: Race, Labor, and Citrus in the Making of Greater Los Angeles, 1900–1970*. Chapel Hill: University of North Carolina Press, 2001.

Garnett, Tara, and Andreas Wilkes. "Appetite for Change: Social, Economic and Environmental Transformations in China's Food System." Food Climate Research Network, University of Oxford, February 2014. www.fcrn.org.uk

Genpan, Li. "Thought and Practice of Sustainable Development in Chinese Traditional Agriculture." *China Agricultural Economic Review* 1 (1) (2009): 97–109.

Gerson, Jacob E., and Benjamin I. Sachs. "Protect Those Who Protect Our Food." *New York Times*, November 12, 2014. http://www.nytimes.com/2014/11/13/opinion/protect-those-who-protect-our-food.html.

Gilmour, Brad, and Fred Gale. "A Maturing Food Retail Sector: Wider Channels for Food Imports?" In "China's Food and Agriculture: Issues for the 21st Century." US Department of Agriculture Economic Research Service, Agriculture Information Bulletin No. AIB-775, April 2002. http://www.ers.usda.gov/publications/aib-agricultural-information-bulletin/aib775.aspx.

Good to China. "Riding the Green Wave into Shanghai: Sustainable Shanghai, the Market." December 2, 2013. http://goodtochina.com/riding-the-green-wave-into-shanghai-sustainable-shanghai-the-market/.

Gottlieb, Robert, and Anupama Joshi. *Food Justice*. Cambridge, MA: MIT Press, 2013.

Gottlieb, Robert, and James Rojas. "Los Angeles Should Cultivate This Rare Urban Seed." *Los Angeles Times*, March 23, 2004. http://articles.latimes.com/2004/mar/23/opinion/oe-gottlieb23.

Gough, Neil. "China's Cofco to Buy Majority Stake in Noble Agricultural Unit." *New York Times*, April 2, 2014. http://dealbook.nytimes.com/2014/04/02/chinas-cofco-to-buy-majority-stake-in-noble-agriculture-unit/?_php=true&_type=blogs&_php=true&_type=blogs&emc=edit_tnt_20140402&nlid=42244359&tntemail0=y&_r=1.

Green, Emily. "Green Dreams." *Los Angeles Times*, October 31, 2004. http://articles.latimes.com/2004/oct/31/magazine/tm-urbangarden44.

Greenhouse, Steven. "Study Finds Federal Contracts Given to Flagrant Violators of Labor Laws." *New York Times*, December 10, 2013. http://www.nytimes.com/2013/12/11/business/study-finds-federal-contracts-given-to-flagrant-violators-of-labor-laws.html.

Guangwei, He. "In China's Heartland: A Toxic Trail Leads from Factories to Fields to Food." *Yale Environment 360*, July 7, 2014. http://e360.yale.edu/feature/chinas _toxic_trail_leads_from_factories_to_food/2784/.

Guzman, Richard. "As Wal-Mart Opening Nears, Division Remains in Chinatown." *DT News*, August 26, 2013. http://www.ladowntownnews.com/news/as-wal-mart -opening-nears-division-remains-in-chinatown/article_8e9ce710-0c4b-11e3-adc5 -0019bb2963f4.html.

Hairong, Yan. *New Masters, New Servants: Migration, Development, and Women Workers in China*. Durham: Duke University Press, 2008.

Hairong, Yan, and Yiyuan Chen. "Debating the Rural Cooperative Movement in China, the Past and the Present." *Journal of Peasant Studies* 40 (6) (2013): 955–981.

He, Guangwei. "The Soil Pollution Crisis in China: A Cleanup Presents Daunting Challenge." Part III of "Tainted Harvest." *Yale Environment 360*, July 14, 2014. http://e360.yale.edu/feature/the_soil_pollution_crisis_in_china_a_cleanup_presents _daunting_challenge/2786/.

He, Na. "Green Fingers and Green Houses in China." *City Farmer News*, September 19, 2012. http://www.cityfarmer.info/2012/09/19/green-fingers-and-green-houses -in-china/.

He, Zhixiong, Lanying Zhang, and Wen Tiejun. "New Rural Regeneration in Contemporary China." In *Touching the Heart, Taking Root* (Hong Kong: Partnerships for Community Development, 2015), 101–118.

Henderson, Elizabeth. "CSA in Taiwan and China—a Tantalizing Glimpse." *Chelsea Green Publishing*, January 16, 2012. http://chelseagreen.com/blogs/elizabethhender son/2012/01/16/csa-in-taiwan-and-china-%E2%80%93-a-tantalizing-glimpse/.

Herring, Louise, Daniel Hui, Paul Morgan, and Caroline Tufft. "Inside China's Hypermarkets: Past and Prospects." McKinsey & Co., May 2012. http://www.mckin seyonmarketingandsales.com/inside-chinas-hypermarkets.

Holdaway, Jennifer, and Lewis Husain. "Food Safety in China: A Mapping of Problems, Governance and Research." Forum on Health, Environment and Development (FORHEAD) Working Group on Food Safety, February 2014. http://webarchive.ssrc .org/cehi/PDFs/Food-Safety-in-China-Web.pdf.

Hong Kong Consumer Council. "Grocery Market Study: Market Power of Supermarket Chains under Scrutiny." December 19, 2013. https://www.consumer.org.hk/ sites/consumer/files/competition_issues/20131219/GMSReport20131219.pdf.

Hong Kong Heritage Museum. "Hong Kong's Food Culture." December 2, 1999. http://www.heritagemuseum.gov.hk/documents/2199315/2199693/Hong_Kong _Food_Culture-E.pdf.

Hong Kong Legislative Council, Panel on Economic Services. "Consumer Council's Report on Wet Markets versus Supermarkets: Competition in the Retailing Sector." LC Paper No. CB(1)1017/03–04(03), February 23, 2004. http://www.legco.gov.hk/yr03-04/english/panels/es/papers/es0223cb1-1017-3e.pdf.

Hong Kong Legislative Council, Panel on Food Safety and Environmental Hygiene. "Prevention and Control of Avian Influenza." November 19, 2013. LC Paper No. CB(2)277/13–14(03). http://www.legco.gov.hk/yr13-14/english/panels/fseh/papers/fe1119cb2-277-3-e.pdf.

Hong Kong Special Administrative Region, Agriculture, Fisheries, and Conservation Department. "Agriculture in HK [Hong Kong]." March 10, 2016. http://www.afcd.gov.hk/english/agriculture/agr_hk/agr_hk.html.

Hong Kong Special Administrative Region, Agriculture, Fisheries, and Conservation Department. "Organic Farming in Hong Kong." 2014. http://www.afcd.gov.hk/textonly/english/agriculture/agr_orgfarm/agr_orgfarm.html.

Hong Kong Special Administrative Region, Department of Health. "Action Plan to Promote Healthy Diet and Physical Activity Participation in Hong Kong." 2013. http://www.change4health.gov.hk/filemanager/common/image/strategic_framework/action_plan/action_plan_e.pdf.

Hong Kong Special Administrative Region, Food and Health Bureau and Agriculture, Fisheries and Conservation Department. "Consultation Document: The New Agricultural Policy: Sustainable Agricultural Development in Hong Kong." December 2014. http://www.afcd.gov.hk/english/whatsnew/what_agr/files/consultation_on_agricultural_policy.pdf.

Hsu, Tiffany. "Fresh & Easy Fail: Tesco Exits US after Profit Tanks 96%." *Los Angeles Times*, April 17, 2013. http://articles.latimes.com/2013/apr/17/business/la-fi-mo-fresh-easy-tesco-us-20130417.

Hu, Dinghuan, et al. "The Emergence of Supermarkets with Chinese Characteristics: Challenges and Opportunities for China's Agricultural Development." *Development Policy Review* 22 (5) (2004): 557–586.

Huang, Jikun, Xiaobing Wang, and Huanguang Qiu. "Small-Scale Farmers in China in the Face of Modernization and Globalization." International Institute for Environment and Development (IIED/Hivos), May 2012. http://pubs.iied.org/16515IIED.html.

Huang, Philip C. C. "China's New-Age Small Farms and Their Vertical Integration: Agribusiness or Coops?" *Modern China* 37 (2) (2011): 107–134.

Huang, Philip C. C., Gao Yuan, and Yusheng Peng. "Capitalization without Proletarianization in China's Agricultural Development." *Modern China* 38 (2) (2012): 139–173.

Huanxin, Zhao, and Wencong Wu. "Tainted Farmland to Be Restored." *China Daily USA*, December 31, 2013. http://usa.chinadaily.com.cn/china/2013-12/31/content_17206152.htm.

Huddleston, Tom Jr. "Chinese Meat Scandal Hurts Sales at Fast-Food Giant Yum Brands." *Fortune*, October 7, 2014. http://fortune.com/2014/10/07/chinese-meat-scandal-hurts-sales-at-fast-food-giant-yum-brands/.

Hui, Mary. "In Organic-Hungry Hong Kong, Corn as High as an Elevator's Climb." *New York Times*, October 3, 2012. http://www.nytimes.com/2012/10/04/world/asia/fearing-tainted-imports-hong-kong-squeezes-in-farms.html.

Institute for Agriculture and Trade Policy. "China's Pig Industry Has Global Impact, New Report Finds." Institute for Agriculture and Trade Policy, May 17, 2011. Web, April 1, 2013.

Iratini, Evelyn. "Retail Giant Remarks That a British Rival Might Be Too Big Raises Critics' Eyebrows." *Los Angeles Times*, September 5, 2005.

Johnson, Ian. "Picking Death over Eviction." *New York Times*, September 8, 2013. http://www.nytimes.com/2013/09/09/world/asia/as-chinese-farmers-fight-for-homes-suicide-is-ultimate-protest.html?_r=0.

Johnson, Ian. "Pitfalls Abound in China's Push from Farm to City." *New York Times*, July 13, 2013.

Jopson, Barney. "Walmart Slows International Expansion." *Financial Times*, August 16, 2012. https://www.ft.com/content/876f65a4-e79d-11e1-86bf-00144feab49a.

Joshi, Anupama, Moira Beery, and Marion Kalb. "Going Local: Paths to Success for Farm to School Programs." Los Angeles: Urban & Environmental Policy Institute and the Community Food Security Coalition, December 2006. http://mda.maryland.gov/farm_to_school/Documents/goinglocal.pdf.

Kaufman, Francine. *Diabesity: A Doctor and Her Patients on the Front Lines of the Obesity-Diabetes Epidemic*. New York: Bantam, 2006.

Kesmodel, David, and Laurie Burkitt. "Inside China's Supersanitary Chicken Farms: Looking to Capitalize on Food Safety Concerns, Tyson Shifts from Using Independent Breeders." *Wall Street Journal*, December 9, 2013. http://online.wsj.com/news/articles/SB10001424052702303559504579197662165181956.

Kessler, Rebecca. "Urban Gardening: Managing the Risk of Contaminated Soil." *Environmental Health Perspectives* 121 (11–12) (2013): A327–A333. http://ehp.niehs.nih.gov/121-a326/.

King, F. H. *Farmers of Forty Centuries: Or Permanent Agriculture in China, Korea and Japan*. London: Jonathan Cape, 1926.

Klein, Jakob A. "Creating Ethical Food Consumers? Promoting Organic Foods in Southwest China." *Social Anthropology* 17 (1) (February 2009): 74–89. http://onlineli brary.wiley.com/doi/10.1111/j.1469-8676.2008.00058.x/abstract.

Kroll, Andy. "Are Walmart's Chinese Factories as Bad as Apple's?" *Mother Jones*, March-April 2012. http://www.motherjones.com/print/161316.

Kuo, Lily. "With Wealth Comes Fat, China Finds." *Los Angeles Times*, October 31, 2010. http://articles.latimes.com/2010/oct/31/world/la-fg-china-fat-20101031.

LaFraniere, Sharon. "In China, Fear of Fake Eggs and 'Recycled' Buns." *New York Times*, May 7, 2011. http://www.nytimes.com/2011/05/08/world/asia/08food .html?pagewanted=all.

Lam, Angus. "Farmers' Market: Manifesting the Spirit of Everyday Life." In *Touching the Heart, Taking Root* (Hong Kong: Partnerships for Community Development, 2015), 36–45. http://www.pcd.org.hk/sites/default/files/publications/Final_pdf_csa _eng_book.pdf.

Lang, Graeme, and Bo Miao. "Food Security for China's Cities." *International Planning Studies* 18 (1) (2013): 5–20. http://dx.doi.org/ 10.1080/13563475.2013.750940.

Lang, Tim, and Michael Heasman. *Food Wars: The Global Battle for Mouths, Minds and Markets*. London: Earthscan, 2004.

Lee, Chang-won, et al. "H5N2 Avian Influenza Outbreak in Texas in 2004: The First Highly Pathogenic Strain in the United States in 20 Years?" *Journal of Virology* 79 (17) (September 2005): 11412–21. http://www.ncbi.nlm.nih.gov/pmc/articles/ PMC1193578/?tool=pubmed.

Leonard, Christopher. *The Meat Racket: The Secret Takeover of America's Food Business*. New York: Simon & Schuster, 2014.

Levinson, Marc. *The Great A&P and the Struggle for Small Business in America*. New York: Hill and Wang, 2011.

Levitt, Tom. "China Facing Bigger Dietary Health Crisis than the U.S." Interview with Barry Popkin. *China Dialogue*, April 7, 2014. https://www.chinadialogue.net/ article/show/single/en/6880-China-facing-bigger-dietary-health-crisis-than-the-US.

Levitt, Tom. "Urban Farming on Brink of Corporate Era." *China Dialogue*, April 3, 2013. https://www.chinadialogue.net/article/show/single/en/5854-Urban-farming -on-brink-of-corporate-era.

Levitt, Tom. "U.S.-Style Intensive Farming Is Not the Solution to China's Meat Problem." *Guardian*, March 3, 2014. http://www.theguardian.com/environment/blog/ 2014/mar/03/us-intensive-farming-chinas-meat-problem.

Lichtenstein, Nelson. *The Retail Revolution: How Wal-Mart Created a Brave New World of Business*. New York: Metropolitan Books, 2009.

Longstreth, Richard. *The Drive-In, the Supermarket, and the Transformation of Commercial Space in Los Angeles, 1914–1941.* Cambridge, MA: MIT Press, 1999.

Los Angeles Food Policy Council. "Los Angeles Food System Snapshot 2013: A Baseline Report on the Los Angeles Regional Foodshed." October 2013. www.goodfoodla .org.

Lui, Haiyan, et al. "Household Composition and Food Away from Home Expenditures in Urban China." Paper presented at the Agricultural and Applied Economics Association (formerly the American Agricultural Economics Association), 2012 Annual Meeting, August 12–14, 2012, Seattle, Washington. http://ageconsearch .umn.edu/bitstream/131057/3/Household%20Composition%20and%20Food%20 Away%20From%20Home%20Expenditures%20in%20Urban%20China.pdf.

Maduri Frank, J. "China's Ownership of an Iconic American Food Company." *United Press International,* October 25, 2013. http://www.upi.com/Top_News/Analysis/ Outside-View/2013/10/25/Chinas-ownership-of-an-iconic-American-food-com pany/UPI-75821382673840/.

Manchester, Alden. "The Transformation of U.S. Food Marketing." In *Food and Agricultural Markets: The Quiet Revolution,* ed. Lyle P. Schertz and Lynn M. Daft. Washington, DC: USDA Economic Research Service, National Planning Association, 1994.

Marion, Bruce W. "Concentration-Relationship in Food Retailing." In *Concentration and Price,* ed. Leonard W. Weiss, 183–194. Cambridge: MIT Press, 1989.

Masunaga, Samantha, and Ivan Penn. "Wal-Mart to Shut 269 Stores." *Los Angeles Times,* January 16, 2016.

Matuszak, Sascha. "China's Organic Food Cooperatives Must Overcome Trust Deficit." *Tea Leaf Nation,* April 23, 2013.

McGee, Patrick. "KKR Buys into China Pork Producer." *Financial Times,* June 6, 2014. https://www.ft.com/content/ad4e7df4-ed28-11e3-8963-00144feabdc0.

Melillo, Edward D. "The First Green Revolution: Debt Peonage and the Making of the Nitrogen Fertilizer Trade, 1840–1930." *American Historical Review* 117 (4) (October 2012): 1028–1060. http://ahr.oxfordjournals.org/content/117/4/1028.extract.

Ming, Chow Sung. "Sharing Hong Kong: From Social and Solidarity Economy to Sharing Economy, and from Fair Trade to Community Supported Agriculture." Paper submitted to the Asian Solidarity Economy Council, 5th RIPESS International Meeting, Manila, Philippines, October 15–18, 2013. http://www.google.com/url?sa =t&rct=j&q=&esrc=s&source=web&cd=1&ved=0CCQQFjAA&url=http%3A%2F%2F www.ripess.org%2Fwp-content%2Fuploads%2F2013%2F06%2FChow-Sung-Ming -Sharing-Hong-Kong-1.doc&ei=Q3U-VfLHEILlsAWWioGgAg&usg=AFQjCNGs7 -wBBcqFFT-HDVIJSNpbz3geXQ&sig2=jhgcB_t2gcomLfIgQIGQkw.

Minghe, Lu. "Shanghai's Dead Pig Story Stretches Back Upstream." *China Dialogue*, March 25, 2013. https://www.chinadialogue.net/article/show/single/en/5820 -Shanghai-s-dead-pig-story-stretches-back-upstream.

Mitchell, Tom, and Barney Jopson. "Official China Union Raises Stakes in Walmart Closure Programme." *Financial Times*, March 23, 2014. https://www.ft.com/ content/2038fd78-b262-11e3-b891-00144feabdc0.

Moses, H. Vincent. "G. Harold Powell and the Corporate Consolidation of the Modern Citrus Enterprise, 1904–1922." *Business History Review* 69 (2) (1995): 119–155. http://www.jstor.org/discover/10.2307/3117097?sid=21105684136323&uid=70 &uid=4&uid=3739560&uid=3739256&uid=2129&uid=2.

Moss, Michael. *Salt, Sugar, Fat: How the Food Giants Hooked Us*. Rev. ed. New York: Random House, 2014.

Mukhija, Vinit, and Anastasia Loukaitou-Sideris, eds. *The Informal American City: Beyond Taco Trucks and Day Labor*. Cambridge, MA: MIT Press, 2014.

Nan, Xu. "A Decade of Food Safety in China." *Chinadialogue*, August 6, 2012. https:// www.chinadialogue.net/article/show/single/en/5083-A-decade-of-food-safety -in-China.

National Health and Family Planning Commission of the PRC. "China's National Program for Food and Nutrition (2014–2020)." May 16, 2014. http://www.china daily.com.cn/m/chinahealth/2014-05/16/content_17514010.htm.

Ng, Marie, et al. "Global, Regional, and National Prevalence of Overweight and Obesity in Children and Adults during 1980–2013: A Systematic Analysis for the Global Burden of Disease." *Lancet*, published online May 29, 2014. http://dx.doi .org/ 10.1016/SO140-6736(14)60460-8.

Nip, Amy. "Grocery Giants ParknShp, Wellcome Accused of Pressuring Suppliers." *South China Morning Post*, December 20, 2013. http://www.scmp.com/news/hong -kong/article/1386263/grocery-giants-parknshop-wellcome-accused-pressuring -suppliers?page=all.

Ogden, Cynthia L., et al. "Prevalence of Childhood and Adult Obesity in the United States, 2011–2012." *Journal of the American Medical Association* 311 (8) (2014): 806–814. http://jama.jamanetwork.com/article.aspx?articleid=1832542.

Oxfam Hong Kong. "Survey on the Impact of Soaring Food Prices on Poor Families in Hong Kong." August 2011. http://www.oxfam.org.hk/filemgr/1630/FoodSurvey ReportAug2011_revised.pdf.

Partnerships for Community Development. "Community Supported Agriculture." http://pcd.org.hk/csa/gb/index.html.

Partnerships for Community Development. *Touching the Heart, Taking Root: CSA in Hong Kong, Taiwan, and Mainland China.* English edition. Hong Kong: Partnerships for Community Development, 2015. http://www.pcd.org.hk/sites/default/files/publications/Final_pdf_csa_eng_book.pdf.

Paull, John. "China's Organic Revolution." In *Marketing of Organic Products: Global Experiences,* ed. S. Bhaskaran and Suchitra Mohanty, 260–275. Hyderabad: Icfai University Press, 2008. http://orgprints.org/14846/1/14846.pdf.

Peine, Emelie K. "Trading on Pork and Beans: Agribusiness and the Construction of the Brazil-China-Soy-Pork Commodity Complex." In *The Ethics and Economics of Agrifood Competition,* ed. H. S. James Jr., 193–210. New York: Springer, 2013.

Pepall, Jennifer. "New Challenges for China's Urban Farmers." IDRC Report 21.3 (1997).

Perry, Charles. "The Cafeteria: An L.A. Original." *Los Angeles Times,* November 5, 2003. http://articles.latimes.com/2003/nov/05/food/fo-cafeteria5.

Philpott, Tom. "Are We Becoming China's Factory Farm?" *Mother Jones,* November-December 2013. http://www.motherjones.com/media/2014/03/china-factory-farm-america-pork?google_editors_picks=true.

Philpott, Tom. "Organic-Farming Movement Sprouts in China." *Grist,* November 3, 2010.

Pi, Chendong, with Zhang Rou and Sarah Horowitz. "Fair or Fowl? Industrialization of Poultry Production in China." Institute for Agriculture and Trade Policy, Minneapolis, February 2014. http://www.iatp.org/documents/fair-or-fowl-industrialization-of-poultry-production-in-china.

Pierson, David. "China's Thirst for Milk Gives U.S. Dairy Farms a Boost." *Los Angeles Times,* March 16, 2014. http://www.latimes.com/business/la-fi-feeding-china-dairy-20140315,0,5345274.story?page=2#ixzz2w99PUX8w.

Plottel, Sophie. "Urban Farming Growing in Shanghai, China." Global Site Plans (GSP): Branding for Environmental Design, February 27, 2013. http://globalsiteplans.com/environmental-non-profit/urban-farming-growing-in-shanghai-china/.

Plume, Karl, and Michael Hirtzer. "U.S. Traders Fear Rejection of Corn Byproduct by GMO-Wary China." *Reuters,* December 12, 2013. http://www.reuters.com/article/2013/12/12/usa-china-ddgs-idUSL1N0JR1SF20131212.

Pole, Antoinette, and Margaret Gray. "Farming Alone? What's Up with the 'C' in Community Supported Agriculture." *Agriculture and Human Values* 30 (2013): 85–100.

Popkin, Barry, and Shufa Du. "Dynamics of the Nutrition Transition toward the Animal Foods Sector in China and Its Implications: A Worried Perspective." *Journal of Nutrition* 133 (November 2003): 3898S–3906S.

Raja, Samina, et al. "Beyond Food Deserts: Measuring and Mapping Racial Disparities in Neighborhood Food Environments." *Journal of Planning Education and Research* 27 (4) (2008): 469–482.

Ramzy, Austin. "Chicken Feet: A Symbol of U.S.-China Tension." *Time*, February 8, 2010.   http://content.time.com/time/world/article/0,8599,1960825,00.html#ixzz2 seeJhfzF.

Redmond, Judith, and Thomas Nelson. "Defining a Food Ethic: Common Values for the Sustainable Food and Farm Movement." Full Belly Farm, Guinda, CA, January 2001, 28.

Research and Markets. "China Diabetes Market Outlook 2018." November 2013. http://www.researchandmarkets.com/reports/2694252/china_diabetes_market_out look_2018.

Riggs, Peter. "A Different Growing Season South of the Mountains: Guangdong Province Rethinks Its Agricultural Development Model." Woodrow Wilson International Center for Scholars, China Environment Forum, *China Development Series* 7 (2005): 47–54. http://www.wilsoncenter.org/sites/default/files/chinaenv7.pdf.

"RPT-Hutchison Drops Plans to Sell Parknshop Supermarkets Business." *Reuters*, October 20, 2013. http://www.reuters.com/article/2013/10/20/hutchison-idUSL3N0I A08A20131020.

RUAF Foundation et al. "Urban Agriculture for Resilient Cities (Green, Productive and Social Inclusive)." World Urban Forum IV Nanjing, China, 2009. http://www .ruaf.org/sites/default/files/booklet%20DVD.pdf.

Sacks, Brianna. "Fast Food Titans Sorry over China Scandal." *Los Angeles Times*, July 22, 2014.

Sanders, Richard. "A Market Road to Sustainable Agriculture? Ecological Agriculture, Green Food and Organic Agriculture in China." *Development and Change* 37 (1) (2006): 201–226.

Sanders, Richard. *Prospects for Sustainable Development in the Chinese Countryside: The Political Economy of Chinese Ecological Agriculture*. Aldershot, UK: Ashgate Publishing, 2000.

Sanders, Richard, and Xingji Xiao. "The Sustainability of Organic Agriculture in Developing Countries: Lessons from China." *International Journal of Environmental, Economic & Social Sustainability* 6 (6) (2010): 233–243.

Schell, Orville. "How Walmart Is Changing China." *Atlantic*, December 2011. http:// www.theatlantic.com/magazine/archive/2011/12/how-walmart-is-changing-china/ 308709/.

Schneider, Mindi. "Feeding China's Pigs: Implications for the Environment, China's Smallholder Farmers and Food Security." Institute for Trade and Agricultural Policy, Minneapolis, May 2011. http://www.iatp.org/documents/feeding-china %E2%80%99s-pigs-implications-for-the-environment-china%E2%80%99s-small holder-farmers-and-food.

Scott, Steffanie, et al. "Contradictions in State- and Civil Society-Driven Developments in China's Ecological Agriculture Society." *Food Policy* 45 (April 2014): 158–166. https://ideas.repec.org/a/eee/jfpoli/v45y2014icp158-166.html.

Shaffer, Amanda. *The Persistence of L.A.'s Grocery Gap: The Need for a New Food Policy and Approach to Market Development.* Los Angeles: Urban & Environmental Policy Institute, Occidental College, 2002. http://scholar.oxy.edu/cgi/viewcontent.cgi ?article=1395&context=uep_faculty.

Shaffer, Amanda, et al. "Shopping for a Market: Evaluating Tesco's Entry into Los Angeles and the United States." Los Angeles: Urban & Environmental Policy Institute, Occidental College, August 1, 2007. http://clkrep.lacity.org/onlinedocs/2010/ 10-1537_misc_plum_12-8-10.pdf.

Shi, Tian. "Ecological Agriculture in China: Bridging the Gap between Rhetoric and Practice of Sustainability." *Ecological Economics* 42 (2002): 359–368. http://jyw.znufe .edu.cn/rkzyhj/zyyjfxdt/stjj/P020060513006683730967.pdf.

Shun-hing, Chan. "Understanding Anew the Value of an Everyday Life with Its Roots in *Nong*." In *Touching the Heart, Taking Root* (Hong Kong: Partnerships for Community Development, 2015), 8–19.

Si, Zhenzhong, Theresa Schumilas, and Steffanie Scott. "Characterizing Alternative Food Networks in China." *Agriculture and Human Values* 32 (2) (June 2015): 299–313.

Silk, Richard. "American Ag Firms: China Is Getting Tougher." *Wall Street Journal*, February 27, 2014. http://blogs.wsj.com/chinarealtime/2014/02/27/american-ag -firms-china-is-getting-tougher/.

Silverstein, Amy. "McDonald's in Beijing Sold Expired Food to Customers." *Global-Post*, March 17, 2012, n. pag.

Skinner, G. William. "Vegetable Supply and Marketing in Chinese Cities." *China Quarterly* 76 (December 1978): 733–793.

Souza, Kim. "US to Accept Some China Poultry Imports." *The City Wire*, September 4, 2013. http://talkbusiness.net/2013/09/u-s-to-accept-some-chinese-poultry -imports/#.UuvIyH7TnIU.

Springham, Stephen. "International Analysis: Will Walmart Acquire Hong Kong's Park n Shop?" *Retail Week*, August 16, 2013. https://www.retail-week.com/topics/international/international-analysis-will-walmart-acquire-hong-kongs-park-n-shop/5052015.fullarticle.article.

Stanway, David. "China Government Survey Shows 16 Percent of Its Soil Is Polluted." *Reuters*, April 17, 2014. http://in.reuters.com/article/2014/04/17/china-pollution-soil-idINL3N0N92DX20140417.

Stoll, Steven. *The Fruits of Natural Advantage: Making the Industrial Countryside in California*. Berkeley: University of California Press, 1998.

Striffler, Steve. *Chicken: The Dangerous Transformation of America's Favorite Food*. New Haven: Yale University Press, 2005.

Surls, Rachel, and Judith Gerber. *From Cows to Concrete: The Rise and Fall of Farming in Los Angeles*. Los Angeles: Angel City Press, 2016.

Surls, Rachel, and Judith Gerber. "Los Angeles: A History of Agricultural Abundance." *Los Angeles Agriculture*, July 16, 2010. http://ucanr.edu/blogs/losangelesagriculture/index.cfm?tagname=Los%20Angeles%20City%20seal

"Sustainable Development of Local Agriculture." Hong Kong Legislative Council Q&A, January 21, 2015. http://www.info.gov.hk/gia/general/201501/21/P201501210568.htm.

Szymanski, Mike. "LAUSD Turns Up Heat on the National Chicken Industry." *LA School Report*, March 10, 2016. http://laschoolreport.com/lausd-turns-up-the-heat-on-the-chicken-industry/.

Tan, Debra. "The State of China's Agriculture." *China Water Risk*, April 9, 2014. http://chinawaterrisk.org/resources/analysis-reviews/the-state-of-chinas-agriculture/.

Thiers, Paul. "From Grassroots Movement to State-Coordinated Market Strategy: The Transformation of Organic Agriculture in China." *Environment and Planning C, Government & Policy* 20 (2002): 357–373.

Tobey, Ronald, and Charles Wetherell. "The Citrus Industry and the Revolution of Corporate Capitalism in Southern California, 1887–1944." *California History* 74 (1) (Spring 1995): 6–21.

Tso, Phoenix. "Promoting Organic: The Little Donkey Farm's Shi Yan." *Agenda* (Durban, South Africa), April 27, 2011. http://wiseconference.weebly.com/2133823458-blog/promoting-organic-the-little-donkey-farms-shi-yan.

Tsoi, Grace. "What Will Happen to the New Territories?" *HK Magazine*, September 13, 2012, http://hk-magazine.com/city-living/article/what-will-happen-new-territories.

Tsui, Bonnie. "China's Pollution Crisis Is Also a Food Safety Crisis." *The Atlantic Cities*, January 16, 2014. http://www.theatlanticcities.com/politics./2014/01/chinas-pollution-problem-also-food-safety-crisis/8122/.

UCLA Center for Health Policy Research. "Child and Teen, 2011–2012 Health Profiles: Los Angeles County." UCLA Center for Health Policy Research, Los Angeles, March 2014. http://healthpolicy.ucla.edu/health-profiles/Child_Teen/Documents/2011-2012/Regions/LosAngelesCounty.pdf.

United States Department of Agriculture, Census of Agriculture. "2012 Census Full Report." http://www.agcensus.usda.gov/Publications/2012/.

United States Department of Agriculture, Food Safety and Inspection Service. "Frequently Asked Questions: Equivalence of China's poultry Processing and Slaughter Inspection Systems." March 4, 2016. http://www.fsis.usda.gov/wps/portal/fsis/newsroom/news-releases-statements-transcripts/news-release-archives-by-year/archive/2016/faq-china-030416.

United States Department of Agriculture, Food Safety and Inspection Service. "Frequently Asked Questions: Equivalence of China's Poultry Processing System." September 26, 2013. http://www.fsis.usda.gov/wps/portal/fsis/newsroom/news-releases-statements-transcripts/news-release-archives-by-year/archive/2013/faq-china-08302013.

United States Department of Agriculture, Foreign Agricultural Service. "China's Food Safety Law (2015)." GAIN Report Number CH 15016, May 18, 2015. http://gain.fas.usda.gov/Recent%20GAIN%20Publications/Amended%20Food%20Safety%20Law%20of%20China_Beijing_China%20-%20Peoples%20Republic%20of_5-18-2015.pdf.

United States Department of Agriculture, Foreign Agricultural Service. "Hong Kong Exporter Guide, 2013." Prepared by Chris Li. Global Agricultural Information Network (GAIN) Report No. HK 1317. May 1, 2013. http://gain.fas.usda.gov/Recent%20GAIN%20Publications/Exporter%20Guide_Hong%20Kong_Hong%20Kong_4-30-2013.pdf.

United States Department of Agriculture, Foreign Agricultural Service. "Hong Kong: Retail Foods, 2013." Prepared by Chris Li. Global Agricultural Information Network (GAIN) Report No. HK 1318. May 1, 2013. http://gain.fas.usda.gov/Recent%20GAIN%20Publications/Retail%20Foods_Hong%20Kong_Hong%20Kong_5-1-2013.pdf.

United States Department of Agriculture, Foreign Agricultural Service. "Livestock and Poultry: World Markets and Trade." November 2013. http://apps.fas.usda.gov/psdonline/circulars/livestock_poultry.pdf.

United States Department of Agriculture, Global Agricultural Information Network. "Hong Kong: Organic Products Market." GAIN Report No. 1505, February 26, 2015.

http://gain.fas.usda.gov/Recent%20GAIN%20Publications/Market%20of%20 Organic%20Products_Hong%20Kong_Hong%20Kong_2-26-2015.pdf.

United States Department of Agriculture, Global Agricultural Information Network. "Hong Kong: Retail Foods. Annual 2015." Prepared by Chris Li. GAIN Report No. HK 1532, January 15, 2016. http://gain.fas.usda.gov/Recent%20GAIN%20Publications/ Retail%20Foods_Hong%20Kong_Hong%20Kong_12-9-2015.pdf.

United States Securities and Exchange Commission. Form 10K, Tyson Foods Inc., September 28, 2013. https://www.sec.gov/Archives/edgar/data/100493/000010049 308000060/form10k_092708.pdf.

"Urban Agriculture in Shenzhen, China." Appropedia, December 2009. http://www .appropedia.org/Urban_agriculture_in_Shenzhen,_China.

"Urban Farmers Union in Beijing, China." *City Farmer News*, August 11, 2012. http:// www.cityfarmer.info/2012/09/05/urban-farmers-union-in-beijing-china/.

U.S. Meat Export Federation. "China: Online U.S. Pork Promotion Goes Live on TMall.com." May 5, 2014. http://www.usmef.org/international-markets/china/.

Vallianatos, Mark, and Elizabeth Medrano. "The Transformation of the School Food Environment in Los Angeles: The Link between Grass Roots Organizing and Policy Implementation." Center for Food and Justice/Healthy School Food Coalition, Urban & Environmental Policy Institute, Los Angeles, July 2009. http://scholar.oxy .edu/cgi/viewcontent.cgi?article=1431&context=uep_faculty.

Van Rijnsoever, Martijn. "How to Feed Shanghai." *Change Magazine*. (Floriade Dialogue 2009–2012.) http://www.changemagazine.nl/doc/2012_1/2012_1_Shanghai .pdf.

Vincent Callebaut Architects. "Farmscrapers in Shenzhen, China." *City Farmer News*, 2013. http://www.cityfarmer.info/2013/03/09/farmingscrapers-in-shenzhen -china-by-vincent-callebaut-architects/#more-41390.

Wang, Jeanette. "62m People in China Obese, Sparking Fears of 'Alarming' Financial Burden." *South China Morning Post*, May 29, 2014. http://www.scmp.com/lifestyle/ health/article/1521011/21-billion-people-are-obese-china-and-us-among-worlds -fattest-new.

Watanabe, Teresa. "L.A. Healthful Lunch Menu Panned by Students." *Los Angeles Times*, December 17, 2011. http://articles.latimes.com/2011/dec/17/local/la-me -food-lausd-20111218.

Watanabe, Teresa. "Local Food Push Healthy All Around." *Los Angeles Times*, November 25, 2013. http://articles.latimes.com/2013/nov/24/local/la-me-lausd-food -20131124.

Watts, Jonathan. "Chinese Villagers Driven Off Land Fear Food May Run Out." *Guardian*, May 19, 2011. http://www.theguardian.com/environment/2011/may/19/china-food-illegal-land-grab-protests.

Web editor Zhang. "'Leisure Agriculture' Is in Vogue in China." *Crienglish.com*, May 28, 2012. Reprint from *City Farmer News*, http://www.cityfarmer.info/2012/05/28/leisure-agriculture-is-in-vogue-in-china/.

Williams, Mark. "The Supermarket Sector in China and Hong Kong: A Tale of Two Systems." *Competition Law Review* 3 (2) (March 2007): 251–268. http://new.clasf.org/CompLRev/Issues/Vol3Issue2Art4Williams.pdf.

Wingfield, Brian, and Shruti Date Singh. "Chicken Processed in China Triggers U.S. Food Safety Protests." *Bloomberg*, September 26, 2013. http://www.bloomberg.com/news/2013-09-27/chicken-processed-in-china-triggers-u-s-food-safety-protests.html.

Wong, Edward. "One-Fifth of China's Farmland Is Polluted, State Study Finds." *New York Times*, April 17, 2014. http://www.nytimes.com/2014/04/18/world/asia/one-fifth-of-chinas-farmland-is-polluted-state-report-finds.html.

Wong, Julian L. "The Food-Energy-Water Nexus: An Integrated Approach to Understanding China's Resource Challenges." *Harvard Asia Quarterly* (Spring 2010): 15–19. Reprint by Center for American Progress. http://www.americanprogress.org/wp-content/uploads/issues/2010/07/pdf/haqspring2010final.pdf.

World Bank. June 2014. "China Economic Update. Special Topic: Changing Food Consumption Patterns in China: Implications for Domestic Supply and International Trade." World Bank Office, Beijing. http://www.worldbank.org/content/dam/Worldbank/document/EAP/China/China_Economic_Update_June2014.pdf.

World Poultry. "Cobb-Vantress Signs Joint Venture Agreement in China." *World Poultry*, July 31, 2013. http://www.worldpoultry.net/Genetics/Articles/2013/7/Cobb-Vantress-signs-joint-venture-agreement-in-China-1323164W/.

World Poultry. "US Poultry, Egg Exports Set Records in 2011." *World Poultry*, February 15, 2012. http://www.worldpoultry.net/Broilers/Markets--Trade/2012/2/US-poultry-egg-exports-set-records-in-2011-WP009994W/.

Wu, Jing, and Catherine Vasilikas-Scaramozza. "Type 2 Diabetes in China." *Decision Resources*, August 2013. http://decisionresources.com/Products-and-Services/Report?r=emermd0113.

Wu, Yangfeng. "Overweight and Obesity in China." *British Medical Journal* 333 (7564) (2006): 362–363. http://archive.oxha.org/knowledge/publications/china_overweightandobesity_bmjed_aug06.pdf.

Xiaochen, Sun, and Lei Lei. "Obesity Rate on the Rise." *China Daily*, August 6, 2013. http://usa.chinadaily.com.cn/china/2013-08/06/content_16872878.htm.

Xu, Yu, et al. "Prevalence and Control of Diabetes in Chinese Adults." *Journal of the American Medical Association* 310 (9) (2013): 948–958. http://jama.jamanetwork .com/article.aspx?articleid=1734701.

Yan, Shi. Web log. 石嫣—食在当地,食在当季. http://blog.sina.com.cn/usashiyan.

Yan, Shi, et al. "Chinese Sustainable Agriculture and the Rising Middle Class: Analysis From Participatory Research In Community Supported Agriculture (CSA) at Little Donkey Farm." URGENCI Resource Center, 2011. http://www.google.com/url?sa=t &rct=j&q=&esrc=s&frm=1&source=web&cd=1&ved=0CB4QFjAA&url=http%3A%2F %2Furgenci.net%2Fwp-content%2Fuploads%2F2015%2F03%2FThe-Development -of-Sustainable-Urban-Agriculture-and-the-Rise-of-the-Urban-Middle-Class .doc&ei=RbpbVaC_HIzItQWfkoHQBg&usg=AFQjCNF1d0iSHdkXBlDlSpsu2T _nbYWQcw&bvm=bv.93756505,d.b2w.

Yan, Shi, et al. "Safe Food, Green Food, Good Food: Chinese Community Supported Agriculture and the Rising Middle Class." *International Journal of Agricultural Sustainability* 9 (4) (2011): 551–558. http://www.tandfonline.com/doi/pdf/10.1080/147359 03.2011.619327.

Yan, Yunxiang. "Food Safety and Social Risk in China 2011." *Journal of Asian Studies* 71 (3) (August 2012): 705–729.

Yang, Yang. "Pesticides and Environmental Health Trends in China." Wilson Center China Environment Forum, February 28, 2007. http://www.wilsoncenter.org/sites/ default/files/pesticides_feb28.pdf.

Yang, Zhengzhou. "Demographic Changes in China's Farmers: The Future of Farming in China." *Asian Social Science* 9 (7) (2013): 136–143.

Yongmin, Bian. "The Challenges for Food Safety in China." *China Perspectives* 53 (May-June 2004). https://chinaperspectives.revues.org/819.

Yoo, Yungsuk Karen. "Tainted Milk: What Kind of Justice for Victims' Families in China?" *Hastings International and Comparative Law Review* 33 (2) (2010): 555–575. https://litigation-essentials.lexisnexis.com/webcd/app?action=DocumentDisplay&cr awlid=1&doctype=cite&docid=33+Hastings+Int%27l+%26+Comp.+L.+Rev.+555&sr ctype=smi&srcid=3B15&key=1696686a61fab15ef83c6190c4246d78.

Zhang, Qian Forrest, and John A. Donaldson. "The Rise of Agrarian Capitalism with Chinese Characteristics: Agricultural Modernization, Agribusiness and Collective Land Rights." *China Journal* (Canberra) 60 (July 2008): 25–47. http://ink .library.smu.edu.sg/cgi/viewcontent.cgi?article=1631&context=soss_research.

Zhangjin, ed. "American Model Adapted for Chinese Soil." *CRI English*, February 11, 2012, n. pag.

Zhao, Yongjun. *China's Disappearing Countryside: Towards Sustainable Land Governance for the Poor*. Surrey, UK: Ashgate Publishing, 2013.

Zheng, Jinran. "Balcony Farmers Are Taking Root." *ChinaDaily*, June 23, 2012. http://usa.chinadaily.com.cn/china/2012-06/23/content_15519100.htm.

Zhouqiong, Wang. "Carrefour to Expand Presence in China." *ChinaDaily.com*, December 5, 2013. http://www.chinadaily.com.cn/china/2013-12/05/content _17155071.htm.

Chapter 6

Aeppel, Timothy. "Bringing Manufacturing to the U.S. after a World Tour." *Wall Street Journal*, July 17, 2014. http://www.wsj.com/articles/bringing-manufacturing-to -the-u-s-after-a-world-tour-1405629164.

Allaire, Julian. "China and Bicycle: The End of the Story?" Velo-city conference, June 2007. http://webcom.upmf-grenoble.fr/edden/spip/IMG/pdf/JA_poster-Veloc ity-juin2007.pdf.

Avila, Eric. *The Folklore of the Freeway: Race and Revolt in the Modernist City*. Minneapolis: University of Minnesota Press, 2014.

Axelrod, Jeremiah B.C. *Inventing Autotopia: Dreams and Visions of the Modern Metropolis in Jazz Age Los Angeles*. Berkeley: University of California Press, 2009.

Baijie, An. "Putting the Brakes on Car Owners." *Asia Weekly*, February 6, 2015. http://epaper.chinadailyasia.com/asia-weekly/article-3962.html.

Banham, Reyner. *Los Angeles: The Architecture of Four Ecologies*. Berkeley: University of California Press, 2009.

Bartholomew, Dana. "Officer Cites 82-Year-Old Woman for Being Too Slow to Negotiate Busy Street." *Los Angeles Daily News*, April 10, 2006.

Burnham, John Chynoweth. "The Gasoline Tax and the Automobile Revolution." *Mississippi Valley Historical Review* 48 (3) (December 1961): 435–459.

California Department of Transportation (Caltrans). "Technical Memorandum: I 710 Corridor Project EIR/EIS Travel Demand Modeling Methodology." February 26, 2010. http://www.dot.ca.gov/dist07/resources/envdocs/docs/710corridor/docs/ I-710_Travel%20Demand%20Modeling_Report_Revise_Feb262010_FINAL.pdf.

Carter, Vanessa, Manuel Pastor, and Madeline Wander. "An Agenda for Equity: A Framework for Building a Just Transportation System in Los Angeles County." USC Program for Environmental and Regional Equity, November 2013. https://dornsife .usc.edu/assets/sites/242/docs/Agenda_Equity_Full_Report_Web02.pdf.

Cervero, Robert, and Jennifer Day. "Suburbanization and Transit-Oriented Development in China." *Transport Policy* 15 (5) (2008): 315–323. http://www.worldtransi tresearch.info/research/1401/.

Chan, L. Y., et al. "Commuter Exposure to Particulate Matter in Public Transportation Modes in Hong Kong." *Atmospheric Environment* 36 (2002): 3363–3373.

Crump, Spencer. *Ride the Big Red Cars: How Trolleys Helped Build Southern California.* Los Angeles: Trans-Anglo Press, 1970.

Cullinane, Sharon. "Attitudes of Hong Kong Residents to Cars and Public Transport: Some Policy Implications." *Transport Reviews* 23 (1) (2003): 21–34.

de Boom, Annemarie, Richard Walker, and Rob Goldup. "Shanghai: The Greatest Cycling City in the World?" *World Transport Policy and Practice* 7 (3) (2001): 53–59.

de Jong, Martin, et al. "Introducing Public-Private Partnerships for Metropolitan Subways in China: What Is the Evidence?" *Journal of Transport Geography* 18 (2013): 301–313.

Denham, T. D. "California's Great Cycle-way." *Good Roads Magazine*, November 1901.

Dingding, Xin, et al. "Subway Costs Feared to Go Off the Rails." *China Daily*, July 31, 2012. http://usa.chinadaily.com.cn/china/2012-07/31/content_15633448.htm.

Dreyfuss, Robert. "China: A 'Kingdom of Bicycles' No Longer." National Public Radio, November 25, 2009. http://www.npr.org/templates/story/story.php?storyId=120811453.

Eidlin, Eric. "What Density Doesn't Tell Us About Sprawl." *Access*, Fall 2010. http://www.accessmagazine.org/articles/fall-2010/density-doesnt-tell-us-sprawl/.

Elkind, Ethan. *Railtown: The Fight for the Los Angeles Metro Rail and the Future of the City.* Berkeley: University of California Press, 2014.

Fischer, Elizabeth. "China's High Speed Rail Revolution." *Railway-technology.com*, November 21, 2012. http://www.railway-technology.com/features/feature124824/.

Freeman, Fox, Wilbur Smith, and Associates. *Hong Kong Mass Transport Study: Report Prepared for the Hong Kong Government.* Hong Kong: Hong Kong Government, 1967.

Friedricks, William B. *Henry E. Huntington and the Creation of Southern California.* Columbus: Ohio State University Press, 1992.

Gallagher, Kelly Sims. *China Shifts Gears: Automakers, Oil, Pollution, and Development.* Cambridge, MA: MIT Press, 2006.

Gallagher, Kelly Sims. *The Globalization of Clean Energy Technology: Lessons from China.* Cambridge, MA: MIT Press, 2014.

Gottlieb, Robert. *Reinventing Los Angeles: Nature and Community in the Global City.* Cambridge, MA: MIT Press, 2007.

Grigg, Angus. "Chinese High Speed Rail Should Confine the XPT to History." *Financial Revie*, June 30, 2015. http://www.afr.com/business/infrastructure/rail/chinese-high-speed-rail-should-confine-the-xpt-to-history-20150629-gi11pb.

Harris, Robert E. G., "Have the Freeways Failed Los Angeles?" *Frontier*, January 1957, 7–11.

Hawthorne, Christopher. "Fast Lane to Gridlock." *Los Angeles Times*, July 12, 2011. http://articles.latimes.com/2011/jul/12/entertainment/la-et-carmageddon -20110712.

Henstell, Bruce. *Sunshine and Wealth: Los Angeles in the Twenties and Thirties*. San Francisco: Chronicle Books, 1984.

Hilton, George W., and John F. Due. *The Electric Interurban Railways in America*. Stanford: Stanford University Press, 1960.

Hise, Greg, and William Deverell. *Eden by Design: The 1930 Olmsted-Bartholomew Plan for the Los Angeles Basin*. Berkeley: University of California Press, 2000.

Hong Kong Mass Transit InfoCenter. "The History." July 2004. http://www.hkmtr. net/Past_History.htm.

Hong Kong Special Administrative Region, Transport Department. 2015 Annual Transport Digest. "Section 3: Registration and Licensing of Vehicles and Drivers," and "Section 6: Vehicle Parking." http://www.td.gov.hk/mini_site/atd/2015/tc/index.html.

Hong Kong Special Administrative Region, Transport Department. "Hong Kong: The Facts." http://www.gov.hk/en/about/abouthk/factsheets/docs/transport.pdf.

Hong Kong Special Administrative Region, Transport Department. "Road Traffic Casualties by Age, 2004–2014." Figure 1.1. 2015. http://www.td.gov.hk/filemanager/en/content_4717/f1.1.pdf.

Hook, Walter. "The Sound of China's Bicycle Industry? One Hand Clapping." *Sustainable Transport*, Fall 2002, 20–22.

Institute for Transportation and Development Policy. "Improving Access for Guangzhou's Urban Villages." Blog, August 24, 2015. https://www.itdp.org/improving -access-for-guangzhous-urban-villages/.

Jeff, Gloria J., General Manager, Department of Transportation, City of Los Angeles. Subject of Pedestrians (CF 06–0762). Inter-Departmental Memo to the Los Angeles City Council, May 4, 2006.

Jiao, Jingjuan, et al. "Impacts on Accessibility of China's Present and Future HSR Network." *Journal of Transport Geography* 40 (October 2014): 123–132. http://www .sciencedirect.com/science/article/pii/S0966692314001525.

Jie, Liu. "Official Data Confirm China as World's Biggest Auto Producer, Consumer, Challenges Remain." *Xinhua*, January 11, 2010. http://www.china-embassy.org/eng/xw/t650869.htm.

Johnson, Ian. "Train Wreck in China Heightens Unease about Safety Standards." *New York Times*, July 24, 2011. http://www.nytimes.com/2011/07/25/world/asia/25train.html?_r=1.

Jones, David W. "California's Freeway Era in Historical Perspective." University of California Institute of Transportation Studies, Berkeley; California Department of Transportation, Sacramento, June 1989.

Lam, Kin-che et al. "Annoyance Response to Mixed Transportation Noise in Hong Kong." *Applied Acoustics* 70 (2009): 1–10.

Lau, Justine. "Veolia Climbs Aboard to Run HK Tram Group." *Financial Times*, April 7, 2009. https://www.ft.com/content/134cc4a2-238b-11de-996a-00144feabdc0.

Laumbach, R. J., et al. "Sickness Response Symptoms among Healthy Volunteers after Controlled Exposures to Diesel Exhaust and Psychological Stress." *Environmental Health Perspectives* 119 (7) (July 2011): 945–950. http://www.ncbi.nlm.nih.gov/pubmed/21330231.

Leung, C. K. "The Growth of Public Passenger Transport." In *Asian Urbanization: A Hong Kong Casebook*, ed. D. J. Dwyer, 137–154. Hong Kong: Hong Kong University Press, 1971.

Linton, Joe. "L.A. City Council Approves New Mobility Plan, Including Vision Zero." *Streetsblog LA*, August 11, 2015. http://la.streetsblog.org/2015/08/11/l-a-city-council-approves-new-mobility-plan-vision-zero/.

Loo, Becky P. Y., and Alice S. Y. Chow. "Changing Urban Form in Hong Kong: What Are the Challenges on Sustainable Transportation?" *International Journal of Sustainable Transportation* 2 (3) (2008): 177–193. http://www.tandfonline.com/doi/pdf/10.1080/15568310701517331#.VInbVH7TnIU.

Loo, Becky P. Y., and Donggen Wang. "Changing Landscapes of Transport and Logistics in China." *Journal of Transport Geography* 40 (October 2014): 1–2. http://www.sciencedirect.com/science/journal/09666923/40.

Lopez, Steve. "Guilty of 'Crossing While Elderly.'" *Los Angeles Times*, April 15, 2006. http://articles.latimes.com/2006/apr/15/local/me-lopez15.

Los Angeles Department of City Planning. "Mobility Plan 2035: An Element of the General Plan." August 2015. http://planning.lacity.org/documents/policy/mobilityplnmemo.pdf.

Los Angeles Department of Transportation. "Great Streets for Los Angeles: Strategic Plan." 2014. http://www.smartgrowthamerica.org/documents/cs/impl/ca-losangeles-dot-strategicplan2014.pdf.

Loukaitou-Sideris, Anastasia, and Robert Gottlieb. "The Day That People Filled the Freeway: Re-Envisioning the Arroyo Seco Parkway and the Urban Environment in Los Angeles." *DISP Journal* (Zurich) 159 4/2004.

Loukaitou-Sideris, Anastasia, and Robert Gottlieb. "Putting the Pleasure Back in the Drive: Reclaiming Urban Parkways for the 21st Century." *Access* 22 (Spring 2003). http://scholar.oxy.edu/cgi/viewcontent.cgi?article=1427&context=uep_faculty.

Lovett, Ian. "A Los Angeles Plan to Reshape the Streetscape Sets off Fears of Gridlock." *New York Times*, September 7, 2015. http://www.nytimes.com/2015/09/08/us/a-los-angeles-plan-to-reshape-the-streetscape-sets-off-fears-of-gridlock.html?&hp&action=click&pgtype=Homepage&module=second-column-region&region=top-news&WT.nav=top-news&_r=0.

Markowitz, Gerald E., and David Rosner. *Lead Wars: The Politics of Science and the Fate of America's Children*. Berkeley: University of California Press, 2014.

Melosi, Martin. "The Automobile and the Environment in American History." Essay produced for the Automobile in American Life and Society Project, sponsored by the University of Michigan-Dearborn Science and Technology Studies Program and the Benson Ford Research Center Program, 2004–2010. http://www.autolife.umd.umich.edu/Environment/E_Overview/E_Overview4.htm.

Millard-Ball, Adam. "Phantom Trips." *Access*, Fall 2014. http://www.accessmagazine.org/articles/fall-2014/phantom-trips/.

Millard-Ball, Adam. "Phantom Trips: Overestimating the Traffic Impacts of New Development." *Journal of Transport and Land Use* 8 (1) (2015): 31–49.

Murphy, Colum. "Southern Chinese City of Shenzhen to place Restrictions on Car Purchases." *Wall Street Journal*, December 30, 2014. http://www.wsj.com/articles/southern-chinese-city-of-shenzhen-to-place-restrictions-on-car-purchases-1419918102.

Murphy, Deborah. "Sunland Incident with Mavis Coyle, Pedestrian Safety Forum." Memo to Los Angeles City Council member Wendy Gruel and Laurie Pollner, May 2, 2006.

National Bicycle Dealers Association. "Industry Overview 2013." Costa Mesa, CA, 2013. http://nbda.com/articles/industry-overview-2013-pg34.htm.

Nelson, Laura J., Armand Emamdjomeh, and Joseph Serna. "Crossing Examination." *Los Angeles Times*, July 12, 2015. http://www.pressreader.com/usa/los-angeles-times/20150712/281479275090678/TextView.

Newton, Damien. "Mayvis Coyle Redux." *Streetsblog*, February 13, 2008. http://la.streetsblog.org/2008/02/13/mayvis-coyle-redux/.

Newton, Damien. "Vision Zero or Zero Vision: LA Needs to Change the Way It Thinks about Safety." *Streetsblog LA*, March 24, 2014.

Newton, Damien, and Joe Linton. "LA DOT's Bold New Strategic Vision: Eliminate Traffic Deaths by 2025." *Streetsblog LA*, September 30, 2014. http://la.streetsblog.org/2014/09/30/ladots-bold-new-strategic-vision-eliminate-l-a-traffic-deaths-by-2025/.

Ng, Jeffrey. "Hong Kong's Subway System Wants to Run the World." *Wall Street Journal*, September 18, 2013. http://www.wsj.com/articles/SB10001424127887323342404579080682338101534.

Ng, Simon. "Walking: The New Mobility." *DutchCham Magazine* 170 (July/August 2014): 28–29. http://www.dutchchamber.hk/ebook/DCM_170/DCM_170.html#p=32.

Norton, Peter D. *Fighting Traffic: The Dawn of the Motor Age in the American City.* Cambridge, MA: MIT Press, 2008.

Padukone, Neil. "The Unique Genius of Hong Kong's Transportation System." *Atlantic,* September 10, 2013. http://www.theatlantic.com/china/archive/2013/09/the-unique-genius-of-hong-kongs-public-transportation-system/279528/.

Pan, Philip P. "Bicycle No Longer King of the Road in China." *Washington Post*, March 12, 2001. https://www.washingtonpost.com/archive/politics/2001/03/12/bicycle-no-longer-king-of-the-road-in-china/f9c66880-fcab-40ff-b86d-f3db13aa1859/.

Rae, John B. *The American Automobile Industry.* Boston: Twayne Publishers, 1984.

Rasmussen, Cecilia. "Bikeway Was Ahead of Its Time." *Los Angeles Times*, November 29, 1998. http://articles.latimes.com/1998/nov/29/local/me-48885.

Riedel, Hans-Ulrich. "Chinese Metro Boom Shows No Sign of Abating." *International Railway Journal*, November 19, 2014. http://www.railjournal.com/index.php/metros/chinese-metro-boom-shows-no-sign-of-abating.html.

Roberts, Ken. "Imports of Bicycles Topping $1 Billion Annually." *Miami Herald*, February 1, 2015. http://www.miamiherald.com/news/business/biz-monday/article8855768.html.

Rowangould, Gregory M. "A Census of the US Near-Roadway Population: Public Health and Environmental Justice Considerations." *Transportation Research Part D, Transport and Environment* 25 (2013): 59–67.

Ruan, Grace S. "Profile of the Chinese Bicycle Market." 2014. Bike Market/CBES 2013–2014, special report, 58–61. http://biketaiwan.com/resource/article/6/157/article-03.pdf.

Sanchez, George J. "'What's Good for Boyle Heights Is Good for the Jews': Creating Multiculturalism on the Eastside during the 1950's." *American Quarterly* 56 (3) (2004): 633–661. http://www.havenscenter.org/files/BoyleHeights.pdf.

Schwartz, Gary T. "Urban Freeways and the Interstate System." *Southern California Law Review* 49 (March 1976): 406–513.

Schwartz, Sam. "Why Mobility Is Overrated." *Los Angeles Times*, September 2, 2015. http://www.pressreader.com/usa/los-angeles-times/20150902/281775627922693/TextView.

Sham, Wai Chi. "The History of Hong Kong and Yaumati Ferry Company Limited, 1923 to the1970s." Master's thesis, Lingnan University, 2007. http://commons.ln.edu.hk/cgi/viewcontent.cgi?article=1001&context=his_etd.

"Shenzhen's Pre-Universiade Eviction of 80,000 'High Risk People' Sparks Controversy." *Xinhua*, April 13, 2011. http://news.xinhuanet.com/english2010/china/2011-04/13/c_13826681.htm.

Shoup, Donald. *The High Cost of Free Parking*. Washington, DC: American Planning Association, 2011.

Shoup, Donald C. "The Trouble with Minimum Parking Requirements." *Transportation Research Part A, Policy and Practice* 33 (1999): 549–574. http://shoup.bol.ucla.edu/Trouble.pdf.

Smith, Steven. "Why China's Subway Boom Went Bust." *CityLab*, September 10, 2012. http://www.citylab.com/commute/2012/09/why-chinas-subway-boom-went-bust/3207/.

South Coast Air Quality Management District. "MATES-IV: Multiple Air Toxics Exposure Study in the South Coast Air Basin." Final Report. May 2015. http://www.aqmd.gov/docs/default-source/air-quality/air-toxic-studies/mates-iv/mates-iv-final-draft-report-4-1-15.pdf?sfvrsn=7.

Suwei, Feng, and Li Qiang. "Car Ownership Control in China's Mega-cities: Shanghai, Beijing, and Guangzhou." *Journeys*, 2013. http://www.lta.gov.sg/ltaacademy/doc/13Sep040-Feng_CarOwnershipControl.pdf.

Tang, Bo-sin et al. "Study of the Integrated Rail-Property Development Model in Hong Kong." Hong Kong Polytechnic University, November 2004. http://www.reconnectingamerica.org/assets/Uploads/mtrstudyrpmodel2004.pdf.

Tang, Rachel. "The Rise of China's Auto Industry and Its Impact on the U.S. Motor Vehicle Industry." Congressional Research Service, November 16, 2009. https://www.fas.org/sgp/crs/row/R40924.pdf.

Tang, Siman, and Hong K. Lo. "The Impact of Public Transport Policy on the Viability and Sustainability of Mass Railway Transit—the Hong Kong Experience." *Transportation Research Part A, Policy and Practice* 42 (2008): 563–576.

Taylor, Brian. "When Finance Leads Planning: Urban Planning, Highway Planning, and Metropolitan Freeways in California." *Journal of Planning Education and Research* 20 (2000): 196–214.

Tillemann, Levi. "China's Electric Car Boom: Should Tesla Motors Worry?" *Fortune*, February 19, 2015. http://fortune.com/2015/02/19/chinas-electric-car-boom-should-tesla-motors-worry/.

Transport Advisory Committee. "Report on Study of Road Traffic Congestion in Hong Kong." December 2014. http://www.thb.gov.hk/eng/boards/transport/land/Full_Eng_C_cover.pdf.

United States Environmental Protection Agency. "Milestones in Mobile Source Air Pollution Control and Regulations." 2012. http://www3.epa.gov/otaq/consumer/milestones.htm.

United States Public Roads Administration. "Toll Roads and Free Roads." 76th Cong., 1st Sess. House document no. 272. Washington, DC: US Government Printing Office, 1939. http://babel.hathitrust.org/cgi/pt?id=wu.89090507302;view=1up;seq=5.

Wachs, Martin. "Autos, Transit, and the Sprawl of Los Angeles: The 1920s." *Journal of the American Planning Association* 50 (3) (1984): 297–310. http://www.tandfonline.com/doi/abs/10.1080/01944368408976597?journalCode=rjpa20.

Wachs, Martin. "The Evolution of Transportation Policy in Los Angeles: Images of Past Policy and Future Prospects." In *The City: Los Angeles and Urban Theory at the End of the Twentieth Century*, ed. Allen J. Scott and Edward Soja, 106–159. Berkeley: University of California Press, 1996.

Wachs, Martin, and Margret Crawford. *The Car and the City: The Automobile, the Built Environment and Daily Life*. Ann Arbor: University of Michigan Press, 1992.

Wang, James J., and Qian Liu. "Understanding the Parking Supply Mechanism in China: A Case Study of Shenzhen." *Journal of Transport Geography* 40 (October 2014): 77–88. http://www.sciencedirect.com/science/article/pii/S0966692314000866.

Watts, Jonathan. "China Restores Bike Lanes Lost to Car Boom." *Guardian*, June 16, 2006. http://www.theguardian.com/world/2006/jun/16/china.transport.

Weber, Christopher. "Carmageddon 2: 405 Freeway Ahead of Schedule: Closure Deemed A Success." *Huffington Post*, October 1, 2012.

Webster, Martin. "Transportation: A Civic Problem." *Engineering and Science* 13 (3) (December 1949): 11–15.

Weikel, Dan. "Two Options Considered for Reconstructing Part of Congested 710 Freeway." *Los Angeles Times*, March 17, 2015. http://www.latimes.com/local/california/la-me-california-commute-20150317-story.html.

Wong, Lai Tim, and Gerald Erick Fryxell. "Stakeholder Influences on Environmental Management Practices: A Study of Fleet Operations in Hong Kong (SAR), China." *Transportation Journal* 43 (4) (Fall 2004): 22–35.

Yan, Wu. "Shenzhen Explains Abrupt Limit on Car Purchase." *China Daily*, December 30, 2015. http://www.chinadaily.com.cn/china/2014-12/30/content_19203728 .htm.

Yeung, Rikki. *Moving Millions: The Commercial Success and Political Controversies of Hong Kong's Railways*. Hong Kong: Hong Kong University Press, 2008.

Yiu, Chan Wing. "A Review of the Public Transport Policy in Hong Kong." University of Hong Kong, June 2001. http://hub.hku.hk/bitstream/10722/28199/1/FullText .pdf?accept=1.

Zacharias, John, and Yuanzhou Tang. "Restructuring and Repositioning Shenzhen, China's New Mega City." *Progress in Planning* 73 (2010): 209–249.

Zahniser, David. "L.A. Rethinks Its Roads." *Los Angeles Times*, August 12, 2015.

Zamichow, Nora. "52,800 Ride the Rails of History as Subway Rolls." *Los Angeles Times*, January 31, 1993. http://articles.latimes.com/1993-01-31/news/mn-1209_1 _metro-red-line.

Zhao, Pengjun. "The Impact of the Built Environment on Bicycle Commuting: Evidence from Beijing." *Urban Studies* (Edinburgh) 51 (5) (April 2014): 1019–1037. http://usj.sagepub.com/content/early/2013/07/12/0042098013494423.

Zhao, Pengjun. "Private Motorized Urban Mobility in China's Large Cities: The Social Causes of Change and an Agenda for Future Research." *Journal of Transport Geography* 40 (2014): 53–63. http://www.sciencedirect.com/science/article/pii/ S0966692314001598.

Zhao, Pengjun, and Phillipa Howden-Chapman. "Social Inequalities in Mobility: The Impact of the *Hukou* System on Migrants' Job Accessibility and Commuting Costs in Beijing." *International Development Planning Review* 32 (3/4) (2010): 363–384.

Chapter 7

Abcarian, Robin. "Long Overdue: Malibu Elitists Who Impede Public Access Now Face Fines." *Los Angeles Times*, June 23, 2014. http://www.latimes.com/local/abcar ian/la-me-ra-malibu-public-access-fines-20140623-column.html#page=1.

Arellano, Gustavo. *Taco USA: How Mexican Food Conquered America*. New York: Scribner, 2012.

Bach, Jonathan. "'They Come In Peasants and Leave Citizens': Urban Villages in the Making of Shenzhen, China." *Cultural Anthropology* 25 (August 2010): 421–458. http://onlinelibrary.wiley.com/wol1/doi/10.1111/j.1548-1360.2010.01066.x/full.

Blakely, Edward J., and Mary Gail Snyder. "Divided We Fall: Gated and Walled Communities in the United States." In *Architecture of Fear*, ed. Nan Ellin, 85–100.

Princeton, NJ: Princeton Architectural Press, 1997. http://www.asu.edu/courses/aph294/total-readings/blakely%20--%20dividedwefall.pdf.

Bontje, Marco. "Creative Shenzhen? A Critical View on Shenzhen's Transformation from a Low-Cost Manufacturing Hub to a Creative Megacity." *International Journal of Cultural and Creative Industries* 1 (2) (March 2014): 52–67.

Brilliant, Ashleigh E. "Some Aspects of Mass Motorization in Southern California, 1919–1929." *Southern California Quarterly* 47 (2) (1965): 191–208.

Campanella, Thomas J. *The Concrete Dragon: China's Urban Revolution and What It Means for the World.* New York: Princeton Architectural Press, 2008.

Chan, Kam Wing. *Cities with Invisible Walls: Reinterpreting Urbanization in Post-1949 China.* Oxford: Oxford University Press, 1994.

Chan, Kam Wing. "Crossing the 50 Percent Population Rubicon: Can China Urbanize to Prosperity?" *Eurasian Geography and Economics* 53 (1) (2012): 63–86. http://www.tandfonline.com/doi/abs/10.2747/1539-7216.53.1.63.

Chan, Yannie. "Yes. We Can Have a Walkable Central." *HK Magazine*, November 20, 2014. http://hk-magazine.com/article/12894/yes-we-can-have-walkable-central.

Chase, John Leighton, Margaret Crawford, and John Kaliski, eds. *Everyday Urbanism.* Expanded ed. New York: Monacelli Press, 2008.

Cheung, Darren Man Wai. "Land Supply and Land-Use Planning of Public Open Space in Hong Kong." PhD thesis, University of Hong Kong, 2015. http://hub.hku.hk/bitstream/10722/209500/2/FullText.pdf?accept=1.

China Dialogue. "Reimagining China's Cities: Towards a Sustainable Urbanization." *China Dialogue*, November 2013. https://s3.amazonaws.com/cd.live/uploads/content/file_en/6480/37_urbanisation_journal_single_file-new_1_.pdf.

Chowkwanyun, Merlin, and Jordan Segall. "How an Exclusive Los Angeles Suburb Lost Its Whiteness." *City Lab*, August 27, 2012. http://www.citylab.com/politics/2012/08/how-exclusive-los-angeles-suburb-lost-its-whiteness/3046/.

Chung, Him. "The Planning of 'Villages-in-the-City' in Shenzhen, China: The Significance of the New State-Led Approach." *International Planning Studies* 14 (3) (August 2009): 253–273.

Chung, M. L., and K. W. Chow. *Pictures of Hong Kong and Its People, 1950s–1970s.* Hong Kong: Commercial Press, 1997.

City of Los Angeles. "Mobility Plan 2035: An Element of the General Plan." August 2015. http://planning.lacity.org/documents/policy/mobilityplnmemo.pdf.

Cosulich-Schwartz, Paolo. "Spatial Injustice in Los Angeles: An Evaluation of Downtown L.A.'s Privately Owned Public Space." Occidental College, 2009. https://www

.oxy.edu/sites/default/files/assets/UEP/Comps/2009/cosulich%20master%20doc _final.pdf.

Cox, Wendell. "China: Urbanizing and Moving East: 2010 Census." *New Geography*, April 5, 2011. http://www.newgeography.com/content/002218-china-urbanizing -and-moving-east-2010-census.

Crawford, Margaret, and Jiong Wu. "The Beginning of the End: The Destruction of Guangzhou's Urban Villages." In *Villages in the City: A Guide to South China's Informal Settlements*, ed. Stefan Al, 19–28. Hong Kong: Hong Kong University Press, 2014.

Culver, Lawrence. *The Frontier of Leisure: Southern California and the Shaping of Modern America*. New York: Oxford University Press, 2010.

Culver, Lawrence. "Race, Recreation, and Conflict between Public and Private Nature in Twentieth-Century Los Angeles." In *Greening the City: Urban Landscapes in the Twentieth Century*, ed. Dorothee Brantz and Sonja Dumpelmann, 95–111. Charlottesville: University of Virginia Press, 2011.

Davis, Mike. "The Case for Letting Malibu Burn." *Environmental History Review* 19 (2) (Summer 1995): 1–36. http://envihistrevi.oxfordjournals.org/content/19/2/1.full .pdf+html.

Davis, Mike. *City of Quartz: Excavating the Future in Los Angeles*. London: Verso, 1990.

Davis, Mike. *Planet of Slums*. London: Verso, 2006.

Doulet, Jean-Francois. "Where Are China's Cities Heading? Three Approaches to the Metropolis in Contemporary China." *China Perspectives* 4 (2008): 4–14.

Flamming, Douglas. *Bound for Freedom: Black Los Angeles in Jim Cow America*. Berkeley: University of California Press, 2005.

Friedmann, John. "Four Theses in the Study of China's Urbanization." *International Journal of Urban and Regional Research* 30 (2) (June 2006): 440–451.

Garcia, Robert, Erica Flores Baltodano, and Edward Mazzarella. "Free the Beach! Public Access, Equal Justice, and the California Coast." Policy Report, Center for Law in the Public Interest and the Surfrider Foundation, 2005. http://www.surfrider.org/ images/uploads/publications/Free_the_Beach.pdf.

Generre, Sam. "The Gated City of Rolling Hills." *South Bay Daily Breeze*, July 14, 2010. http://blogs.dailybreeze.com/history/2010/07/14/rolling-hills/.

Groves, Martha. "Malibu App Celebrates Beach Access—Right Next to David Geffen's House." *Los Angeles Times*, June 7, 2014. http://www.latimes.com/local/lanow/ la-me-ln-malibu-beach-app-20140607-story.html.

Haila, Anne. "Real Estate in Global Cities: Singapore and Hong Kong as Property States." *Urban Studies* (Edinburgh) 37 (12) (2000): 2241–2256.

Hao, Pu, et al. "Spatial Analyses of the Urban Village Development Process in Shenzhen China." *International Journal of Urban and Regional Research* 37 (6) (November 2013): 2177–2197.

Hawthorne, Christopher. "For Gehry, the L.A. River Is All About the Water." *Los Angeles Times*, August 9, 2015. http://www.latimes.com/local/california/la-et-la -river-notebook-20150809-story.html#page=1.

Hawthorne, Christopher. "'Latino Urbanism' Influences a Los Angeles in Flux." *Los Angeles Times*, December 6, 2014. http://www.latimes.com/entertainment/arts/ architecture/.

Hou, Jeffrey, ed. *Insurgent Public Space: Guerrilla Urbanism and the Remaking of Contemporary Cities*. London: Routledge, 2010.

Houkai, Wei. "China's Urban Transformation Strategy in New Period." *Chinese Journal of Urban and Environmental Studies* 1 (1) (2013): 1–28. http://www.worldscientific .com/doi/abs/10.1142/S2345748113500036.

Hsing, You-tien. *The Great Urban Transformation: Politics of Land and Property in China*. New York: Oxford University Press, 2010.

Institute for Transportation and Development Policy. "Parking Guidebook for Chinese Cities (2013)." http://www.itdp.org/documents/Parking_Guidebook_for _Chinese_Cities.pdf.

Jackson, Kenneth. *Crabgrass Frontier: The Suburbanization of the United States*. New York: Oxford University Press, 1983.

Jao, Karen. "Lingering Thoughts on the Los Angeles River: A Q&A with Jenny Price." *KCET*, May 29, 2013. http://www.kcet.org/socal/departures/lariver/confluence/river -notes/lingering-thoughts-on-the-los-angeles-river-a-qa-with-jenny-price.html.

Jie, Fan, and Wolgang Taubmann. "Migrant Enclaves in Large Chinese Cities." In *The New Chinese City: Globalization and Market Reform*, ed. John Logan, 183–197. New York: Wiley-Blackwell, 2002.

Kaiman, Jonathan. "China's 'Eco-cities': Empty of Hospitals, Shopping Centres, and People." *Guardian*, April 14, 2014. http://www.theguardian.com/cities/2014/apr/14/ china-tianjin-eco-city-empty-hospitals-people.

Lau, Joyce. "Art Spawned by Protest: Now to Make It Live On." *New York Times*, November 16, 2014. http://www.nytimes.com/2014/11/15/world/asia/rescuing -protest-artwork-from-hong-kongs-streets.html.

Liu, Ye., et al. "Growth of Rural Migrant Enclaves in Guangzhou, China: Agency, Everyday Practice, and Social Mobility." *Urban Studies* (Edinburgh) 52 (16) (December 2015): 3086–3105.

Low, Setha. "How Private Interests Take Over Public Space: Zoning, Taxes, and the Incorporation of Gated Communities." In *The Politics of Public Space*, ed. Setha Low and Neil Smith, 81–103. New York: Routledge, 2006.

Low, Setha, and Neil Smith, eds. *The Politics of Public Space*. New York: Routledge, 2006.

Ma, Laurence. "Anti-urbanism in China." *Proceedings of the Association of American Geographers* 8 (1976): 114–118.

Ma, Laurence J. C., and Fulong Wu. "The Chinese City in Transition: Towards Theorizing China's Urban Restructuring." In *Restructuring the Chinese City: Changing Society, Economy and Space*. London: Routledge, 2005.

Makinen, Julie. "Polo and Pomp (Ponies for China's 1%)." *Los Angeles Times*, January 28, 2014. http://www.latimes.com/world/la-fg-c1-china-snow-polo-pictures,0,6558574.photogallery#axzz2rimqjgcJ.

Miao, Pu. "Deserted Streets in a Jammed Town: The Gated Community in Chinese Cities and Its Solution." *Journal of Urban Design* 8 (1) (2003): 45–66.

Miller, Tom. *China's Urban Billion: The Story Behind the Biggest Migration in Human History*. London: Zed Books, 2012.

Ming, Xiaodong. "Infrastructure's Central Role in China's New Urbanization." McKinsey & Co., 2014. file:///C:/Users/gottlieb/Downloads/InfrastructuresCentral Role.pdf.

Ng, Simon. "Walking: The New Mobility." *Dutchcham Magazine* 170 (July/August 2014). http://www.dutchchamber.hk/ebook/DCM_170/DCM_170.html#p=1.

Ng, Simon, et al. "Walkable City, Living Streets." Civic Exchange, Hong Kong, October 2012.

O'Donnell, Mary Ann. "Maillen Hotel and Apartments in Shenzhen, China." *Architectural Review*, August 28, 2012. http://www.architectural-review.com/buildings/maillen-hotel-and-apartments-in-shenzhen-china-by-c/8634622.article.

Opinion Internationale. "How Innovation Helps Large Cities Like Shenzhen Face the Challenges of Urbanization." Interview with Huang Weiwen. *Opinion Internationale Newsletter*, March 24, 2014. http://www.opinion-internationale.com/en/2014/03/24/how-innovation-helps-large-cities-like-shenzhen-face-the-challenges-of-urbanization_23792.html.

Plan for a Healthy L.A. Draft Plan. http://healthyplan.la/wordpress/wp-content/uploads/2014/02/PUBLIC-RELEASE-DRAFT-PLAN-FOR-A-HEALTHY-LOS-ANGELESv2.pdf.

Pow, Choon-Piew, and Lily Kong. "Marketing the Chinese Dream Home: Gated Communities and Representations of the Good Life in (Post-) socialist Shanghai."

Urban Geography 28 (2) (2007): 129–159. http://www.tandfonline.com/doi/abs/ 10.2747/0272-3638.28.2.129.

Roberts, Dexter. "Premier Li Keqiang Wants More Chinese in the Cities." *Bloomberg Businessweek*, June 6, 2013, http://www.businessweek.com/printer/articles/123194 -premier-li-keqiang-wants-more-chinese-in-the-cities.

Rojas, James. "Latino Urbanism in Los Angeles: A Model for Urban Improvisation and Reinvention." In *Insurgent Public Space: Guerrilla Urbanism and the Remaking of Contemporary Cities*, ed. Jeffrey Hou, 36–44. London: Routledge, 2010.

Sanchez, George J. "'What's Good for Boyle Heights Is Good for the Jews': Creating Multiculturalism on the Eastside during the 1950's." *American Quarterly* 56 (3) (2004): 633–661. http://www.havenscenter.org/files/BoyleHeights.pdf.

Sassen, Saskia. "The Global Street: Making the Political." *Globalizations* 8 (5) (2011): 573–579.

Schiffman, R., et al., eds. *Beyong Zuccotti Park: Freedom of Assembly and the Occupation of Public Space*. Oakland, CA: New Village Press, 2012.

Sides, Josh. *L.A. City Limits: African American Los Angeles from the Great Depression to the Present*. Berkeley: University of California Press, 2003.

Siu, Helen. "Grounding Displacement: Uncivil Urban Spaces in Postreform South China." *American Anthropologist* 34 (2) (2007): 329–350.

Smith, Nick. "City in the Village: Huanggang and China's Urban Renewal." In *Villages in the City: A Guide to South China's Informal Settlements*, ed. Stefan Al, 29–41. Hong Kong: Hong Kong University Press, 2014.

Sorkin, Michael. "Bull in China's Shop." *Nation*, February 24, 2014, 33–35.

Wang, Jun, and Stephen S. Y. Lau. (2013) "Hierarchical Production of Privacy: Gating in Compact Living in Hong Kong." *Current Urban Studies* 1 (2) (June 2013): 11–18. http://www.scirp.org/journal/PaperInformation.aspx?PaperID=32971#.VV4 U4UZcK4o.

Wassener, Bettina, and Grace Tsoi. "Have Nots Squeezed and Stacked in Hong Kong." *New York Times*, September 27, 2013. http://www.nytimes.com/2013/09/28/ business/international/have-nots-squeezed-and-stacked-in-hong-kong.html.

Wu, Fulong. "Gated and Packaged Suburbia: Packaging and Branding Chinese Suburban Residential Development." *Cities* 27 (5) (2010): 385–396.

Wu, Fulong. "Neighborhood Attachment, Social Participation, and Willingness to Stay in China's Low-Income Communities." *Urban Affairs Review* 48 (4) (2012): 547–570.

Yip, Stanley C.T. "Planning for Eco-cities in China: Visions, Approaches and Challenges." 44th ISOCARP Congress, 2008. http://www.isocarp.net/data/case_studies/1162.pdf.

Zhang, Li. "Migrant Enclaves and Impacts of Redevelopment Policy in Chinese Cities." In *Restructuring the Chinese City: Changing Society, Economy and Space*, ed. Laurence J. C. Ma and Fulong Wu, 243–259. London: Routledge, 2005.

Chapter 8

Beder, Sharon. "Activism versus Negotiation: Strategies for the Environmental Movement." *Social Alternatives* 10 (4) (December 1991): 53–56. http://www.uow.edu.au/~sharonb/activism.html#fn13.

Boyce, James K., and Manuel Pastor. "Clearing the Air: Incorporating Air Quality and Environmental Justice into Climate Policy." *Climatic Change* 120 (4) (October 2013): 801–814. http://www.peri.umass.edu/fileadmin/pdf/other_publication_types/magazine___journal_articles/Boyce__Pastor_-_Climatic_Change_2013.pdf.

Cai, Yongshun. *Collective Resistance in China: Why Popular Protests Succeed or Fail.* Stanford: Stanford University Press, 2010.

Chiu, Stephen Wing-Kai, Hung Ho-fung, and On-Kwok Lai. "Environmental Movements in Hong Kong." In *Asia's Environmental Movements: Comparative Perspectives*, ed. F. Lee Yok-shiu and Alvin Y. So, 55–89. London: Routledge, 1999.

Chiu, Stephen Wing-Kai, and Tai Lok Lui. *The Dynamics of Social Movements in Hong Kong.* Hong Kong: Hong Kong University Press, 2000.

Clifford, Mark L. *The Greening of Asia: The Business Case for Solving Asia's Environmental Emergency.* New York: Columbia Business School Publishing, Columbia University Press, 2015.

Congressional-Executive Commission on China. "Regulations on the Registration and Management of Social Organizations." (CECC full translation.) Issued October 25, 1998. http://translate.google.com/translate?hl=en&sl=zh-CN&u=http://www.cecc.gov/resources/legal-provisions/regulations-on-the-registration-and-management-of-social-organizations&prev=/search%3Fq%3Dchina%2Bregulations%2Bon%2Bthe%2Bregistration%2Band%2Bmanagement%2Bof%2Bsocial%2Borganizations%26biw%3D784%26bih%3D396.

Farber, Daniel A. "Pollution Markets and Social Equity: Analyzing the Fairness of Cap and Trade." *Ecology Law Quarterly* 39 (1) (2012): 1–56. http://scholarship.law.berkeley.edu/cgi/viewcontent.cgi?article=3047&context=facpubs.

FitzPatrick, Liam. "Heroes of the Environment: Christine Loh." *Time*, October 17, 2007. http://content.time.com/time/specials/2007/article/0,28804,1663317_1663320_1669919,00.html.

Garrett, Daniel. *Counter-hegemonic Resistance in China's Hong Kong: Visualizing Protest in the City*. Singapore: Springer, 2015.

Geall, S., ed. *China and the Environment: The Green Revolution*. London: Zed Books, 2013.

Gottlieb, Robert, et al. *The Next Los Angeles: The Struggle for a Livable City*. Berkeley: University of California Press, 2006.

Green Beagle Environmental Institute. http://bjep.org.cn/Home/IndexEn.

Harvey, David. *Rebel Cities: From the Right to the City to the Urban Revolution*. London: Verso, 2012.

Ho, Peter. "Greening without Conflict: Environmentalism, NGOs and Civil Society in China." *Development and Change* 32 (2001): 893–921.

Ho, Peter, and Richard Louis Edmonds. *China's Embedded Activism: Opportunities and Constraints of a Social Movement*. London: Routledge, 2008.

Hsing, You-tien and Ching Kwan Lee, eds. *Reclaiming Chinese Society: The New Social Activism*. London: Routledge, 2010.

Hsu, Carolyn. "Beyond Civil Society: An Organizational Perspective on State-NGO Relations in the People's Republic of China." *Journal of Civil Society* 6 (3) (2010): 259–277.

Hsu, Jennifer. "Layers of the Urban State: Migrant Organisations and the Chinese State." *Urban Studies* (Edinburgh) 49 (16) (December 2012): 3513–3530.

Hui, Wang. *The End of the Revolution: China and the Limits of Modernity*. London: Verso, 2009.

Jie, Feng, and Wang Tao. "Officials Struggling to Respond to China's Year of Environmental Protests." *China Dialogue* (reprint from *Southern Weekend*), December 6, 2012. https://www.chinadialogue.net/article/show/single/en/5438-Officials-struggling-to-respond-to-China-s-year-of-environment-protests-.

Lai, On-Kwok. "Greening of Hong Kong? Forms of Manifestation of Environmental Movements." In *The Dynamics of Social Movement in Hong Kong*, ed. Stephen Wing-Kai Chiu and Tai Lok Lui, 259–296. Hong Kong: Hong Kong University Press, 2000.

Lam Wai-man. *Understanding the Political Culture of Hong Kong: The Paradox of Activism and De-politicization*. London: Routledge, 2004.

Lang, Graeme, and Ying Xu. "Anti-incinerator Campaigns and the Evolution of Protest Politics in China." *Environmental Politics* 22 (5) (2013): 832–848.

Lee, Kingsyhon, and Ming-Sho Ho. "The Maoming Anti-PX Protest of 2014." *China Perspectives*. 2014, no. 3, pp. 33–39. http://chinaperspectives.revues.org/6537?file=1+.

Lee, Yok-shui, and Alvin Y. So, eds. *Asia's Environmental Movements: Comparative Perspectives*. London: Routledge, 1999.

Lo, Carlos Wing-Hung, and Sai Wing Leung. "Environmental Agency and Public Opinion in Guangzhou: The Limits of a Popular Approach to Environmental Governance." *China Quarterly* (163) (September 2000): 677–704.

Loh, Christine. "Time for an Open Mind and Cool Head." *South China Morning Post*, May 5, 1997. http://www.scmp.com/article/194690/time-open-mind-and-cool -head.

Loh, Christine, and Civic Exchange. *Building Democracy: Creating Good Government for Hong Kong*. Hong Kong: Hong Kong University Press, 2003.

Loh, Christine, and Civic Exchange, eds. *Functional Constituencies: A Unique Feature of the Hong Kong Legislative Council*. Hong Kong: Hong Kong University Press, 2006.

Loh, Christine, and Civic Exchange. *Getting Heard: A Handbook for Hong Kong Citizens*. Hong Kong: Hong Kong University Press, 2002.

Lopez, Steve. "Power Grab Topples Another Defender of California's Environment." *Los Angeles Times*, March 4, 2016. http://www.latimes.com/local/california/la-me -lopez-district-20160306-column.html.

Ma, Ngok. "Social Movement, Civil Society, and Democratic Development in Hong Kong." Paper presented at the Conference on Emerging Social Movements in China, University of Hong Kong, March 23, 2005.

Ma, Ngok. "Social Movements and State-Society Relationship in Hong Kong." In *Social Movements in China and Hong Kong: The Expansion of Protest Space*, ed. Khun Eng Kuah-Pearce and Gilles Guiheux, 45–64. Amsterdam: Amsterdam University Press, 2009.

Ma, Shu-Yun. "The Chinese Discourse on Civil Society." *China Quarterly* (137) (March 1994): 180–193. http://unpan1.un.org/intradoc/groups/public/documents/ un-dpadm/unpan041405.pdf.

Maohong, Bao. "Environmental NGOs in Transforming China." *Nature and Culture* 4 (1) (Spring 2009): 1–16. http://www.ingentaconnect.com/content/berghahn/ natcult/2009/00000004/00000001/art00001.

Morello-Frosch, Rachel. "Environmental Policies Must Tackle Social Inequities." *Environmental Health News*, June 21, 2012. http://www.environmentalhealthnews .org/ehs/news/2012/pollution-poverty-people-of-color-op-ed-morello-frosch.

Purcell, Mark. "Excavating Lefebvre: The Right to the City and Its Urban Politics of the Inhabitants." *GeoJournal* 58 (2002): 99–108. http://faculty.washington.edu/ mpurcell/geojournal.pdf.

Reaves, Joseph A. "Nuclear Plant Jolts Hong Kong out of Apathy." *Chicago Tribune*, August 17, 1986. http://articles.chicagotribune.com/1986-08-17/news/8603010439 _1_hong-kong-daya-bay-nuclear-plant-build.

Roosevelt, Margot. "A New Crop of Eco-Warriors Take to Their Own Streets." *Los Angeles Times*, September 24, 2009. http://www.latimes.com/local/la-me-air-pollu tion24-2009sep24-story.html#page=2.

Shieh, Shawn. "On the Eve of the Third Plenum, Are We Seeing a Depoliticization of The NGO Sector?" *NGOs in China* blog, November 6, 2013. http://ngochina. blogspot.com/2013/11/on-eve-of-third-plenum-are-we-seeing.html.

Simon, Karla W. *Civil Society in China: The Legal Framework from Ancient Times to the "New Reform Era."* New York: Oxford University Press, 2013.

Simon, Karla W. "Two Steps Forward, One Step Back: Developments in the Regulation of Civil Society Organizations." *International Journal of Civil Society Law* 7 (4) (October 2009): 51–56.

So, Alvin Y. "The Development of Post-Modernist Social Movements in the Hong Kong Special Administrative Region." In *East Asian Social Movements: Power, Protest and Change in a Dynamic Region*, ed. Jeffrey Broadbent and Vicky Brockman, 365–378. Springer, 2010.

Spires, Anthony J. "Lessons from Abroad: Foreign Influences on China's Emerging Civil Society." *China Journal* (Canberra) 68 (July 2012): 125–146. https://search.infor mit.com.au/documentSummary;dn=725986067901337;res=IELHSS.

Spires, Anthony J., Lin Tao, and Kin-man Chan, "Societal Support for China's Grassroots Ngos: Evidence from Yunnan, Guangdong, and Beijing." *China Journal* 71 (January 2014): 65–90.

Tang, Shui-Yan, and Xueyong Zhan. "Environmental NGOs." In *Institutions, Regulatory Styles, Society and Environmental Governance in China*, ed. Carlos Wing-Hung Lo and Shui-Yan Tang. London: Routledge, 2014.

Unger, J., ed. *Associations and the Chinese State: Contested Spaces*. Armonk, NY: M. E. Sharpe, 2008.

Wang, Alex. "The Role of Law in Environmental Protection in China: Recent Developments." *Vermont Journal of Environmental Law* 8 (2) (2007): 195–223. https://www .law.ucla.edu/~/media/Files/UCLA/Law/Pages/Publications/CEN_ICLP_PUB%20 Role%20Law%20China%20Environment.ashx.

Wang, Stephanie. "NGOs Tread Lightly on China's Turf." *Asia Times*, September 12, 2009. http://www.atimes.com/atimes/China/KI12Ad02.html.

Wasserstrom, Jeffrey. "The Pollution Crisis and Environmental Activism in China: A Q&A with Ralph Litzinger." *Dissent*, May 15, 2013. http://www.dissentmagazine

.org/online_articles/the-pollution-crisis-and-environmental-activism-in-china-a-qa -with-anthropologist-ralph-litzinger.

Wen, Dale. "Alternative Voices and Actions from within China." China Study Group, April 3, 2006. http://chinastudygroup.net/2006/04/alternative-voices-and -actions-from-within-china/.

Yang, Guobin. "Civic Environmentalism." In *Reclaiming Chinese Society: The New Social Activism*, ed. You-tien Hsing and Ching Kwan Lee, 119–139. London: Routledge, 2010.

Yee, Herbert S., and Wong Yiu-chung. "Hong Kong: The Politics of the Daya Bay Nuclear Plant Debate." *International Affairs* 63 (4) (Autumn 1987): 617–630.

Zhang, Joy Y., and Michael Barr. *Green Politics in China: Environmental Governance and State-Society Relations*. London: Pluto Press, 2013.

Zhang, Joy Y., and Michael Barr. "Recasting Subjectivity through the Lenses: New Forms of Environmental Mobilisation in China." *Environmental Politics* 22 (5) (2013): 849–865. http://dx.doi.org/10.1080/09644016.2013.817761.

Index

Urban and Industrial Environments

Series editor: Robert Gottlieb, Henry R. Luce Professor of Urban and Environmental Policy, Occidental College

Maureen Smith, *The U.S. Paper Industry and Sustainable Production: An Argument for Restructuring*

Keith Pezzoli, *Human Settlements and Planning for Ecological Sustainability: The Case of Mexico City*

Sarah Hammond Creighton, *Greening the Ivory Tower: Improving the Environmental Track Record of Universities, Colleges, and Other Institutions*

Jan Mazurek, *Making Microchips: Policy, Globalization, and Economic Restructuring in the Semiconductor Industry*

William A. Shutkin, *The Land That Could Be: Environmentalism and Democracy in the Twenty-First Century*

Richard Hofrichter, ed., *Reclaiming the Environmental Debate: The Politics of Health in a Toxic Culture*

Robert Gottlieb, *Environmentalism Unbound: Exploring New Pathways for Change*

Kenneth Geiser, *Materials Matter: Toward a Sustainable Materials Policy*

Thomas D. Beamish, *Silent Spill: The Organization of an Industrial Crisis*

Matthew Gandy, *Concrete and Clay: Reworking Nature in New York City*

David Naguib Pellow, *Garbage Wars: The Struggle for Environmental Justice in Chicago*

Julian Agyeman, Robert D. Bullard, and Bob Evans, eds., *Just Sustainabilities: Development in an Unequal World*

Barbara L. Allen, *Uneasy Alchemy: Citizens and Experts in Louisiana's Chemical Corridor Disputes*

Dara O'Rourke, *Community-Driven Regulation: Balancing Development and the Environment in Vietnam*

Brian K. Obach, *Labor and the Environmental Movement: The Quest for Common Ground*

Peggy F. Barlett and Geoffrey W. Chase, eds., *Sustainability on Campus: Stories and Strategies for Change*

Steve Lerner, *Diamond: A Struggle for Environmental Justice in Louisiana's Chemical Corridor*

Jason Corburn, *Street Science: Community Knowledge and Environmental Health Justice*

Peggy F. Barlett, ed., *Urban Place: Reconnecting with the Natural World*

David Naguib Pellow and Robert J. Brulle, eds., *Power, Justice, and the Environment: A Critical Appraisal of the Environmental Justice Movement*

Eran Ben-Joseph, *The Code of the City: Standards and the Hidden Language of Place Making*

Nancy J. Myers and Carolyn Raffensperger, eds., *Precautionary Tools for Reshaping Environmental Policy*

Kelly Sims Gallagher, *China Shifts Gears: Automakers, Oil, Pollution, and Development*

Kerry H. Whiteside, *Precautionary Politics: Principle and Practice in Confronting Environmental Risk*